Magnesium-based Syntactic, Metastable and Nano-composites

Sankaranarayanan Seetharaman
Dhivya Sankaranarayanan • Eugene Wong
Manoj Gupta

Magnesium-based Syntactic, Metastable and Nano-composites

Springer

Sankaranarayanan Seetharaman
Advanced Remanufacturing and Tech Cnt
Singapore, Singapore

Dhivya Sankaranarayanan
College of Design and Engineering
National University of Singapore
Singapore, Singapore

Eugene Wong
Engineering Cluster
Singapore Institute of Technology
Singapore, Singapore

Manoj Gupta
Department of Mechanical Engineering
National University of Singapore
Singapore, Singapore

ISBN 978-3-031-96513-5 ISBN 978-3-031-96514-2 (eBook)
https://doi.org/10.1007/978-3-031-96514-2

This Springer imprint is published by the registered company Springer Nature Switzerland AG
The registered company address is: Gewerbestrasse 11, 6330 Cham, Switzerland

Preface

Magnesium, the lightest structural metal, has emerged as a cornerstone in the development of lightweight, high-performance materials. Its exceptional strength-to-weight ratio, recyclability, and abundance make it indispensable in industries, such as aerospace, automotive, and biomedical engineering. However, magnesium's inherent limitations in strength, ductility, and thermal stability necessitate innovations to unlock its full potential.

This book, *Magnesium-Based Syntactic, Metastable and Nano-composites*, provides a comprehensive examination of the evolution of magnesium composites, structured to guide readers from foundational concepts to advanced applications. The journey begins with an introduction to magnesium, emphasizing its properties and advantages as a structural material. This is followed by an overview of metal matrix composites (MMCs) and their fabrication methods, which serve as the building blocks for magnesium-based composites.

This book then transitions to magnesium composites with micron-sized reinforcements, exploring their ability to enhance mechanical properties through the incorporation of traditional ceramic and metallic particulates. A detailed section on magnesium nanocomposites follows, showcasing how nano-scale reinforcements—such as carbon nanotubes, graphene, and ceramic nanoparticles—enable significant advancements in strength, ductility, and thermal stability. This section is particularly comprehensive, reflecting the critical role of nanotechnology in revolutionizing magnesium composites.

Next, the focus shifts to magnesium-based composites with hybrid reinforcements and subsequently to the use of novel reinforcements. The unique capabilities of syntactic foams, renowned for their lightweight structure and energy-absorbing capabilities are introduced. These materials, reinforced with hollow microspheres, are ideal for crash-resistant and impact-sensitive applications. The final section also examines magnesium composites reinforced with amorphous metallic glasses and metastable particles, including high-entropy alloys. These cutting-edge reinforcements deliver remarkable properties, such as enhanced toughness, thermal stability, and corrosion resistance, expanding the horizons of magnesium composites.

Throughout this book, readers will find in-depth discussions of fabrication techniques—spanning liquid-state processing, solid-state methods, and in situ synthesis—alongside insights into the interplay between microstructure, reinforcement type, and composite performance. By addressing challenges such as reinforcement dispersion, interfacial bonding, and scalability, this book not only presents the current state of magnesium composites but also charts a path forward for research and innovation.

This work is designed for materials scientists, engineers, and industry professionals, offering both foundational knowledge and advanced insights into magnesium-based composites. We hope it serves as a valuable resource, inspiring future developments and applications in the field of lightweight, high-performance materials.

Singapore, Singapore Sankaranarayanan Seetharaman
Singapore, Singapore Dhivya Sankaranarayanan
Singapore, Singapore Eugene Wong
Singapore, Singapore Manoj Gupta

Acknowledgments

We would like to take this opportunity to express our heartfelt thanks to all the people who have helped and contributed to the publication of this book, in particular to our families for their continual support and understanding and to co-workers, friends, and students for their assistance and encouragement.

Competing Interests The authors have no competing interests to declare that are relevant to the content of this manuscript.

Contents

About the Authors

Sankaranarayanan Seetharaman is a Senior Scientist and Technical Lead at the Advanced Remanufacturing and Technology Centre (ARTC) in Singapore, specializing in additive manufacturing and high-performance materials. He holds a Ph.D. in Mechanical Engineering (Materials and Manufacturing) from the National University of Singapore (NUS). His expertise spans advanced high-performance alloys, metal matrix composites, nano-structured materials, microstructural engineering, advanced materials characterization, and applied machine learning. At ARTC, he leads innovation in metal additive manufacturing, with a focus on next-generation alloys and composites, building on previous experience as a Senior Application Engineer at ANSYS, where he spearheaded global technical initiatives in additive manufacturing simulations. He also held visiting researcher roles at the Indian Institute of Science, Bangalore (2011–2012) and Helmholtz Zentrum Geesthacht, Germany (2014–2015). A professional member of The Minerals, Metals and Materials Society with active contributions to the Light Metals Division, he has published over 70 research articles and conference papers, with a current h-index of 29.

Dhivya Sankaranarayanan is a graduate student at the College of Design and Engineering, National University of Singapore, pursuing an M.Sc. in Biomedical Engineering with expected graduation in December 2025. With a robust foundation in biotechnology, Dhivya has expertise in molecular biology techniques, cell culture, biomaterials, and clinical research. She is also skilled in regulatory affairs and clinical trials management, reflecting her interdisciplinary approach to advancing biomedical science.

Eugene Wong is currently the Director of Programmes for the Engineering Cluster at Singapore Institute of Technology (SIT), where he oversees the management of the engineering programmes offered by SIT. Prior to joining SIT, Eugene was with Newcastle University International Singapore. He obtained his Ph.D. and B. Eng in Mechanical Engineering with a specialization in materials engineering in design from the National University of Singapore in 2008 and 2003, respectively. In addition to his role at SIT, Eugene is the Deputy Head for the National Additive Manufacturing Innovation Cluster (NAMIC) Hub at SIT. His research expertise lies in the area of additive manufacturing, lightweight composite materials, and microwave processing of materials. He was a recipient of NUS President Graduate Fellowship (2006–2007), The Institute of Engineers Singapore (IES) Outstanding Volunteer and Best Committee (2014), and Singapore Armed Forces National Serviceman of the Year (2015).

Manoj Gupta was a former Head of Materials Division of the Mechanical Engineering Department. He earned his Ph.D. from the University of California, Irvine, USA (1992), and postdoctoral research at the University of Alberta, Canada (1992). He has worked for about 40 years in the area of metallic materials with a particular focus on metallic composites of different types. He has taken multiple consultancy work related to failure of materials besides teaching many people in corporate sectors on the same topic. He is currently among the top 1% researchers as per Stanford' List, The Universal Scientific Education and Research Network, and ResearchGate. He has published over 730 peer-reviewed journal papers and owns two US patents and two trade secrets. His h-index is 89, with over 30,000 citations. He has also co-authored eight books published by John Wiley, Springer, and MRF—USA. A multiple award winner, he actively collaborates and visits Japan, France, Saudi Arabia, Qatar, China, the USA, and India as a visiting researcher, professor, and chair professor.

Chapter 1
Introduction

1.1 Introduction to Magnesium

Magnesium is a technologically significant element with a rich history and diverse applications. As the sixth most abundant element in the Earth's crust, magnesium is not only plentiful but also essential for various biological and industrial processes. Its unique status as the lightest structural metal makes it valuable in applications ranging from aerospace to medicine.

This comprehensive write-up delves into the multifaceted world of magnesium, exploring its historical significance, intrinsic properties, and the diverse alloys that enhance its utility. This chapter also examines the myriad applications of magnesium across different industries, the limitations that challenge its use, and the promising future outlook driven by ongoing research and technological advancements.

1.2 History of Magnesium

Magnesium's history is deeply rooted in both ancient civilizations and modern scientific breakthroughs. The element derive its name from "Magnesia," a region in Thessaly, Greece, where its compounds were first identified. The Greeks and Romans recognized the medicinal properties of magnesium compounds, using them as laxatives and antacids. These early uses reflects the ancient knowledge of magnesium compounds, long before the element itself was isolated [1, 2].

S. Seetharaman et al., *Magnesium-based Syntactic, Metastable and Nano-composites*, https://doi.org/10.1007/978-3-031-96514-2_1

1.2.1 Early Scientific Discoveries

The first significant steps toward isolating magnesium began in the mid-eighteenth century. In 1755, Scottish chemist Joseph Black identified magnesium carbonate, although he did not isolate magnesium as an element. This discovery laid the foundation for further research. The turning point came in 1808 when Sir Humphry Davy, an English chemist, successfully isolated magnesium through electrolysis, using a mixture of magnesium oxide and mercuric oxide. Davy's isolation of magnesium was a groundbreaking achievement, as it marked the first time the metal had been obtained in its pure form. This discovery set the stage for magnesium's role in various scientific and industrial applications [3].

Building upon Davy's work, Antoine Bussy, a French chemist, made further progress in 1831 by producing magnesium in a coherent form. Bussy's achievement allowed for more practical use of the metal, though magnesium remained difficult to extract from its ores, which delayed its widespread application [4].

1.2.2 Industrial Production and Early Applications

Magnesium's industrial journey began in the late nineteenth century. In 1886, Germany pioneered the electrolytic production of magnesium, marking the beginning of its commercial-scale production. However, early industrial use was limited due to the complexities and costs associated with extracting magnesium. One notable early application was in photography, where magnesium powder was ignited to create a bright flash of light used in flash photography. Although effective, this application remained niche due to the high production costs [5].

1.2.3 Wartime Importance and Expansion

The significance of magnesium dramatically increased during the two World Wars, particularly due to its lightweight properties, which made it ideal for military applications. During World War I and even more so in World War II, magnesium was heavily utilized in the production of aircraft components, bombs, and incendiary devices. Germany, especially, invested substantially in magnesium production during World War II, recognizing its strategic value. The U.S. similarly prioritized magnesium as a lightweight material, employing it to reduce the weight of military equipment and improve aircraft performance.

During World War II, magnesium was not only used for its low density but also its ability to burn at high temperatures, which made it valuable for incendiary bombs. The metal's importance as a military resource underscored the need for efficient extraction processes, which continued to evolve throughout the war [6].

1.2.4 Post-War Development and Peacetime Uses

After World War II, magnesium production shifted toward civilian applications. The growing automotive and aerospace industries quickly recognized the metal's potential to reduce vehicle weight, thereby enhancing fuel efficiency and performance. By the 1950s and 1960s, magnesium was being used extensively in automotive parts, further aided by technological advancements in magnesium extraction processes, such as the Pidgeon process. This process, developed in the 1940s, allowed for more cost-effective production of magnesium, which enabled its broader industrial use [7].

In this period, magnesium alloys began to be developed, which expanded the metal's utility in various industries. As lightweight materials became increasingly important in transportation and engineering sectors, magnesium and its alloys were positioned as a key component of future innovations.

1.2.5 Magnesium in Modern Industry

Today, magnesium continues to be a critical material in various industries, including aerospace, automotive, and electronics. Ongoing research and development into magnesium alloys have significantly expanded its applications, allowing the metal to maintain its relevance in modern engineering and manufacturing. Technological advances have improved the efficiency of magnesium extraction, making it more accessible for industrial use while reducing production costs. Apart from engineering sectors, magnesium-based materials are also evolving as temporary implant materials due to their bioresorbability. This is likely to open multi-billion-dollar market for magnesium-based materials should the materials are successfully introduced with tailored properties.

1.3 Properties of Magnesium

Magnesium is a remarkable metal, known for its unique combination of physical, chemical, mechanical, and thermal properties. As the lightest structural metal, magnesium's exceptional characteristics make it suitable for various applications, particularly in industries that prioritize weight reduction, such as aerospace, space, rail, road, maritime, automotive, sports defense and electronics sectors.

1.3.1 Physical Properties

Magnesium's most notable property is its low density of 1.74 g/cm^3, nearly two-thirds the density of aluminum and one quarter of steel. This lightness makes it an ideal material for applications where weight reduction is crucial, such as in transportation and aerospace. Its hexagonal close-packed (HCP) crystal structure strongly influences its mechanical properties, particularly its limited ductility at room temperature. The HCP structure restricts the number of slip systems, limiting the metal's ability to deform plastically under stress, unlike metals with more slip systems like aluminum or copper.

Magnesium's melting point is 650°C and its boiling point is 1090°C. These relatively low values make magnesium easier to cast and form, allowing for the creation of complex shapes and components in industrial applications. This ability to be cast efficiently is highly valued in the manufacturing sector, especially in automotive and aerospace components where both precision and lightness are critical.

1.3.2 Mechanical Properties

Magnesium is prized for its excellent strength-to-weight ratio, which makes it a favorable material for structural components that must be both strong and lightweight. The tensile strength of pure magnesium is approximately 190 MPa, but alloying can significantly improve this, with tensile strengths ranging from 170 to 600 MPa depending on the alloy and processing method. The elastic modulus of magnesium is around 45 GPa, lower than both aluminum and steel, meaning that magnesium is less stiff and more flexible than these materials.

Despite its relatively modest mechanical properties in its pure form, alloying magnesium with elements such as aluminum, zinc, or rare earth elements can greatly enhance its strength, ductility, and impact resistance. This versatility allows magnesium alloys to be tailored for a wide range of applications, from structural components to mechanical systems requiring flexibility and strength.

A key challenges for magnesium is its lower ductility compared with metals that possess cubic crystal structures. Due to its HCP structure, magnesium is more prone to cracking under tensile stress, which limits its formability at room temperature. However, when processed at elevated temperatures, magnesium can undergo dynamic recrystallization, allowing for enhanced ductility and formability. This behavior makes magnesium suitable for high-temperature applications and processes such as hot forming and extrusion.

1.3.3 Chemical Properties

Magnesium is highly reactive in its pure form. When exposed to air, it quickly forms a thin oxide layer of magnesium oxide (MgO), which acts as a passivation layer that protects the metal from further oxidation. While this oxide layer helps to prevent rapid corrosion, magnesium remains more susceptible to corrosion than many other metals, particularly in environments exposed to saltwater or other corrosive agents.

The high reactivity of magnesium is also evident in its flammability, particularly when in powdered or finely divided form. Magnesium powder can ignite easily and burns with a bright white flame, a property that is exploited in pyrotechnics, flares, and incendiary devices. Due to this flammability, safety measures are necessary when handling magnesium, especially in molten form or during machining processes where sparks or heat may be generated.

1.3.4 Thermal Properties

Magnesium exhibits good thermal conductivity, with a thermal conductivity of approximately 156 W/mK. This makes it an effective material for applications where heat dissipation is essential, such as heat sinks in electronics or components exposed to high temperatures. Its ability to conduct heat efficiently helps prevent the buildup of heat in mechanical systems, reducing the risk of failure due to thermal stress.

Magnesium also has a high coefficient of thermal expansion (25.2 μm/mK), meaning it expands more than many other metals when subjected to heat. This property is particularly important in applications that undergo significant temperature changes, as the expansion and contraction of the material must be accounted for in design. Additionally, magnesium has excellent vibration damping properties, which help reduce noise and mechanical stress in systems such as automotive and aerospace components.

1.3.5 Electrical Properties

While magnesium is not as conductive as copper or aluminum, it still exhibits reasonable electrical conductivity, approximately 40% that of copper. This makes magnesium suitable for use in electrical components, particularly where weight savings are important, such as in automotive or aerospace wiring. Magnesium's combination of electrical conductivity and light weight allows for the development of components that balance performance with efficiency.

The properties of Mg are listed in Table 1.1.

Table 1.1 Properties of magnesium compared to other structural materials [1, 8–12]

Properties	Mg	Al	Ti	Steel	HDPE	PEEK
Atomic number	12	13	22	26	–	–
Atomic weight	24.32	26.98	47.87	55.85	–	–
Crystal structure	HCP	FCC	HCP	BCC/FCC	–	–
Density (g/cm^3)	1.74	2.70	4.50	7.85	0.93–0.97	1.30
Melting point (°C)	650	660	1668	1370–1510	130–137	343
Boiling point (°C)	1105	2520	3287	2861	–	–
Thermal expansion coefficient (μm/mK)	25.2	23.6	8.6	11–13	100–200	47
Thermal conductance (cal/cm^2/cm/°C/s)	0.38	0.5	0.22	0.12–0.16	0.44	0.25
Electrical conductance (x10^{-7} S/m)	2.34	3.5	2.38	6–10	–	–
Specific heat (J/g·K)	1.02	0.897	0.523	0.49	2.3	1.3
Latent heat of fusion (KJ/Mol)	8.954	10.79	15.451	13.812	188.63	1304
Latent heat of vaporization (KJ/Mol)	127.4	293.4	4211	349.65	7324	7324
Elastic modulus (GPa)	45	69	116	200–210	0.8	3.6
Yield strength (MPa)	15–140	35–150	275–580	250–1000	20–30	90–100
Tensile strength (MPa)	60–220	75–170	434–1400	400–2000	25–40	90–100
Specific strength (kNm/kg)	35–260	7–200	50–300	25–100	40–436	69–771
Specific stiffness (MNm/kg)	21–29	25–38	25–30	25–30	0.41–1.246	2.77–3.087
Thermal conductivity (W/m·K)	–	–	21.9	50.28	0.449	0.254
Hardness (Rockwell)	–	–	36	80–1008	45–705	85–1094
Poisson's ratio	–	–	0.34	0.27–0.308	0.40–0.455	0.364

1.4 Common Magnesium Alloys

In modern engineering and manufacturing, the demand for materials that combine lightweight properties with high strength and durability is ever-increasing. Magnesium alloys fulfill this need exceptionally well. They are the lightest structural metals available, with a density approximately two-thirds that of aluminum and one-quarter that of steel. This makes them ideal for applications where weight reduction is crucial, such as in the automotive, aerospace, defense, and electronics industries. Reducing weight in vehicles and aircraft leads to improved fuel efficiency, lower emissions, and enhanced performance. Additionally, magnesium alloys offer excellent castability, good machinability, and a high strength-to-weight ratio, making them versatile for various manufacturing processes and applications [7, 13–17].

1.4.1 *Role of Alloying Elements in Magnesium Alloys*

Alloying elements play a crucial role in determining the mechanical and physical properties of magnesium alloys [18]. Figure 1.1 illustrates the historical development of various magnesium alloy systems [19]. The following section is an overview of the most common alloying elements used in magnesium alloys and their effects on the properties of magnesium:

- **Aluminum (Al)** enhances castability and strength without impairing weldability. However, Mg-Al alloys suffer from creep and limited high-temperature mechanical properties. When the Al addition exceeds 6%, the alloy becomes heat treatable. Magnesium alloys containing more than 1.5% Al are susceptible to stress corrosion and must be stress relieved after welding.
- **Zinc (Zn)** improves fluidity during casting and is mostly used in combination with aluminum for strength enhancement. The amount of zinc addition is limited to less than 2% to prevent hot shortness. Zinc combined with aluminum also helps circumvent the harmful corrosive effects of iron and nickel impurities.
- **Zirconium (Zr)** acts as a grain refiner due to its lattice parameters being close to those of Mg. It is commonly used with Zn and rare earths but not in Al-containing alloys.

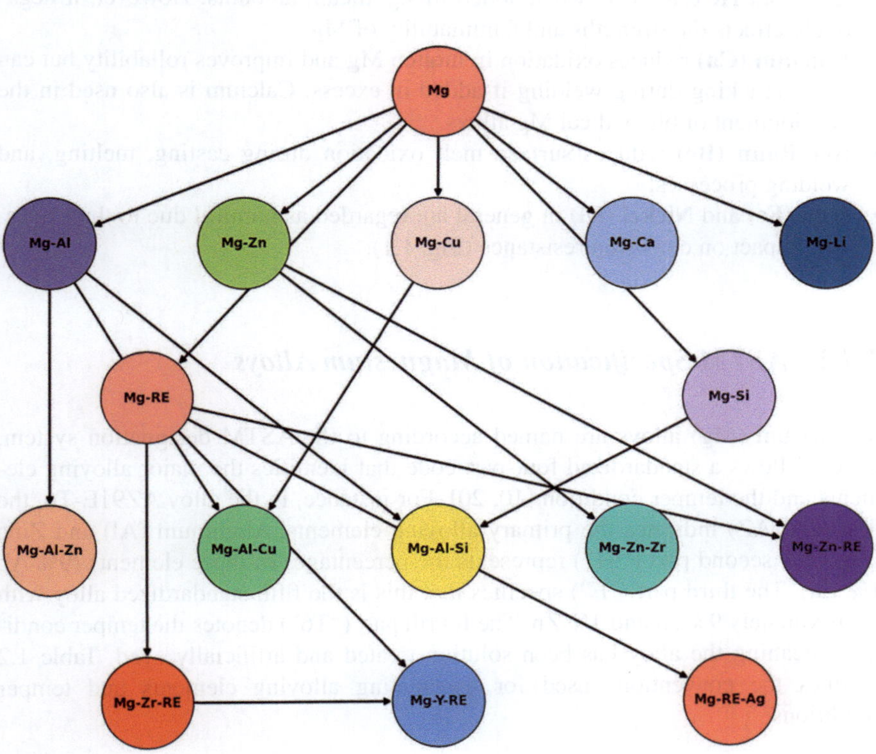

Fig. 1.1 Development of magnesium alloys

- Manganese (Mn) enhances corrosion resistance by removing heavy metals like iron from the melt. A small quantity of Mn addition improves the saltwater corrosion resistance of Mg-Al and Mg-Al-Zn alloys.
- **Silicon (Si)** increases fluidity in the molten state but decreases corrosion resistance in Mg alloys containing iron. It is widely used in the development of die-cast Mg alloys.
- **Copper (Cu)** improves room temperature and high-temperature strengths but adversely affects corrosion resistance if added in excess.
- **Strontium (Sr)** is used in combination with other alloying elements to improve the creep performance of Mg alloys.
- **Thorium (Th)** is known to increase the creep strength of Mg-Zn alloys to up to 370 °C. It has also been shown to improve the weldability.
- **Silver (Ag)** enhances mechanical properties through age hardening and is recommended for biomedical alloys due to its antibacterial properties.
- **Tin (Sn)** improves the ductility of Mg alloys by reducing the hot cracks during thermo-mechanical processing.
- **Rare Earth Elements (RE)** improve high-temperature creep and corrosion resistance. For example, cerium (Ce) improves ductility, and yttrium (Y) enhances creep resistance up to 300°C.
- **Lithium (Li)** enhances ductility and formability by changing the crystal structure from HCP to BCC when added in significant amounts. However, it negatively affects the strengths and flammability of Mg.
- **Calcium (Ca)** reduces oxidation in molten Mg and improves rollability but can cause cracking during welding if added in excess. Calcium is also used in the development of biomedical Mg alloys.
- **Beryllium (Be)** reduces surface melt oxidation during casting, melting, and welding processes.
- **Iron (Fe)** and **Nickel (Ni)** in general are regarded as harmful due to their negative impact on corrosion resistance (Fig. 1.1).

1.4.2 ASTM Specification of Magnesium Alloys

Magnesium (Mg) alloys are named according to the ASTM designation system, which follows a standardized four-part code that identifies the major alloying elements and the temper condition [10, 20]. For instance, in the alloy AZ91E-T6, the first part (AZ) indicates the primary alloying elements, Aluminum (Al) and Zinc (Zn). The second part ("91") represents the percentages of these elements (9% Al, 1% Zn). The third part ("E") specifies that this is the fifth standardized alloy with approximately 9% Al and 1% Zn. The fourth part ("T6") denotes the temper condition, meaning the alloy has been solution treated and artificially aged. Table 1.2 outlines the conventions used for designating alloying elements and temper conditions.

Table 1.2 ASTM designation for magnesium alloys [11]

Part	Description	
Part I	Two letters indicating the primary alloying elements in decreasing order of concentration.	
	Alloying element	*Letter*
	Aluminum	A
	Copper	C
	Thorium	H
	Strontium	J
	Zirconium	K
	Lithium	L
	Manganese	M
	Silver	Q
	Silicon	S
	Tin	T
	Gadolinium	V
	Yttrium	W
	Calcium	X
	Zinc	Z
Part II	Two numbers representing the percentage of the primary alloying elements (rounded to whole numbers)	
Part III	A letter distinguishing different alloys with the same element percentages	
Part IV	A letter-number combination indicating the temper condition	
	Temper condition	*Symbol*
	As fabricated	F
	Annealed and recrystallized (wrought products)	O
	Strain hardened	H
	Strain hardened only	H1
	Strain hardened and partially annealed	H2
	Strain hardened and stabilized	H3
	Solution treated (unstable temper)	W
	Thermally treated (stable temper)	T
	Naturally aged	T1
	Annealed (cast products)	T2
	Solution heat treated and cold worked	T3
	Solution heat treated and naturally aged	T4
	Artificially aged	T5
	Solution heat treated and artificially aged	T6
	Solution heat treated and stabilized	T7
	Solution heat treated, cold worked, and artificially aged	T8
	Solution heat treated, artificially aged, and cold worked	T9
	Artificially aged and cold worked	T10

1.4.3 Classification of Magnesium Alloys

Magnesium alloys can be broadly classified into two major categories—(i) cast and (ii) wrought alloy forms. The cast Mg alloys can be further classified into—(a) aluminum-containing alloys and (b) aluminum-free alloys [15, 20].

Magnesium-aluminum (Mg-Al) alloys are widely used in noncritical applications for magnesium castings. These alloys often display micro-shrinkage during sand casting, leading to a preference for high-pressure die-casting as a favored fabrication method. The strength of AZ (Mg-Al-Zn-Mn) alloys is enhanced by the addition of Al and Zn, along with heat treatment. However, this improvement can come at the expense of ductility, creep resistance, and high-temperature behavior. One of the primary shortcomings of aluminum-containing alloys is the formation of the $Mg_{17}Al_{12}$ phase. This phase is brittle and thermally unstable, rapidly dissolving upon exposure to temperatures above 180 °C. This dissolution can significantly degrade the mechanical properties and thermal stability of the alloy, making it unsuitable for high-temperature applications. In contrast, AM (Mg-Al-Mn) alloys offer higher toughness and elongation while maintaining comparable corrosion resistance. The AS (Mg-Al-Si-Mn) and AE (Mg-Al-RE) series alloys demonstrate good creep resistance up to 150 °C. Particularly noteworthy are the benefits of silicon and rare earth (RE) additions for low-Al-containing Mg alloys like AS21 and AE42, which exhibit superior mechanical strength and creep resistance within their respective series. For applications requiring high temperatures and resistance to creep, the development of AJ (Mg-Al-Sr) series alloys is underway, leveraging the addition of strontium to promote the formation of thermally stable Al_4Sr intermetallics [11, 21, 22].

Aluminum-free alloys are developed to overcome the drawbacks associated with aluminum-containing alloys, that is, to avoid the formation of the brittle and thermally unstable $Mg_{17}Al_{12}$ phase, thereby enhancing their thermal stability and mechanical properties. Aluminum-free Mg-alloy castings integrate small amounts of zirconium as a grain refiner, along with zinc (ZK), rare earth (ZE), yttrium (WE), and silver (QE) additions. These alloys are suitable for parts operating at temperatures up to 300 °C for extended durations. ZK (Mg-Zn-Zr) and ZE (Mg-Zn-RE) alloys offer enhanced strength, yet they are prone to hot-shortness due to higher zinc levels. The QE series alloys demonstrate improved mechanical properties at both room and elevated temperatures, attributed to substantial grain refinement facilitated by Zr addition and subsequent heat treatment. Notably, yttrium-containing alloys like WE43 and WE54 exhibit comparable corrosion resistance to AZ91D but possess superior high-temperature properties. Recent advancements include the use of alloying additions such as Gd, La, Sm, and Yb to enhance the high-temperature performance and corrosion resistance of cast Mg alloys [13, 15, 23–25]. Table 1.3 presents the general characteristics and properties of few important cast Mg alloys.

Wrought magnesium (Mg) alloys, known for their fine microstructure, generally offer superior mechanical properties compared to cast alloys. These alloys are commercially available in various forms such as rolled plates, forgings, and extruded

Table 1.3 Composition and properties of few cast Mg-alloys [8, 11, 18, 22–24, 26]

Alloy	Process[a]	Density (g/cc)	Heat capacity (kJ/kg K)	CTE (μ/K)	Thermal conduct (W/mK)	Temper[a]	Tensile strength (MPa)	Yield strength (MPa)	Elongation (%)	Remarks
AE42 (Al: 4%, RE: 2.4%, Mn: 0.25%)	DC	1.79	1020	16.1	84	F	230	145	6	
AE44 (Al: 4%, RE: 4%)	HPDC	1.82	1020	25.2	85	F	245	142	10	
AJ62 (Al: 6%, Sr: 2.4%)	HPDC	1.8	1150	27.2	77	F	240	143	3	
AM100A (Al:10%, Mn: 0.1%)	SPC	1.83	1050	25	73	F	150	83	2	Aging: 232 °C/5 h; solution at 432 °C/16–24 h; age hardening: 232 °C/5 h
						T6	275	110	4	
AM20 (Al: 2%, Zn: 0.2%, Mn: 0.2%, Si: 0.05%)	DC	1.75	1020	26	94	F	210	90	20	
AM50A (Al: 5%, Mn: 0.35%)	DC	1.77	1020	26	65	F	230	125	8	
AM60A, B (Al: 6%, Mn: 0.3%)	DC	1.8	1000	25.6	67	F	220	130	6	
AS21 (Al: 2.2%, Si: 1%, Mn: 0.1%)	DC	1.78	1020	26.1	84	F	220	120	7	
AS41A & B (Al: 4.25%, Mn: 0.35%)	DC	1.78	1010	26.1	68	F	240	140	8	

(continued)

Table 1.3 (continued)

Alloy	Process[a]	Density (g/cc)	Heat capacity (kJ/kg K)	CTE (μ/K)	Thermal conduct (W/mK)	Temper[a]	Tensile strength (MPa)	Yield strength (MPa)	Elongation (%)	Remarks
AZ63A (Al: 6%, Zn: 3%, Si: 0.3, Mn: 0.1%)	SPC	1.83	–	26.1	77	F	200	97	6	Aging: 260 °C/4 h; solution at 391 °C/10–14 h; age hardening: 218 °C/5 h
						T4	275	97	12	
						T5	200	105	4	
						T6	275	130	5	
AZ81A (Al: 7.6%, Zn: 0.7%, Mn: 0.13%)	SPC	1.8	800	27.2	51.1	T4	275	85	15	Solution: 418 °C/16–24 h
AZ91A, C, C, D & E (Al: 9%, Zn: 0.7%, Mn: 0.3%)	DC	1.81	1050	26	72	F	185	150	3	Compositions A,B,D are typically DC while C, E are SPC processed
						T4	250	80	4.5	
						T6	230	130	2.3	
	SPC					F	165	95		Solution at 418 °C for 16–24 h; Age hardening at 168 °C for 16 h
						T4	275	85		
						T6	275	195		
EQ21A (Nd: 2.5%, Ag: 1.5%, Zr: 0.7%)	SPC	1.81	1000	26.7	113	T6	234	172	2	Solution at 530 °C for 4–8 h; age hardening at 200 °C for 16 h
EZ33A (Nd: 3.5%, Zr: 0.7%)	SPC	1.8	1040	26.4	99.5	T5	160	105	3	Aging: 175 °C for 16 h
HK31 (Th: 3%, Zr: 0.7%)	DC	1.8	1000	26.8	92					
QE22A (Ag: 2.5%, RE: 2%, Zr: 0.6%)	SPC	1.81	1000	26.7	113	T6	275	205	4	Solution at 530 °C for 4–8 h; age hardening at 204 °C for 8 h

WE43A (Y: 4.75%, Nd: 3%, Zr: 0.7%, Zn: 0.2%)	SPC	1.8	965	26.7	51.3	T6	235	190	4	Solution at 535 °C for 4–8 h; age hardening at 250 °C for 16 h
WE54A (Y:5.25%, RE: 1.75%, Nd: 1.75)	SPC	1.8	–	24.6	52	T6	270	190	4	Solution at 535 °C for 4–8 h; age hardening at 250 °C for 16 h
ZC63A (Zn: 6.5%, Cu: 3%, Mn: 0.75%)	SPC	1.87	960	26	122	T6	210	125	4	Solution at 445 °C for 4–8 h; age hardening at 200 °C for 16 h
ZE41A (Zn: 4.2%, RE: 1.2%, Zr: 0.7%)	SPC	1.84	–	15.1	124	T5	205	140	3.5	Aging: 329 °C for 2 h
ZE63A (Zn: 5.7%, RE: 2.5%, Zr: 0.7%)	SPC	1.87	960	–	109	T6	295	190	7	Solution at 491 °C for 10–72 h; age hardening at 141 °C for 48 h
ZK51A (Zn: 4.5%, Zr: 0.7%)	SPC	1.83	1000	–	110	T5	276	165	3	Aging: 177 °C for 12 h

[a] Process: *DC* die casting, *SC* sand casting, *SPC* sand and permanent moud casting

[b] Temper Conditions: *F* fabricated, *T4* solution heat treated and naturally aged, *T5* artificially aged, *T6* solution heat treated and artificially aged

bars. Due to the hexagonal close-packed (HCP) crystal structure of magnesium, processing wrought alloys through methods like extrusion, forging, and rolling requires temperatures above 300 °C to activate multiple slip systems. Common wrought Mg alloy compositions include AZ31, AZ61, AZ80, ZC71, and ZK60. Their general characteristics and mechanical properties with respect to wrought processing conditions are summarized in Table 1.4. Alloys such as AZ31, AZ61, and AZ80 have similar zinc content, resulting in almost identical densities and thermal properties. However, AZ80A, with its higher aluminum content, exhibits superior mechanical properties. In comparison, ZC71, which contains zinc and copper, surpasses AZ80A in strength but has lower ductility. ZK60A, with zinc and zirconium, shows superior mechanical properties [11, 23, 27, 29].

Recently, rare earth (RE) additions, which help in modifying texture and refining grains, have been incorporated to achieve high strength and formability. Research has focused on developing custom heat treatment profiles to create ultra-high-strength Mg-RE alloys with very fine long-period stacking ordered (LPSO) structures. These precipitates act as barriers to dislocation motion, enhancing the alloy's yield strength. However, these alloys typically require significant amounts of rare earth additions, which can be costly. One notable RE-containing magnesium alloy is the ZE series, which includes zinc and rare earth elements such as neodymium. The presence of 14-H LPSO phases in these alloys contributes to a combination of high strength, ductility, and corrosion resistance, making them suitable for structural applications in demanding environments [20, 25].

Another strategy to improve the formability of Mg is the addition of lithium (Li). Increased Li content reduces the axial ratio (c/a) of HCP-Mg, enhancing the activation of non-basal prismatic planes. Li additions exceeding 5.5 wt.% transform Mg-Li alloys into a duplex (HCP + BCC) structure, and beyond 11.5 wt.% Li, they transition entirely into a BCC structure. Despite the mechanical benefits, Mg-Li alloys exhibit poor corrosion behavior and flammability concerns. Ongoing research explores other alloying additions such as calcium (Ca), strontium (Sr), tin (Sn), silver (Ag), and yttrium (Y) to further develop wrought Mg alloys. Notably, Sr addition has shown remarkable similarity to REs in producing a two-component texture that enhances formability in Mg-Zn alloys [27, 30–34].

Building on these advancements, high-entropy alloys (HEAs) represent a new class of materials that contain five or more principal elements in near-equimolar ratios. These alloys are characterized by their complex microstructures, high strength, and excellent thermal stability. Recently, researchers have explored the potential of incorporating magnesium into high- and medium-entropy alloys to create lightweight structural materials with enhanced properties. In Mg-based HEAs, magnesium is combined with other lightweight elements such as lithium (Li), zinc (Zn), and aluminum (Al) to create alloys with exceptional strength-to-weight ratios. The high configurational entropy of these alloys leads to the formation of simple solid-solution phases, which contribute to their remarkable mechanical properties. Mg-based HEAs have shown promise for applications where lightweight materials with high strength and corrosion resistance are critical. Their unique

Table 1.4 Composition and properties of few wrought magnesium alloys [8, 11, 18, 22, 23, 27, 28]

Alloy name	Density (g/cc)	CTE (μ/K)	Specific heat capacity (J/kg*K)	Ther. conduc-tivity (W/m*K)	Form	Tensile strength (MPa)	Yield strength (MPa)	Ductility (%)	Comp. yield strength (MPa)	Hardness (HB500)
AZ31B (Al = 2.5–3.5%; Zn = 0.6–1.4%; Mn = 0.2% min)	1.77	26	1000	96	Extrusion	260 (F)	200 (F)	15 (F)	95 (F)	49 (F)
					Plates & Sheets	255 (T4)	150 (T4)	21 (T4) 15 (T6)	110 (T4)	73 (T6)
						290 (T6)	220 (T6)		180 (T6)	
					Forging	260 (F)	195 (F)	9 (F)	85 (F)	–
AZ61A (Al = 2.5–3.5%; Zn = 0.6–1.4%; Mn = 0.2% min)	1.77	26	1000	96	Extrusion	310 (F)	230 (F)	16 (F)	130 (F)	60 (F)
					Forging	195 (F)	180 (F)	12 (F)	115 (F)	–
AZ80A (Al = 7.8–9.2%; Zn = 0.2–0.8%; Mn = 0.12% min)	1.8	26	1050	76	Extrusion	340 (F)	250 (F)	11 (F)	140 (F)	82 (T5)
						380 (T5)	275 (T5)	7 (T5)	240 (T5)	
	1.8	26	1050	76	Forging	315 (F)	215 (F)	8 (F)	170 (F)	–
						345 (T5)	235 (T5)	6 (T5)	195 (T5)	
						345 (T6)	250 (T6)	5 (T6)	185 (T6)	
ZC71 (Zn = 6–7%; Cu= 1–1.5%; Mn = 0.5–1%)	1.83	26	–	122	Extrusion	360 (F)	340 (F)	5 (F)	–	80 (F)
ZK60A (Zn = 4.8–6.2%; Zr = 0.45% min)	1.83	26	1000	120	Extrusion	340 (F)	250 (F)	14 (F)	185 (F)	88 (T6)
						365 (T5)	305 (T5)	11 (T5)	250 (T5)	
	1.83	26	1000	120	Forging	305 (T5)	220 (T5)	16 (T5)	195 (T5)	–
						325 (T6)	270 (T6)	11 (T6)	170 (T6)	
HM31 (Th = 2.5–3.5%; Mn = 1.2% min)	1.8	26	1000	104	Extrusion	290 (F)	230 (F)	10 (F)	–	–
						300 (T5)	270 (T5)	10 (T5)		
HK31A (Th = 2.5–3.5%; Mn = 1.2% min)	1.8	26	1000	104	Plates & Sheets	255 (T6)	200 (T6)	9 (T6)	160 (T6)	68 (T6)

microstructures, combined with the benefits of magnesium, make them ideal candidates for next-generation structural materials in aerospace and defense [35, 36].

1.5 Applications of Magnesium

Magnesium alloys are renowned for their exceptional properties, making them highly suitable for a wide range of applications across various industries. These alloys are characterized by their lightweight nature, being approximately 33% lighter than aluminum and 75% lighter than steel, which significantly contributes to improved fuel efficiency and performance in automotive and aerospace sectors. Additionally, magnesium alloys exhibit high specific strength, excellent castability, and good damping capacity, making them ideal for structural and mechanical applications. Their biocompatibility and biodegradability further enhance their suitability for medical devices, while their excellent electromagnetic interference (EMI) shielding properties make them valuable in the electronics industry. The combination of these advantageous characteristics ensures that magnesium alloys are indispensable in driving innovation and efficiency across diverse domains [2, 6, 7, 22, 37–39].

In the automotive industry, Mg alloys are highly valued due to their structural lightweight capabilities to reduce the vehicle weight, which enhances fuel efficiency and performance. Key applications include engine and transmission parts such as engine blocks, intake manifolds, lower crankcases, transmission cases, gearboxes, and cylinder heads (Table 1.5). Interior parts like steering wheel frames, instrument panels, seat frames, and airbag retainers also benefit from magnesium's properties. Additionally, chassis components such as wheel rims, lift gates, valve covers, and ABS mounting brackets are commonly made from magnesium alloys. The use of magnesium in automobiles dates back to the 1930s, with Volkswagen and Porsche being early adopters. Modern applications focus on engine and transmission parts, interior components, and chassis elements, contributing to significant weight savings and improved fuel efficiency (Fig. 1.2).

Magnesium alloys are becoming increasingly significant in the aerospace industry due to their unique properties, including high specific strength, stiffness, excellent damping properties, electromagnetic shielding, and thermal conductivity. These characteristics make Mg alloys particularly valuable for manufacturing critical components in aircraft, missiles, and spacecraft, where reducing weight is crucial for enhancing performance and efficiency.

One of the primary advantages of Mg alloys in aerospace applications is their ability to significantly reduce the weight of components. This weight reduction leads to improved maneuverability of aircraft and lower launch costs for spacecraft. Common Mg alloys used in aerospace include AZ91E, QE22 (MSR), ZE41 (RZ5), EQ21 (ZRE1), and WE43. Each of these alloys has specific properties that make them suitable for various parts and operating conditions. For instance, WE43 is

Table 1.5 Mg alloys in automotive applications [7, 22, 39, 40]

Component	Car models
Engine and transmission parts	
Engine block	BMW (N52), Ford, Porsche (911)
Intake manifold	BMW (V8 motor), GM (V8 North Star motor), Chrysler
Lower crankcase	Chrysler (jeep), Alfa Romeo (GTV), GM (Oldsmobile), McLaren Motors (F1-V12)
Transmission case	AutoZAZ-Daewoo (Tavria, Slavuta, Daewoo-Sens), Volvo Motors (LCP), Porshe AG (911 series), Volkswagen (Volkswagen Passat), Audi (A4, A6), Mercedes-Benz, Ford (Bronco, Aerostar 1994)
Gear box and controls housing	AutoZAZ-Daewoo (Tavria, Slavuta, Daewoo-Sens)
Cylinder block	GM (Pontiac Gran AM, Corvette)
Cylinder head	Dodge (Dodge Raw), Honda Motors (City Turbo), Alfa Romeo (GTV), AutoZAZ-Daewoo (Tavria, Slavuta, Daewoo-Sens), Honda, BMW, Ford, Isusu, Volvo Motors (LCP), Chrysler
Camshaft drive chain case (AZ91D)	Porsche AG (911 series), Chrysler (jeep 1993, Viper)
Clutch case (AZ91B, AZ91D, AZ91HP)	Alfa Romeo (GTV), AutoZAZ-Daewoo (Tavria, Slavuta, Daewoo-Sens), Volvo Motors (LCP), Ford (Ranger), GM (Corvette)
Oil pan and pump housing	Ford (Ranger), Chrysler (jeep), Chrysler (Viper), McLaren Motors (F1-V12)
Engine cradle	Ford, GM (Corvette)
Brackets (brake pedal, clutch, and accelerator)	Ford, GM, Chrysler
Interior parts	
Steering wheel frame	Ford (Thunderbird, Cougar, Taurus, Sable, Ranger), Chrysler (Chrysler Plymouth, LH midsize 1993), Toyota, BMW (MINI), Lexus (Lexus LS430), GM (Oldsmobile, Pontiac, Buick)
Steering link bracing	GM (LH Midsize)
Instrument panel	GM, Chrysler (jeep), Ford, Audi (A8), Toyota (Toyota Century)
Seat frame	GM (Impact), Mercedes-Benz (Mercedes Roadster 300/400/500 SL), Lexus (Lexus LS430);
Brake and clutch pedal	GM Oldsmobile, Pontiac, Buick
Door inner	Aston Martin (DB9 2004)
Console bracket	Ford
Air bag retainer	Chrysler
Chassis components	
Wheel rims	Toyota (Toyota 2000GT, Toyota Supra), Alfa Romeo (GTV), Porshe AG (944 and 911)
Lift gate	Chrysler (Pacifica 2017), Lincoln (MKT luxury crossover 2010), Mercedes (E-Class T-Model 2009), Aston Martin (Vanquish S 2017)
Valve cover	GM (Corvette)
ABS mounting bracket	Ford, Chrysler

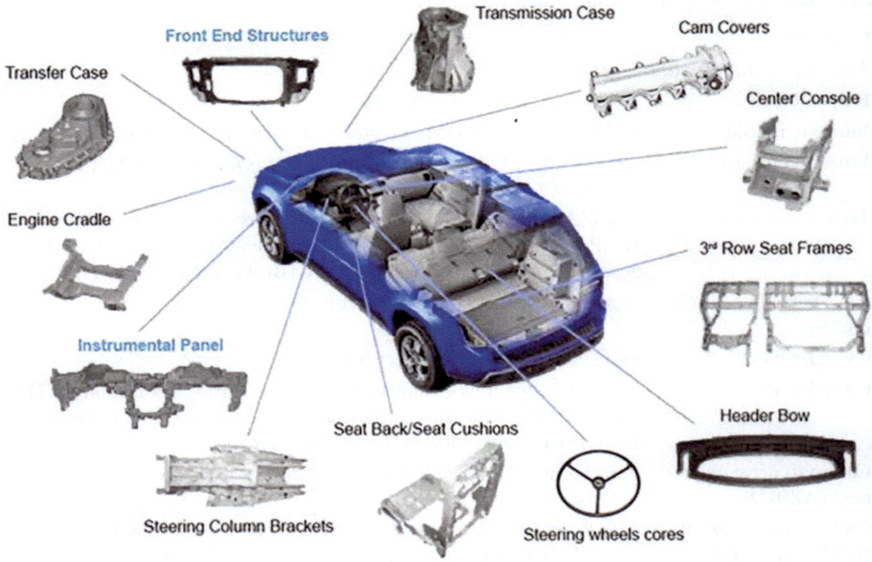

Fig. 1.2 Applications of Mg alloys in automobile industry [38]

known for its excellent corrosion resistance and is commonly used in manufacturing aircraft propeller housings [41].

Despite their advantages, Mg alloys face several challenges in aerospace applications. These include poor corrosion resistance, low strength at high temperatures, and difficulties in casting. To address these issues, ongoing research focuses on developing Mg-rare earth (Mg-RE) alloys, which offer better strength and high-temperature resistance. These advancements are crucial for overcoming the current limitations and enabling broader use of Mg alloys in aerospace.

The historical context of Mg alloys in aerospace dates back to the early twentieth century, when they were employed in military equipment such as aircraft, bombers, and missiles. However, the use of Mg alloys in aerospace applications declined in the 1970s and 1980s due to poor corrosion resistance. Now, there is a resurgence in interest driven by the strong needs toward lightweighting of aircrafts to achieve fuel energy savings. For example, Mg alloys are currently used in the making of aircraft structures, transmission systems, rocket and missile parts, and spacecraft shells (Fig. 1.3) as mentioned below.

- *Aircraft Components*: Magnesium alloys are used in structural components like aircraft frames, engine casings, and wheel assemblies. They contribute to significant weight reductions, improving fuel efficiency and performance. Common magnesium alloys used in aircraft manufacturing include AZ91E, QE22, ZE41, EQ21, and WE43. For example, WE43 is known for its excellent corrosion resistance and is used in aircraft propeller housings.

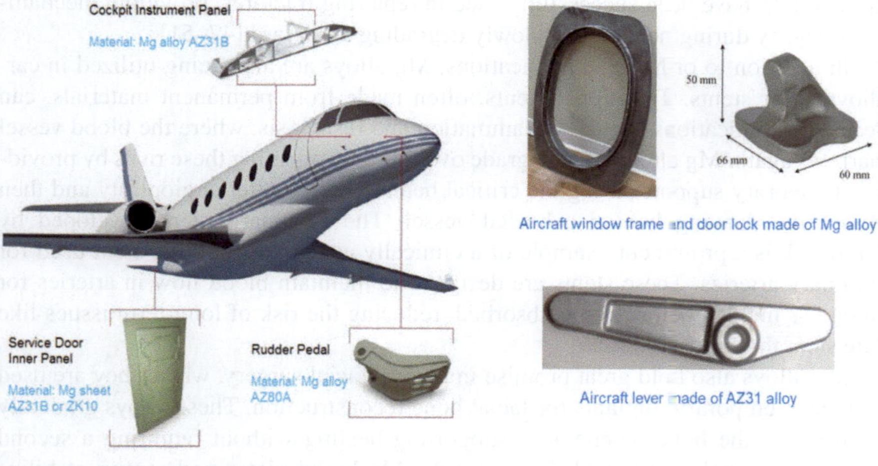

Fig. 1.3 Aerospace applications of Mg alloys [38]

- *Helicopter Transmission Systems*: High-strength heat-resistant magnesium alloys like EV21 and EV31 are used in helicopter gearbox housings. For example, the military aircrafts such as the Eurofighter Typhoon and F16 utilize RE containing Mg alloys for making transmission casings as they offer excellent performance under high heat and loading conditions and corrosion resistance.
- *Missiles and Spacecraft*: Magnesium alloys are employed in the production of essential mechanical equipment parts such as chassis, transfer boxes, and power supplies. They are also used in the manufacture of satellite components and rocket propellants. The Titan intercontinental ballistic missile, for instance, utilized about 900 kg of wrought magnesium alloy.

Magnesium (Mg) alloys are rapidly emerging as a key material in biomedical applications, particularly due to their biodegradable and biocompatible nature. Their use in medical devices is driven by the ability to degrade in the human body at a controlled rate, making them an excellent alternative to traditional metallic implants. These properties, combined with their mechanical strength and a density similar to human bone, position Mg alloys as a prime candidate for temporary medical implants [17, 28, 42–46].

One of the most significant areas of application for Mg alloys is in orthopedic implants. Traditional implants made from metals like titanium or stainless steel often require removal once the bone has healed, necessitating a second surgery (revision surgery). In contrast, Mg alloy implants gradually degrade, allowing the bone to heal and adapt to mechanical stresses, eliminating the need for additional surgeries. Screws, plates, and pins made from Mg alloys are widely tested for applications in bone fixation, particularly for fractures, where they provide the necessary support during the healing process. For instance, Mg alloy screws like those made from WE43 have been successfully used in repairing fractures, providing mechanical integrity during healing and slowly degrading afterward [47–51].

In addition to orthopedic applications, Mg alloys are also being utilized in cardiovascular stents. Traditional stents, often made from permanent materials, can lead to complications such as inflammation and restenosis, where the blood vessel narrows again. Mg alloy stents degrade over time, eliminating these risks by providing temporary support during the critical healing period after angioplasty and then dissolving, leaving behind a healed vessel. The Magmaris stent developed by Biotronik is a prominent example of a clinically approved Mg alloy stent used for coronary arteries. These stents are designed to maintain blood flow in arteries for about 12 months before being absorbed, reducing the risk of long-term issues like late stent thrombosis [52–56].

Mg alloys also hold great promise in maxillofacial surgery, where they are used to create temporary implants for facial bone reconstruction. These alloys gradually dissolve as the bone regenerates, supporting healing without requiring a second surgery for implant removal. For example, Mg-based plates and screws stabilize jawbones following fractures or surgeries, offering both mechanical support and gradual biodegradation as healing occurs [57, 58]. Similarly, dental applications are another area of interest, where Mg alloys could be used for temporary implants or scaffolds that degrade once the bone or gum tissue has sufficiently regenerated. Mg alloy-based scaffolds for periodontal regeneration are being developed to assist in the healing of damaged gum and bone structures, with the implants dissolving once the restoration is complete.

In tissue engineering, Mg alloys are gaining attention for their use in scaffolds designed to support the regeneration of tissues like bones and cartilage. These scaffolds offer the advantage of providing temporary support and then degrading as the new tissue forms. This eliminates the need for permanent foreign objects in the body and reduces the risk of long-term complications [59–61]. Researchers are also exploring the potential for 3D-printed Mg alloy scaffolds, which can be customized for specific injuries or defects and dissolve once they have fulfilled their purpose [62, 63].

Mg alloys are also being explored for use in drug delivery systems. In these applications, Mg alloys serve as carriers for drugs, gradually releasing therapeutic agents as they degrade. This is particularly useful in localized treatments, such as cancer therapy, where drugs can be delivered directly to the tumor site, minimizing systemic side effects and improving treatment efficacy. For instance, drug-eluting Mg alloy stents have shown promise in both supporting blood vessels and delivering medication to prevent restenosis. Additionally, Mg alloys have potential applications in wound healing devices such as biodegradable staples or clips. These devices degrade as the wound heals, eliminating the need for removal and reducing the risk of infection or scarring [64, 65].

Few examples corresponding to the biomedical applications of Mg alloys are highlighted in Fig. 1.4.

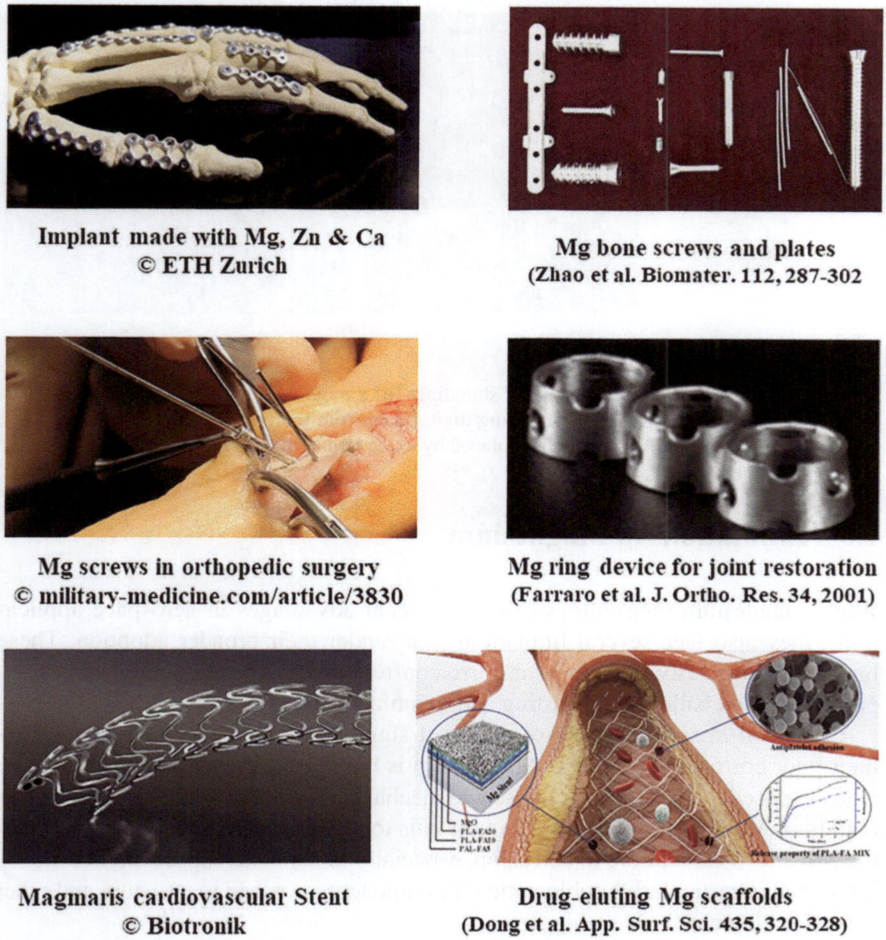

Fig. 1.4 Magnesium alloys for biomedical applications [51, 56, 63, 66]

In the electronics industry, magnesium alloys' lightweight nature and excellent electromagnetic interference (EMI) shielding properties are highly valued. Applications include consumer electronics like smartphone cases, laptop frames, and camera bodies, as well as enclosures for electronic equipment to protect against EMI [67–69]. Magnesium alloys provide durability and heat dissipation, making them suitable for portable electronic devices. The sports industry also utilizes magnesium alloys to manufacture lightweight and high-performance equipment. Applications include bicycle frames and components, tennis racquets, golf clubs, wheelchairs, and snowboards. Magnesium alloys improve user experience by providing smoother and more comfortable rides and better performance in sports activities (Fig. 1.5) [71].

(a) (b) (c)

Fig. 1.5 Cages used for electromagnetic shielding: (**a**) a mobile phone, (**b**) a power plant, (**c**) a metallic fabric flexible pouch for protecting digital data from theft and to improve privacy [70]. Currently used aluminum can easily be replaced by much lighter magnesium especially in portable electronic gadgets

1.6 Limitations of Magnesium

While magnesium (Mg) alloys offer significant advantages in aerospace applications, they also face several limitations that hinder their broader adoption. These limitations primarily include poor corrosion resistance, low strength at high temperatures, and challenges in casting and manufacturing processes.

Corrosion Resistance: One of the most significant drawbacks of Mg alloys is their poor corrosion resistance. Magnesium is highly reactive and tends to oxidize when exposed to air, forming a loose and mechanically weak surface film composed mainly of MgO and $Mg(OH)_2$. This film fails to provide sustained protection, making Mg alloys susceptible to corrosion, especially in harsh aerospace environments. This issue is particularly problematic for components exposed to moisture and other corrosive elements.

High-Temperature Performance: Mg alloys generally exhibit low strength at high temperatures, which limits their use in critical aerospace components that must withstand extreme conditions. As temperature increases, the strength, hardness, and high-temperature creep performance of Mg alloys significantly deteriorate. Conventional Mg alloys typically function well only below 120 °C, whereas aerospace applications often require materials that can perform reliably at much higher temperatures.

Casting and Manufacturing Challenges: Mg alloys are prone to casting defects such as porosity and hot cracking, which result in low yield rates and inconsistent quality. These issues arise due to the difficulties in controlling the plastic deformation conditions during casting, leading to unstable microstructures and mechanical properties. Additionally, the low ignition point of Mg alloys poses a fire hazard during manufacturing processes, requiring stringent safety measures.

Mechanical Properties: Compared to other aerospace materials like aluminum alloys, Mg alloys have relatively low material strength, especially at high temperatures. This limitation affects their ability to bear heavy loads and endure the mechanical stresses encountered in aerospace applications. Furthermore, Mg alloys exhibit poor creep resistance, which is the tendency of a material to deform permanently under constant stress at high temperatures.

Surface Treatment Requirements: To improve the corrosion resistance and mechanical properties of Mg alloys, various surface treatments such as micro-arc oxidation, electrochemical deposition, and physical vapor deposition are necessary. These treatments add complexity and cost to the manufacturing process, making Mg alloys less economically viable compared to other materials.

1.7 Future Outlook

Magnesium (Mg) alloys are increasingly recognized for their potential to revolutionize various industries due to their unique combination of properties, including high specific strength, lightweight characteristics, excellent damping properties, and superior thermal and electrical conductivity. As advancements in material science and engineering continue to evolve, Mg alloys are finding applications in diverse fields such as aerospace, automotive, biomedical, electronics, defense, and sports. This future outlook explores the ongoing innovations and research that are paving the way for broader adoption of Mg alloys, highlighting their transformative impact on next-generation technologies and sustainable development. By addressing current challenges and leveraging new opportunities, Mg alloys are set to play a crucial role in creating lighter, more efficient, and environmentally friendly solutions across multiple sectors.

1. *Advancements in Alloy Composition*: Ongoing research is focused on developing new Mg alloys with enhanced properties. The introduction of rare earth elements (REEs) such as yttrium (Y), gadolinium (Gd), and scandium (Sc) has significantly improved the high-temperature performance, strength, and creep resistance of Mg alloys. These advancements are crucial for expanding the use of Mg alloys in applications that require materials to withstand extreme conditions.
2. *Improved Manufacturing Processes*: Innovations in manufacturing processes, such as asymmetric extrusion, semi-solid casting, and low-pressure casting, are enhancing the quality and performance of Mg alloy components. These processes help overcome traditional challenges like casting defects and poor mechanical properties, making Mg alloys more viable for a wide range of applications.
3. *Enhanced Surface Treatments*: Developments in surface treatment technologies, including micro-arc oxidation, electrochemical deposition, and physical vapor deposition, are addressing the corrosion resistance issues of Mg alloys. These

treatments improve the durability and reliability of Mg alloy components, making them suitable for use in harsh environments.

4. *Sustainable Development and Recycling*: The recycling of Mg alloy scrap is becoming increasingly important for sustainable development. Advanced recycling methods, such as melt refining, solid-phase regeneration, and composite purification, are being developed to recover high-quality Mg alloys from scrap. This not only reduces environmental impact but also enhances the economic viability of Mg alloys.
5. *Potential of Magnesium Matrix Composites*: Magnesium matrix composites, reinforced with materials like silicon carbide (SiC), are showing significant potential for various applications. These composites offer improved strength and elastic modulus, making them ideal for lightweight structural components in diverse fields.
6. *Functional Magnesium Materials*: Research into functional Mg materials, such as hydrogen storage materials and Mg battery materials, is opening new avenues for their application. These materials offer advantages over traditional options and are expected to become key innovation points in the future development of technologies across multiple sectors.
7. *Balancing Strength and Ductility*: The development of high-ductility Mg alloys, guided by the theory of Solid Solution Strengthening and Ductilizing (SSSD), is crucial for applications that require a balance between strength and ductility. Achieving this balance will enhance the performance and safety of Mg alloy components in various uses.
8. *Other Challenges*: Despite the promising advancements, several challenges still remain, such as improving the high-temperature performance, corrosion resistance, and manufacturability of Mg alloys. Continued research and development efforts are essential to overcome these barriers and unlock the full potential of Mg alloys in diverse applications.

1.8 Summary

Magnesium is a lightweight, versatile metal with a rich history and broad industrial significance. Its early use in medicine and subsequent isolation in the eighteenth and nineteenth centuries paved the way for its applications in aerospace, automotive, and military industries. Magnesium's low density, good thermal conductivity, and strength-to-weight ratio make it invaluable for reducing weight in engineering applications. However, its reactivity and flammability pose challenges, necessitating careful handling and surface treatments.

The development of magnesium alloys has significantly expanded its utility. Alloying with elements like aluminum, zinc, and rare earths enhances its strength, corrosion resistance, and high-temperature performance. Magnesium alloys are integral to industries such as automotive (engine blocks, chassis components), aerospace (aircraft frames, helicopter gearboxes), and medicine (biodegradable implants like screws and stents). They also feature prominently in electronics, offering lightweight and EMI-shielding solutions for devices. Despite these benefits, magnesium's susceptibility to corrosion, low strength at high temperatures, and manufacturing defects remain significant limitations.

Looking ahead, advancements in alloy composition, development of composites (using different length scales of ceramic, metallic, hybrid, metastable and hollow reinforcements), microstructure and properties tailoring using innovative heat treatments and cryogenic treatments, surface treatments, and manufacturing technologies are addressing these challenges. Recycling initiatives and eco-friendly production methods further position magnesium as a sustainable material for the future. With ongoing research into applications such as hydrogen storage, batteries, and high-performance composites, magnesium is set to play a pivotal role in next-generation innovations across multiple industries.

References

1. E.F. Emley, *Principles of Magnesium Technology* (Pergamon Press, 1966). https://doi.org/10.1029/2005JF000394
2. H.E. Friedrich, B.L. Mordike, Technology of magnesium and magnesium alloys, in *Magnesium Technology Metallurgy Design Data Application*, (Springer, 2006), pp. 219–430
3. R. Holmes, Humphry davy and the chemical moment. Clin. Chem. **57**(11), 1625 (2011)
4. F. Czerwinski, Oxidation characteristics of magnesium alloys. JOM **64**, 1477–1483 (2012). https://doi.org/10.1007/s11837-012-0477-z
5. K. Flint, *Flash!: Photography, Writing, and Surprising Illumination* (Oxford University Press, 2017)
6. S.N. Mathaudhu, E.A. Nyberg, Magnesium alloys in US military applications: past, current and future solutions, in *Essential Readings in Magnesium Technology*, (Springer, 2016), pp. 71–76
7. D.S. Kumar, C.T. Sasanka, K. Ravindra, K.N.S. Suman, Magnesium and its alloys in automotive applications—a review. Am. J. Mater. Sci. Technol. **4**(1), 12–30 (2015). https://doi.org/10.7726/ajmst.2015.1002
8. M. Gupta, N.M.L. Sharon, *Magnesium, Magnesium Alloys, and Magnesium Composites* (Wiley, 2010). https://doi.org/10.1002/9780470905098
9. T.U. B., K.U. Kainer, *Magnesium Alloys & Technologies* (Wiley, 2003). https://doi.org/10.1002/3527602046
10. J.F. King, Technology of magnesium and magnesium alloys, in *Magnesium Technology: Metallurgy, Design Data, Applications*, (Springer, Berlin, 2006), pp. 219–430. https://doi.org/10.1007/3-540-30812-1_6
11. C. Moosbrugger, *Engineering Properties of Magnesium Alloys* (ASM International, 2017). https://doi.org/10.1002/9780470905098.ch1
12. H. Watarai, Trend of research and development for magnesium alloys—reducing the weight of structural materials in motor vehicles. Sci. Technol. Trends **18**, 84–97 (2004)
13. A.A. Luo, Recent magnesium alloy development for elevated temperature applications. Int. Mater. Rev. **49**(1), 13–30 (2004). https://doi.org/10.1179/095066004225010497
14. H. Hu, A. Yu, N. Li, J.E. Allison, Potential magnesium alloys for high temperature die cast automotive applications: a review. Mater. Manuf. Process. **18**(5), 687–717 (2003). https://doi.org/10.1081/AMP-120024970

15. F. Pan, M. Yang, X. Chen, A review on casting magnesium alloys: modification of commercial alloys and development of new alloys. J. Mater. Sci. Technol. **32**(12), 1211–1221 (2016). https://doi.org/10.1016/j.jmst.2016.07.001
16. R. Radha, D. Sreekanth, Insight of magnesium alloys and composites for orthopedic implant applications—a review. J. Magnes. Alloys **5**(3), 286–312 (2017). https://doi.org/10.1016/j.jma.2017.08.003
17. N. Sezer, Z. Evis, S.M. Kayhan, A. Tahmasebifar, M. Koç, Review of magnesium-based biomaterials and their applications. J. Magnes. Alloys **6**(1), 23–43 (2018). https://doi.org/10.1016/j.jma.2018.02.003
18. M.M. Avedesian, H. Baker, *ASM Specialty Handbook: Magnesium and Magnesium Alloys* (ASM International, Materials Park, OH, 1999)
19. M. Gupta, S. Seetharaman, Magnesium based nanocomposites for cleaner transport, in *Nanotechnology for Energy Sustainability*, (Wiley, 2017). https://doi.org/10.1002/9783527696109.ch33
20. C. Bettles, M. Barnett, *Advances in Wrought Magnesium Alloys: Fundamentals of Processing, Prop Erties and Applications* (Elsevier, 2012). https://doi.org/10.1533/9780857093844
21. A.A. Luo, Magnesium casting technology for structural applications. J. Magnes. Alloys **1**(1), 2–22 (2013). https://doi.org/10.1016/j.jma.2013.02.002
22. M.K. Kulekci, Magnesium and its alloys applications in automotive industry. Int. J. Adv. Manuf. Technol. **39**(9–10), 851–865 (2008). https://doi.org/10.1007/s00170-007-1279-2
23. K.U. Kainer, F. Buch, The current state of technology and potential for further development of magnesium applications, in *Magnesium—Alloys and Technology*, (Wiley, 2003), pp. 1–22. https://doi.org/10.1002/3527602046.ch1
24. F. Von Buch, B. Mordike, High-temperature properties of magnesium alloys, in *Magnesium—Alloys Technology*, (Wiley, Hoboken, 2004), pp. 106–129
25. S. You, Y. Huang, K.U. Kainer, N. Hort, Recent research and developments on wrought magnesium alloys. J. Magnes. Alloys **5**(3), 239–253 (2017). https://doi.org/10.1016/j.jma.2017.09.001
26. J. Song, J. She, D. Chen, F. Pan, Latest research advances on magnesium and magnesium alloys worldwide. J. Magnes. Alloys **8**(1), 1–41 (2020). https://doi.org/10.1016/j.jma.2020.02.003
27. J. Becker, G. Fischer, Manufacturing and potential of extruded and forged magnesium products, in *Magnesium—Alloys and Technology*, (Wiley, 2003), pp. 90–105. https://doi.org/10.1002/3527602046.ch6
28. Z. Yang, J.P. Li, J.X. Zhang, G.W. Lorimer, J. Robson, Review on research and development of magnesium alloys. Acta Metall. Sin. Engl. Lett. **21**(5), 313–328 (2008). https://doi.org/10.1016/S1006-7191(08)60054-X
29. G.E. Dieter, D. Bacon, *Mechanical Metallurgy*, SI Metric Edition, vol 273 (McGraw Hill, 1988). https://doi.org/10.1016/S0016-0032(62)91145-6
30. A. Białobrzeski, K. Saja, K. Hubner, Ultralight magnesium-lithium alloys. Arch. Foundry Eng. **7**(3), 11–16 (2007)
31. R. Wu et al., Recent progress in magnesium-lithium alloys. Int. Mater. Rev. **60**(2), 65–100 (2015). https://doi.org/10.1179/1743280414Y.0000000044
32. C.O. Muga, Z.W. Zhang, Strengthening mechanisms of magnesium-lithium based alloys and composites. Adv. Mater. Sci. Eng. **2016**(12), 1–11 (2016). https://doi.org/10.1155/2016/1078187
33. A. Sadeghi, M. Hoseini, M. Pekguleryuz, Effect of Sr addition on texture evolution of Mg-3Al-1Zn (AZ31) alloy during extrusion. Mater. Sci. Eng. A **528**(7–8), 3096–3104 (2011). https://doi.org/10.1016/j.msea.2010.12.091
34. A. Sadeghi, M. Pekguleryuz, Recrystallization and texture evolution of Mg-3%Al-1%Zn-(0.4-0.8)%Sr a lloys during extrusion. Mater. Sci. Eng. A **528**(3), 1678–1685 (2011). https://doi.org/10.1016/j.msea.2010.10.096
35. A. Babenko et al., Comprehensive insights into recent innovations: magnesium-inclusive high-entropy alloys. J. Magnes. Alloys **12**(4), 1311–1345 (2024)

36. K.S. Tun, V. Charadva, M. Gupta, Lightweight medium entropy magnesium alloy with exceptional compressive strength and ductility combination. J. Mater. Eng. Perform. **30**, 2422–2432 (2021)
37. A. Tharumarajah, P. Koltun, Is there an environmental advantage of using magnesium components for light-weighting cars? J. Clean. Prod. **15**(11–12), 1007–1013 (2007). https://doi.org/10.1016/j.jclepro.2006.05.022
38. A. Dziubińska, A. Gontarz, M. Dziubiński, M. Barszcz, The forming of magnesium alloy forgings for aircraft and automotive applications. Adv. Sci. Technol. Res. J. **10**(31), 158–168 (2016). https://doi.org/10.12913/22998624/64003
39. L. Pawlica, L. Cizek, M. Greger, J. Pavelcová, T. Taski, Mechanical properties of the model casting magnesium alloy AZ 91, in *11th International Scientific Conference: Achievements in Mechanical a nd Materials Engineering*, (Polish Academy of Sciences, 2002), pp. 47–50
40. J. Subramanian, S. Seetharaman, M. Gupta, Processing and properties of aluminum and magnesium based composites C ontaining amorphous reinforcement: a review. Metals **5**(2), 743–762 (2015). https://doi.org/10.3390/met5020743
41. H. Furuya, N. Kogiso, S. Matunaga, K. Senda, Applications of magnesium alloys for aerospace structure systems. Mater. Sci. Forum **5**(2), 189–201 (2000). https://doi.org/10.4028/www.scientific.net/msf.350-351.341
42. M. Haghshenas, Mechanical characteristics of biodegradable magnesium matrix composite s: a review. J. Magnes. Alloys **5**, 189–201 (2017). https://doi.org/10.1016/j.jma.2017.05.001
43. M. Razavi, M.H. Fathi, M. Meratian, Bio-corrosion behavior of magnesium-fluorapatite nanocomposite for bio medical applications. Mater. Lett. **64**(22), 2487–2490 (2010). https://doi.org/10.1016/j.matlet.2010.07.079
44. J. Dong et al., Advances in degradation behavior of biomedical magnesium alloys: a review. J. Alloys Compd. **908**, 164600 (2022)
45. N. Li, Y. Zheng, Novel magnesium alloys developed for biomedical application: a review. J. Mater. Sci. Technol. **29**(6), 489–502 (2013)
46. V.V. Ramalingam, P. Ramasamy, M.D. Kovukkal, G. Myilsamy, Research and development in magnesium alloys for industrial and biomedical applications: a review. Met. Mater. Int. **26**, 409–430 (2020)
47. S. Agarwal, J. Curtin, B. Duffy, S. Jaiswal, Biodegradable magnesium alloys for orthopaedic applications: a review on corrosion, biocompatibility and surface modifications. Mater. Sci. Eng. C **68**, 948–963 (2016)
48. S. Kamrani, C. Fleck, Biodegradable magnesium alloys as temporary orthopaedic implants: a review. Biometals **32**, 185–193 (2019)
49. Y. Lu, S. Deshmukh, I. Jones, Y.-L. Chiu, Biodegradable magnesium alloys for orthopaedic applications. Biomater. Transl. **2**(3), 214 (2021)
50. G.E.J. Poinern, S. Brundavanam, D. Fawcett, Biomedical magnesium alloys: a review of material properties, surface modifications and potential as a biodegradable orthopaedic implant. Am. J. Biomed. Eng. **2**(6), 218–240 (2012)
51. NanoMAG, Craniomaxillofacial applications for bioabsorable magnesium. https://www.nanomag.us/applications.html. Accessed 10 Apr 2020
52. B. Heublein, R. Rohde, V. Kaese, M. Niemeyer, W. Hartung, A. Haverich, Biocorrosion of magnesium alloys: a new principle in cardiovascular implant technology? Heart **89**(6), 651–656 (2003)
53. Z.-Q. Zhang, Y.-X. Yang, J.-A. Li, R.-C. Zeng, S.-K. Guan, Advances in coatings on magnesium alloys for cardiovascular stents—a review. Bioact. Mater. **6**(12), 4729–4757 (2021)
54. N. Zhao, D. Zhu, Application of Mg-based alloys for cardiovascular stents. Int. J. Biomed. Eng. Technol. **12**(4), 382–398 (2013)
55. L. Mao et al., A promising biodegradable magnesium alloy suitable for clinical vascular stent application. Sci. Rep. **7**(1), 46343 (2017)

56. Freesolve® Resorbable coronary scaffolds, Biotronik Resorbable Magnesium Scaffold (RMS), https://www.biotronik.com/en-int/products/vascular-intervention/coronary-vascular-intervention/freesolve
57. O. Charyeva et al., Histological comparison of new biodegradable magnesium-based implants for maxillofacial applications. J. Maxillofac. Oral Surg. **14**, 637–645 (2015)
58. S. Vujović, J. Desnica, D. Stanišić, I. Ognjanović, M. Stevanovic, G. Rosic, Applications of biodegradable magnesium-based materials in reconstructive oral and maxillofacial surgery: a review. Molecules **27**(17), 5529 (2022)
59. S. Dutta, M. Roy, Recent developments in engineered magnesium scaffolds for bone tissue engineering. ACS Biomater Sci. Eng. **9**(6), 3010–3031 (2023)
60. D. Persaud-Sharma, A. McGoron, Biodegradable magnesium alloys: a review of material development and applications. J. Biomim. Biomater. Tissue Eng. **12**, 25–39 (2012)
61. M. Yazdimamaghani, M. Razavi, D. Vashaee, K. Moharamzadeh, A.R. Boccaccini, L. Tayebi, Porous magnesium-based scaffolds for tissue engineering. Mater. Sci. Eng. C **71**, 1253–1266 (2017)
62. A. Lulian, Additive manufactured magnesium-based scaffolds for tissue engineering. Materials **15**(23), 8693 (2022)
63. Fraunhofer Institute of Laser Technology, SLM adapted to work with soft alloys and materials. https://optics.org/news/7/3/8. Accessed 1 May 2020
64. H.R. Bakhsheshi-Rad, Z. Hadisi, E. Hamzah, A.F. Ismail, M. Aziz, M. Kashefian, Drug delivery and cytocompatibility of ciprofloxacin loaded gelatin nanofibers-coated Mg alloy. Mater. Lett. **207**, 179–182 (2017). https://doi.org/10.1016/j.matlet.2017.07.072
65. L.-Y. Li et al., Advances in functionalized polymer coatings on biodegradable magnesium alloys—a review. Acta Biomater. **79**, 23–36 (2018). https://doi.org/10.1016/j.actbio.2018.08.030
66. A. Caggiano, Magnesium alloys for implants. https://www.iom3.org/materials-world-magazine/news/2019/nov/26/magnesium-alloys-implants. Accessed 1 May 2020
67. B. Landkof, Magnesium applications in aerospace and electronic industries, in *Magnesium Alloys and Their Applications*, (Wiley, 2000), pp. 168–172. https://doi.org/10.1002/3527607552.ch28
68. Q.-D. Nguyen, C.-G. Choi, Recent advances in multifunctional electromagnetic interference shielding materials. Heliyon **10**(10), e31118 (2024). https://doi.org/10.1016/j.heliyon.2024.e31118
69. 张志华; Zhang Zhi-hua, 潘复生; Pan Fu-Sheng, 陈先华; Chen Xian-hua, 刘娟; Liu Juan, 镁及其合金的电磁屏蔽性能研究 Electromagnetic shielding properties of magnesium and magnesium alloys. 材料工程, Mater Eng. **1**, 52–57 (2013). https://doi.org/10.3969/j.issn.1001-4381.2013.01.011
70. S. Gupta, M. Gupta, The promise of magnesium based materials in electromagnetic shielding. Juniper Online J. Mater. Sci. **2**(5), 1–6 (2027)
71. J. Tan, S. Ramakrishna, Applications of magnesium and its alloys: a review. Appl. Sci. **11**(15), 6861 (2021). https://doi.org/10.3390/app11156861

Chapter 2
Metal Matrix Composites

Metal matrix composites (MMCs) are advanced materials consisting of a metallic matrix reinforced with ceramic, metallic, or metastable phases, offering superior mechanical properties, wear resistance, thermal stability, and damping characteristics. Common matrix materials include lightweight metals like aluminum, magnesium, and titanium for structural applications, and steel and nickel-based superalloys for high-temperature environments. Reinforcements, such as continuous fibers, particulates, and nanoscale materials, enhance properties like strength, stiffness, and thermal stability. MMCs are categorized into laminate, fiber-reinforced, and particle-reinforced composites, each optimized for specific applications. Innovations in unconventional reinforcements, such as hollow particles for weight reduction and metastable phases for unique mechanical properties, further broaden their scope. The properties of MMCs are influenced by the type, size, and distribution of reinforcements as well as the matrix-reinforcement interface. Strengthening mechanisms, including load transfer, thermal mismatch, dislocation density increases, and grain refinement, contribute to their exceptional performance. These composites are extensively used in aerospace, automotive, electronics, and sports industries, where high strength-to-weight ratios, thermal stability, and wear resistance are critical. This review highlights their classification, properties, and key strengthening mechanisms, providing a comprehensive overview of their potential in modern engineering applications.

2.1 Introduction

Metal matrix composites (MMCs) are advanced materials composed of a continuous metallic matrix reinforced typically with ceramic or metallic phases. These composites offer enhanced mechanical properties, wear resistance, creep resistance, damping characteristics, and reduced thermal expansion. MMCs are suitable for a wide range of applications, including automotive, aerospace, consumer electronics,

S. Seetharaman et al., *Magnesium-based Syntactic, Metastable and Nano-composites*, https://doi.org/10.1007/978-3-031-96514-2_2

packaging, and sports industries. While the common matrix materials include lightweight metals such as aluminum, magnesium, and titanium, steel, cobalt, nickel and cobalt-nickel alloy matrices are also employed for high-temperature applications (Fig. 2.1) [1–8].

The matrix material in MMCs serves as the primary load-bearing component, while the reinforcement materials provide additional strength, stiffness, and targeted properties such as wear resistance and thermal stability, owing to their mechanical stability, chemical compatibility, high strength, hardness, and Young's modulus. Based on the size and shape, reinforcements can be broadly classified into continuous and discontinuous reinforcement forms. The most common continuous reinforcement are the carbon or ceramic fibers. As these fibers are brittle and flaw sensitive, they are often provided protective coatings to avoid any unwanted chemical reactions and to improve the bonding/wetting characteristics. The fiber reinforcements are also known to exhibit size effects (i.e., the strength of these fibers decreases as the length increases), and they can be further classified into long or short fibers. Similarly, discontinuous reinforcement includes ceramic and metallic particulates ranging from a few nanometer to several hundred micrometers, which impart isotropic properties to MMCs. Table 2.1 lists the properties of few matrix materials and ceramic reinforcements.

In general, the properties of MMCs are determined by various factors, including the morphology, type, length scale and properties of the reinforcement, the type of matrix, and the matrix/reinforcement interface. The choice of matrix and reinforcement materials, as well as their volume fractions and distribution, play a crucial role in defining the overall performance of the composite. The interaction between the matrix and the reinforcement, including the bonding at the interface, also plays a crucial role in determining the overall properties of the MMC. Besides, the properties of composite materials for a given composition can further be tailored or enhanced using judicious selection of primary and secondary processing techniques and heat/cryogenic treatments.

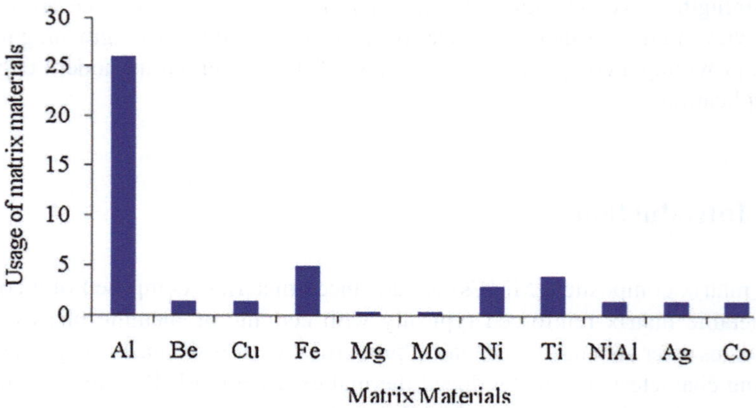

Fig. 2.1 Matrix materials used to develop MMCs [1]

Table 2.1 Properties of matrix materials and selected ceramic reinforcements [8, 9]

Metal	Crystal structure	Density (g/ cm^3)	Melting point (°C)	Thermal conduc-tivity (W/m.K)	Thermal expansion coefficient (µm/m K)	Mohr hardness (Hv)	Modulus (GPa)	Yield strength (MPa)	Ductility (%)
Matrix materials									
Al alloys	FCC	2.5–2.9	580–660	150–235	23–24	2.5–3	68–82	30–500	1–44
Mg alloys	HCP	1.74–1.95	650–700	96–156	25–26	2.5	42–47	70–400	3.5–18
Ti alloys	HCP/BCC	4.4–4.8	< 1670	15–22	8.6–9.8	6	90–120	250–1250	1–40
Ferrous alloys	BCC/FCC	7.1–8.0	< 1538	50–80	11–13	4–5	187–215	240–690	18–31
Cu alloys	FCC	8.93–8.94	1084	330–401	16.5	3	112–148	30–500	3–50
Ni Superalloys	FCC	7.75–8.65	< 1450	11–30	12–16	5–6	150–245	300–1900	0.5–60
Reinforcements									
Al2O3	Hexagonal	3.9	2050	25	8.3	6.5	410	–	–
AlN	Hexagonal	3.25	2300	10	6	–	350	–	–
B4C	Rhombohedral	2.52	2450	29	5–6	9.5	450	–	–
BN	Hexagonal	2.2	3000	25	3.8	1–2	90	–	–
SiC	Hexagonal	3.21	2300	59	4.7–5	9.7	480	–	–
Si3N4	α-Trigonal/β-Hex./γ-cubic	3.29	1900	29	3.3	–	310	–	–
TiB2	Hexagonal	4.5	2900	27	7.4	–	370	–	–
TiC	Cubic	4.93	3140	29	7.4	–	320	–	–
TiN	Cubic	5.24	2950	29	9.4	–	600	–	–
WC	Hexagonal	15.7	2800	110	5.2	9.5	690	–	–

2.2 Classification of Metal Matrix Composites

MMCs can be classified based on the type and form of reinforcement used. The main categories include laminate composites, fiber-reinforced composites, and particle-reinforced composites (Fig. 2.2). They are widely used in applications requiring high strength-to-weight ratios, such as aerospace structures and sporting goods [8, 10–12].

Laminate Composites Laminate composites consist of multiple layers of sheet laminates made of matrix and reinforcement materials. These layers are stacked and bonded in specific patterns to achieve desired strength and properties. Based on the stacking sequence, laminate composites can be classified into:

(a) *Angle-Ply Laminates*: Layers are oriented at specific angles relative to each other.
(b) *Cross-Ply Laminates*: Layers are oriented perpendicular to each other.
(c) *Symmetric Laminates*: Layers are symmetrically arranged about the mid-plane.
(d) *Antisymmetric Laminates*: Layers are arranged in an antisymmetric pattern about the mid-plane.
(e) *Balanced Laminates*: Layers are arranged to balance the in-plane stresses.

Fiber-Reinforced MMCs Fiber-reinforced MMCs can be broadly classified into long-fiber and short-fiber-reinforced composites [13–17].

(a) *Long-Fiber-Reinforced Composites*: These consist of a matrix reinforced by continuous fibers. The fibers provide high strength and stiffness to the composite. Long-fiber composites are typically used in applications requiring high load-bearing capacity and structural integrity. Examples include carbon fiber-reinforced aluminum and silicon carbide fiber-reinforced titanium.

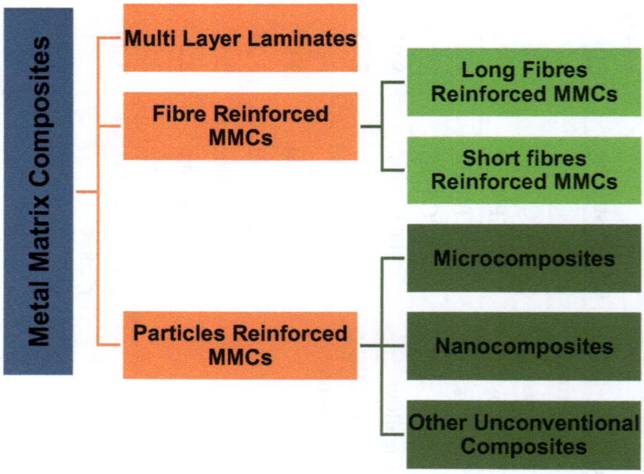

Fig. 2.2 Classification of metal matrix composites

(b) *Short-Fiber-Reinforced Composites*: These consist of a matrix reinforced by discontinuous fibers or whiskers (length < 100*diameter). Short-fiber composites offer improved mechanical properties and are easier to process compared to long-fiber composites. They are used in applications where isotropic properties and ease of manufacturing are important. Examples include aluminum reinforced with short alumina fibers.

Fiber reinforcements can be further classified based on their material composition, such as carbon fibers, ceramic fibers (e.g., silicon carbide, alumina), and metallic fibers (e.g., steel, tungsten).

Particle-Reinforced MMCs Particle-reinforced MMCs consist of a matrix reinforced by particles. The particle dispersion within the matrix can be either random or oriented [3, 18]. These composites can be further classified based on the size of the particulate reinforcement:

(a) *Micron-Sized Reinforcement*: Micron-sized particles are typically in the range of a few micrometers to hundreds of micrometers. They provide significant improvements in mechanical properties such as elastic modulus and wear resistance. Commonly used micron-sized reinforcements include silicon carbide (SiC), aluminum oxide (Al_2O_3), and boron carbide (B_4C) [19].
(b) *Nano-Sized Reinforcement*: When the dimensions of the particulate reinforcement are less than 100 nm, the resulting composite is referred to as a nanocomposite. Nanoscale reinforcements contribute significantly to the strengthening of the matrix material without adversely affecting its ductility. Examples include carbon nanotubes (CNTs) and graphene platelets. Nanocomposites are receiving significant attention due to their potential to achieve superior mechanical properties at lower reinforcement volume fractions. For example, nanocomposites with outstanding properties are obtained through the effective dispersion of hard phases such as multiwalled CNTs (with strengths up to 150 GPa) or nanoscale graphene platelets.
(c) *Other Unconventional Composites:* In addition to the conventional classifications, there are other unconventional composites that utilize unique reinforcement materials which include:

   (i) *Metastable or Metallic Glass Particles*: These composites use metastable phases as reinforcements, which can provide unique mechanical properties and stability under specific conditions. Metastable reinforcements can enhance the strength and toughness of the composite by undergoing phase transformations under stress or temperature changes. These reinforcements can provide a better strength and ductility synergy at micron length scale when compared to traditional ceramic reinforcements such as SiC and Al_2O_3.
   (ii) *Hollow Particle Reinforcements*: Hollow particles are used as reinforcements to reduce the density of the composite while maintaining or enhancing its mechanical properties. These composites are particularly useful in applications where weight reduction is critical, such as in aerospace and

automotive industries. These reinforcements are also used to develop parts where crash worthiness is required such as in collisions and impacts, typically relevant to automobile industry. Hollow particle reinforcements also enhance damping and thermal insulation properties.

2.3 Properties of MMCs

The final properties of composite materials are influenced by various factors, including the reinforcement's form, volume fraction, geometry, distribution, the matrix–reinforcement interface, void content, and the manufacturing process. The presence, amount, and distribution of reinforcement strongly affect dislocation behavior, influencing both the physical and mechanical properties of the composite. Beyond these, reinforcement also impacts properties such as wear resistance and damping capacity.

As indicated earlier, the properties and hence the performance of metal matrix composites (MMCs) is determined by the types of the matrix and the reinforcement as well as the characteristics of the matrix–reinforcement interface. Notably, poor bonding between the two phases can significantly degrade the overall composite performance. The following sections outline key properties of MMCs, beginning with volume and weight fractions, moving on to density, coefficient of thermal expansion, modulus of elasticity, and tensile strength [20, 21].

2.3.1 *Volume and Weight Fractions*

The mechanical properties of a composite are often derived using the rule of mixtures, where the properties are a volume-weighted average of the matrix and reinforcement properties. To calculate these properties, it is essential to know the relative proportions of the matrix and reinforcement, expressed as either volume or weight fractions. Weight fractions are often used because they are easier to measure, while volume fractions are essential for computing the composite's properties. Conversion between these two is therefore necessary.

The weight fractions of matrix and reinforcement can be calculated as:

$$W_M = \frac{w_m}{w_c} \quad \text{and} \quad W_R = \frac{w_r}{w_c} \tag{2.1}$$

where w_m, w_r, and w_c represent the weights of the matrix, reinforcement, and composite, respectively [18].

Similarly, volume fractions of matrix and reinforcements can be calculated by

$$V_M = \frac{v_m}{v_c} \quad \text{and} \quad V_R = \frac{v_r}{v_c} \tag{2.2}$$

where v_m, v_r, and v_c represent the volumes of the matrix, reinforcement, and composite, respectively [18].

Using the relationships between weight, volume, and density, the following expression for weight can be derived:

$$w_c = \rho_c v_c; \quad w_m = \rho_m v_m; \quad w_r = \rho_r v_r \tag{2.3}$$

where ρ_m, ρ_r, and ρ_c represent the densities of the matrix, reinforcement, and composite, respectively [18].

The relationship between volume and weight fractions is then given by:

$$W_m = \frac{\rho_m}{\rho_c} V_m \quad \text{and} \quad W_R = \frac{\rho_r}{\rho_c} V_r \tag{2.4}$$

2.3.2 *Density*

The composite's density can be calculated using the volume fractions and densities of its constituents [18, 21] as follows:

$$\rho_c v_c = \rho_m v_m + \rho_r v_r \tag{2.6}$$

Alternatively, based on weight fractions:

$$\frac{1}{\rho_c} = \frac{W_m}{\rho_m} + \frac{W_r}{\rho_r} \tag{2.7}$$

The composite's density can then be computed as below [18, 21]:

$$\rho_C = \frac{1}{\frac{W_m}{\rho_m} + \frac{W_r}{\rho_r}} \tag{2.8}$$

2.3.3 Coefficient of Thermal Expansion

The coefficient of thermal expansion (CTE) for composites, based on the rule of mixtures [18, 21], is calculated as:

$$\alpha_c = \alpha_m V_m + \alpha_r V_r \tag{2.9}$$

where α_c, α_m and α_r refers to the thermal expansion coefficients of the composite, matrix, and reinforcement materials, respectively.

For particle-reinforced MMCs, the Turner model [22] provides a more precise calculation:

$$\alpha_c = \frac{(\alpha_m V_m K_m + \alpha_r V_r K_r)}{(V_m K_m + V_r K_r)} \tag{2.10}$$

where K_m and K_r represent the bulk moduli of the matrix and reinforcement, respectively.

For continuous fiber-reinforced composites, CTEs along longitudinal and transverse directions can be calculated as follows:

Longitudinal CTE:

$$\alpha_{cl} = \frac{\alpha_m E_m V_m + \alpha_f E_r V_r}{E_m V_m + E_r V_r} \tag{2.11}$$

Transverse CTE:

$$\alpha_{ct} = (1 + V_m)\alpha_m + \alpha_f V_r \tag{2.12}$$

where E_m, E_r are the elastic moduli of the matrix and the fiber reinforcement, respectively, and ν_m refers to the Poisson's ratio of the matrix.

2.3.4 Modulus of Elasticity

The composite's elastic modulus depends on the type and volume fraction of reinforcement [18, 21]:

$$E_C = E_m V_m + E_r V_r \tag{2.13}$$

For discontinuous fiber and particle-reinforced composites, the modulus can be expressed as:

$$E_c = \frac{E_m\left(1+2sqV_r\right)}{1-qV_r} \tag{2.14}$$

where s is the particle aspect ratio and q is a ratio involving the matrix and reinforcement elastic moduli:

$$q = \frac{\left(\frac{E_r}{E_m}-1\right)}{\left(\frac{E_r}{E_m}+2s\right)} \tag{2.15}$$

2.3.5 *Tensile Strength*

The strength of MMCs generally surpasses that of their unreinforced counterparts:

$$\sigma_c = \sigma_m + \Sigma\Delta\sigma_i \tag{2.16}$$

where $\Delta\sigma_i$ refers to strengthening contributions from various mechanisms, including load transfer, enhanced dislocation density, thermal residual stresses, Orowan strengthening, and grain boundary strengthening [18, 21].

Load Transfer: The strengthening contribution from load transfer is given by:

$$\Delta\sigma_{LT} = V_r\sigma_m\frac{\left(L+t\right)AL}{4L} \tag{2.17}$$

CTE and Modulus Mismatch: Strengthening from CTE mismatch and modulus differences can be expressed as:

$$\Delta\sigma_{CTE} = \frac{A\Delta\alpha\Delta TV_r}{bdv_m} \quad \text{and} \quad \Delta\sigma_{EM} = \frac{6V_r}{\pi d} \tag{2.18}$$

where A is a geometric constant, $\Delta\alpha$ is the CTE mismatch, ΔT is the temperature difference, and b is the Burgers vector.

$$\Delta\sigma_{CTE+EM} = \sqrt{3}\beta Gb\left(\sqrt{\Delta\sigma_{CTE}}+\sqrt{\Delta\sigma_{EM}}\right) \tag{2.19}$$

Thermal Residual Stresses: This contribution can be calculated as:

$$\Delta\sigma_{Ts} = \frac{E_r E_m}{E_r V_r + E_m V_m} V_r \Delta\alpha\Delta T \quad (2.20)$$

Orowan Strengthening: This mechanism restricts dislocation motion, calculated as:

$$\Delta\sigma_{OR} = \frac{0.13bG}{d_r\left(\sqrt[3]{\frac{1}{2}V_r} - 1\right)} \ln\left(\frac{d}{2b}\right) \quad (2.21)$$

Grain size strengthening: The reinforcement particles refine the matrix grain size, leading to grain boundary strengthening, as defined below:

$$\Delta\sigma_{HP} = \frac{K_y}{\sqrt{d}} \quad (2.22)$$

where k_y is the strengthening coefficient which is a characteristic constant of each material, and d is the grain size which can be calculated by

$$d_m = \frac{4\alpha d}{3V_r} \quad (2.23)$$

Cumulative Strengthening: The cumulative strengthening contributions from multiple mechanisms can be approximated by summing the squares of each contribution:

$$\sigma_c = \sigma_{m+} \sqrt{\Sigma\Delta\sigma_i^2} \quad (2.24)$$

Another method proposed by Zhang and Chen [23, 24] combines different mechanisms, such as Orowan strengthening, CTE mismatch, and load-bearing effects as follows:

$$\sigma_c = (1 + 0.5V_r)\left(\sigma_m + A + B + \frac{AB}{\sigma_m}\right) \quad (2.25)$$

$$A = 1.25G_m b\sqrt{\frac{12\Delta\alpha\Delta T V_r}{bdV_m}} \quad (2.26)$$

$$B = \frac{0.13G_m b}{d\sqrt[3]{\frac{1}{2V_r}} - 1} \ln\frac{d}{2b} \quad (2.27)$$

2.4 Summary

MMCs are advanced materials composed of a metallic matrix reinforced with ceramic, intermetallic or metastable metallic phases, offering enhanced mechanical properties, wear resistance, thermal stability, and damping characteristics. Common matrix materials include lightweight metals like aluminum, magnesium, and titanium, while steel and nickel-based superalloys are employed for high-temperature applications. The reinforcements, which can be continuous fibers such as carbon or ceramic or particulate forms like silicon carbide and aluminum oxide, play a crucial role in improving properties like strength and stiffness. The selection of reinforcement type, size, and distribution, along with its interaction with the matrix, significantly influences the performance of MMCs, making them suitable for aerospace, automotive, electronics, sports, and packaging applications.

MMCs are classified based on their reinforcements. Laminate composites consist of stacked layers of matrix and reinforcement, strategically arranged to optimize strength and stress distribution. Fiber-reinforced MMCs utilize long fibers for superior load-bearing capacity and stiffness or short fibers for isotropic properties and easier processing. Particle-reinforced MMCs employ micron-sized particles for mechanical strength and wear resistance, while nanoscale reinforcements, such as graphene and carbon nanotubes, offer exceptional strength without sacrificing ductility. Unconventional composites, such as those using hollow particles for weight reduction or metastable reinforcements for unique mechanical properties, further expand the versatility of MMCs.

The properties of MMCs are defined by several factors, including the volume fraction, geometry, and distribution of reinforcements, as well as the quality of bonding at the matrix-reinforcement interface. Key properties such as density, thermal expansion, modulus of elasticity, and tensile strength are often modeled using the rule of mixtures. Strengthening mechanisms, including load transfer, thermal mismatch, increased dislocation density, Orowan strengthening, and grain refinement, contribute cumulatively to their outstanding performance. These characteristics make MMCs indispensable in applications requiring high strength-to-weight ratios, thermal stability, and wear resistance, enabling their extensive use in cutting-edge technological and industrial sectors.

References

1. A. Adetayo, M. Maleque, M.M. Rahman, Metal matrix composite brake rotor: historical development and product life cycle analysis. Int. J. Autom. Mech. Eng. **4**, 471–480 (2011). https://doi.org/10.15282/ijame.4.2011.8.0038
2. K.U. Kainer, Basics of metal matrix composites, in *Metal Matrix Composites: Custom-made Materials for Automotive and Aerospace Engineering*, (Wiley, 2006). https://doi.org/10.1002/3527608117.ch1
3. I.A. Ibrahim, F.A. Mohamed, E.J. Lavernia, Particulate reinforced metal matrix composites—a review. J. Mater. Sci. **26**(5), 1137–1156 (1991). https://doi.org/10.1007/BF00544448
4. R. Casati, M. Vedani, Metal matrix composites reinforced by Nano-particles—a review. Metals **4**(1), 65–83 (2014). https://doi.org/10.3390/met4010065

5. N. Chawla, K.K. Chawla, Metal-matrix composites in ground transportation. JOM **58**(11), 67–70 (2006). https://doi.org/10.1007/s11837-006-0231-5
6. P.K. Rohatgi, Metal matrix composites. Def. Sci. J. **43**(4), 323 (1993)
7. A. Evans, C. San Marchi, A. Mortensen, A. Evans, C. San Marchi, A. Mortensen, *Metal matrix composites* (Springer, 2003)
8. S. Seetharaman, M. Gupta, Fundamentals of metal matrix composites, in *Reference Module in Materials Science and Materials Engineering*, (Elsevier, 2021). https://doi.org/10.1016/B978-0-12-803581-8.00001-1
9. M. Gupta, S. Seetharaman, Magnesium based nanocomposites for cleaner transport, in *Nanotechnology for Energy Sustainability*, (Wiley, 2017). https://doi.org/10.1002/9783527696109.ch33
10. W.C. Harrigan, Commercial processing of metal matrix composites. Mater. Sci. Eng. A **244**, 75–79 (1998). https://doi.org/10.1016/s0921-5093(97)00828-9
11. Z. Gürdal, R.T. Haftka, P. Hajela, *Design and Optimization of Laminated Composite Materials* (John Wiley & Sons, 1999)
12. X.-S. Yi, S. Du, L. Zhang, *Composite Materials Engineering, Volume 1: Fundamentals of Composite Materials* (Springer, 2018)
13. W. Hall, Z. Javanbakht, *Design and Manufacture of Fibre-Reinforced Composites* (Springer, 2021)
14. S. Seshan, A. Guruprasad, M. Prabha, A. Sudhakar, Fibre-reinforced metal matrix composites—a review. J. Indian Inst. Sci. **76**(1), 1 (1996)
15. A. Shrivastava, *Fundamentals of composite materials* (Sankalp Publication, 2021)
16. C. Soutis, Fibre reinforced composites in aircraft construction. Prog. Aerosp. Sci. **41**(2), 143–151 (2005)
17. M. Walker, R. Smith, A methodology to design fibre reinforced laminated composite structures for maximum strength. Compos. Part B **34**(2), 209–214 (2003)
18. D.J. Lloyd, Particle reinforced aluminium and magnesium matrix composites. Int. Mater. Rev. **39**, 1–23 (1994). https://doi.org/10.1179/imr.1994.39.1.1
19. S.F. Hassan, *Creation of New Magnesium-Based Material Using Different Types of Rein Forcements* (PhD thesis, 2006)
20. G.E. Dieter, D. Bacon, Mechanical metallurgy SI metric edition. J. Franklin Inst. **273**, 338 (1988). https://doi.org/10.1016/S0016-0032(62)91145-6
21. W. Callister, D. Rethwisch, *Materials Science and Engineering: An Introduction*, vol 94 (Wiley, 2007). https://doi.org/10.1016/0025-5416(87)90343-0
22. P.S. Turner, Thermal-expansion stresses in reinforced plastics. J. Res. Natl. Bur. Stand. **37**, 239–250 (1946). https://doi.org/10.6028/jres.037.015
23. Z. Zhang, D.L. Chen, Consideration of Orowan strengthening effect in particulate-reinforced metal matrix nanocomposites: a model for predicting their yield strength. Scr. Mater. **54**, 1321–1326 (2006). https://doi.org/10.1016/j.scriptamat.2005.12.017
24. Z. Zhang, D.L. Chen, Contribution of Orowan strengthening effect in particulate-reinforced metal matrix nanocomposites. Mater. Sci. Eng. A **483–484**, 148–152 (2008). https://doi.org/10.1016/j.msea.2006.10.184

Chapter 3
Synthesis Techniques for Magnesium-Based Metal Matrix Composites

This chapter provides a comprehensive overview of the fabrication methods used to produce magnesium-based metal matrix composites (MMCs), focusing on their ability to meet the growing demand for lightweight, high-performance materials in aerospace, automotive, and biomedical industries. Readers will gain insights into the primary fabrication techniques—liquid-state processing, solid-state processing, and in situ synthesis—each evaluated for its suitability in tailoring composite properties for specific applications. Key methods such as melt infiltration, ultrasonic infiltration, squeeze casting, and stir casting are explored within liquid-state processing, highlighting their scalability and effectiveness in distributing reinforcements like nano-scale particles and hollow microspheres. Solid-state approaches, including powder metallurgy, additive manufacturing, and severe plastic deformation, are discussed for their precision in reinforcement placement and minimization of defects. This chapter also delves into in situ synthesis techniques, which enable the formation of reinforcements directly within the magnesium matrix, resulting in improved bonding and microstructural control. Additionally, advanced reinforcements such as high-entropy alloys, metallic glasses, and syntactic foams are introduced, showcasing their potential to enhance properties like strength-to-weight ratios, thermal stability, wear resistance, and energy absorption. This chapter equips readers with a thorough understanding of fabrication methodologies and their impact on the performance and application potential of magnesium-based MMCs, offering guidance for future research and industrial applications.

S. Seetharaman et al., *Magnesium-based Syntactic, Metastable and Nano-composites*, https://doi.org/10.1007/978-3-031-96514-2_3

3.1 Introduction

Magnesium (Mg)-based metal matrix composites (MMCs) have garnered significant attention over the years due to their impressive combination of lightweight and high mechanical strength. With magnesium being one of the lightest structural metals, its alloys and composites offer unparalleled advantages in weight-sensitive applications, such as aerospace, automotive, and biomedical fields. Magnesium's density of about 1.74 g/cm^3 is considerably lower than aluminum (2.7 g/cm^3) and steel (7.85 g/cm^3), making it an excellent candidate for producing weight-efficient materials with enhanced mechanical properties.

The increasing demand for more advanced materials has led to innovations in magnesium-based composites. Traditional composites typically employed larger-scale reinforcements like fibers and particles; however, recent developments have introduced nano-scale reinforcements, metastable phases, and syntactic foams into magnesium matrices. These reinforcements enhance not only the mechanical properties but also the functional attributes of the material, including thermal stability, wear resistance, and energy absorption. Nano-reinforced Mg composites are particularly noteworthy for their exceptional strength-to-weight ratio and the potential to deliver high-performance materials in sectors where weight reduction is critical. Meanwhile, metastable and amorphous metallic reinforcements have introduced new avenues in improving ductility, toughness, and hardness. Syntactic foams, with their hollow microsphere reinforcements, bring exceptional lightweight properties coupled with energy-absorbing capabilities, ideal for crash-resistant applications.

To advance magnesium-based materials, particularly composites, it is important to process them suitably, economically, and in an environmental-friendly manner. It should be noted that processing strongly influences the microstructure, which in turn affects the properties of end material including composites. Accordingly, this section provides an in-depth exploration of the fabrication methods employed in producing magnesium-based nano-, metastable, and syntactic foam composites. Each fabrication method presents its own set of challenges and opportunities in creating materials with tailored properties for specific applications. Whether through liquid-state processing, solid-state techniques, or in situ synthesis, these composites' unique properties depend heavily on the optimization of their fabrication routes.

3.2 Fabrication Methods for Magnesium-Based Composites

The fabrication of magnesium-based composites can be broadly categorized into three primary methods: liquid-state processing, solid-state processing, and in situ synthesis. Each approach has its benefits and limitations depending on the type of reinforcement used and the desired properties of the final material.

3.2.1 *Liquid-State Processing*

Liquid-state processing is one of the most widely adopted techniques for producing magnesium-based composites due to its relatively simple methodology and suitability for large-scale production. In this process, magnesium is melted, and the reinforcements are introduced into the molten metal. The goal is to ensure homogeneous distribution of the reinforcement throughout the magnesium matrix before solidifying the composite into the desired shape.

3.2.1.1 Melt Infiltration

Melt infiltration is a popular technique within the liquid-state processing category, where molten magnesium is infiltrated into a porous reinforcement preform. The porous structure of the preform allows the molten metal to flow into the reinforcement phase, filling the voids and creating a composite material upon solidification. However, one of the main challenges with melt infiltration is achieving good wettability between the molten magnesium and the reinforcement material. Magnesium tends to have poor wettability with many common reinforcement phases, such as ceramic particles, leading to incomplete infiltration or weak bonding at the matrix–reinforcement interface. To address this issue, forced infiltration techniques, such as pressure-assisted infiltration, have been employed. In this method, mechanical or gas pressure is applied to force the molten magnesium into the reinforcement preform, overcoming the poor wettability issue (Fig. 3.1). The pressure helps improve the infiltration efficiency, ensuring complete filling of the porous reinforcement phase. However, pressure-assisted infiltration often requires complex equipment and precise control of processing parameters, which can limit its scalability for industrial applications [1–4].

3.2.1.2 Ultrasonic Infiltration

Ultrasonic infiltration is another liquid-state technique that uses high-frequency ultrasonic waves to improve the infiltration of molten magnesium in reinforcement preform (Fig. 3.2). The ultrasonic vibrations create cavitation bubbles in the molten

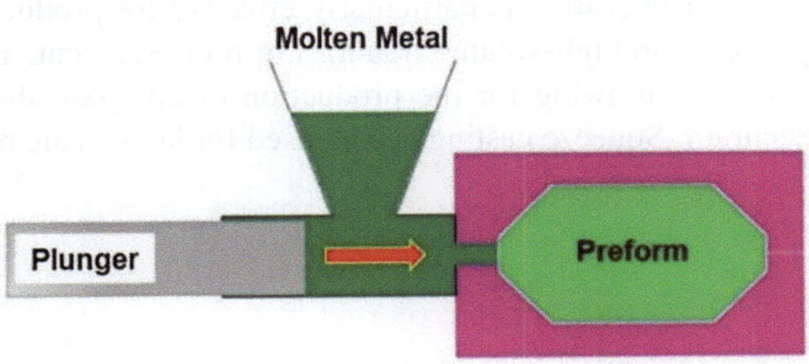

Fig. 3.1 Pressure die infiltration process [5]

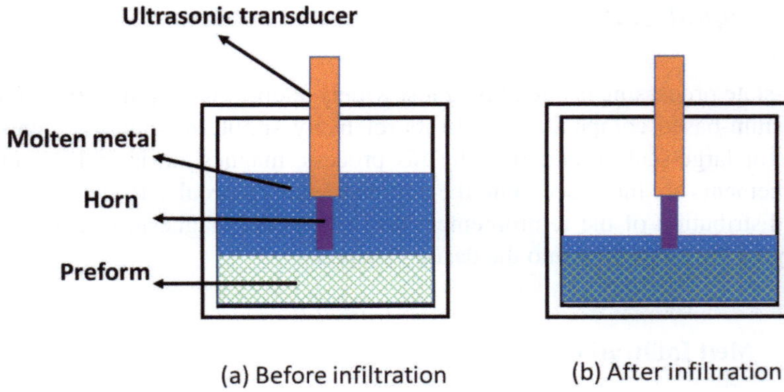

Fig. 3.2 Ultrasonic infiltration [7]

metal, which collapse and generate intense localized pressure, helping break up agglomerated reinforcements and improving their dispersion. Additionally, the acoustic cavitation helps reduce the surface tension of the molten magnesium, enhancing its wettability with the reinforcement phase [6].

Ultrasonic infiltration has proven particularly effective for dispersing nano-scale reinforcements, which tend to agglomerate due to their high surface area-to-volume ratio. The improved distribution of nano-reinforcements achieved through ultrasonic processing leads to composites with superior mechanical properties, such as increased strength and hardness. However, the application of ultrasonic infiltration requires precise control over processing parameters, such as ultrasonic frequency, power, and duration, to avoid damage to the reinforcement phase and ensure uniform distribution [6, 8].

3.2.1.3 Squeeze Casting

Squeeze casting is a liquid-state technique that applies mechanical pressure to the molten metal during the solidification process. In this method, molten magnesium slurry is poured into a mold containing the reinforcement phase, and mechanical pressure is applied (Fig. 3.3). The pressure assists in reducing the formation of defects such as porosity and shrinkage, resulting in a denser and more homogeneous composite structure [9, 10].

Squeeze casting is particularly effective for producing magnesium-based composites with high-volume fractions of reinforcement. The process is also relatively versatile, allowing for the production of complex shapes with high dimensional accuracy. Squeeze casting can be used for large-scale production [5].

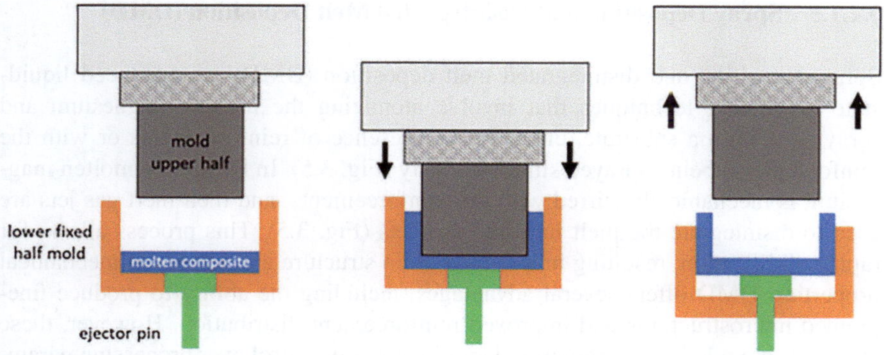

Fig. 3.3 Squeeze casting [5]

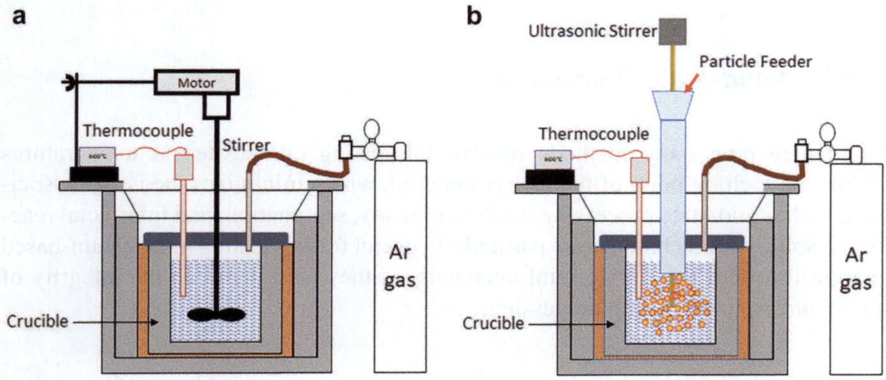

Fig. 3.4 Schematic of (**a**) mechanical stir casting and (**b**) ultrasonic stir casting [13]

3.2.1.4 Stir Casting

Stir casting is one of the most cost-effective liquid-state processing techniques for fabricating magnesium-based composites. In this method, reinforcements are introduced into the molten magnesium while the mixture is mechanically stirred to achieve uniform distribution. The stirring action helps break up agglomerated reinforcements and promotes their dispersion throughout the molten matrix (Fig. 3.4). This method is well-suited for large-scale production due to its simplicity and cost-effectiveness. However, achieving a uniform distribution of reinforcements can be challenging, especially when dealing with nano-scale reinforcements, which tend to agglomerate during processing. To address this issue, ultrasonic-assisted stir casting has been developed, where high-frequency ultrasonic waves are applied during the stirring process to improve the dispersion of reinforcements and enhance the mechanical properties of the resulting composite [1, 7, 11, 12].

3.2.1.5 Spray Deposition and Disintegrated Melt Deposition (DMD)

Spray deposition and disintegrated melt deposition (DMD) are advanced liquid-state processing techniques that involve atomizing the molten magnesium and spraying it onto a substrate, either in the presence of reinforcements or with the reinforcements being sprayed simultaneously (Fig. 3.5). In DMD, the molten magnesium is mechanically stirred with the reinforcements, and then inert gas jets are used to disintegrate the melt into fine droplets (Fig. 3.5). This process allows for rapid solidification, resulting in a refined grain structure and improved mechanical properties. DMD offers several advantages, including the ability to produce fine-grained microstructures and improved reinforcement distribution. However, these processes are relatively complex and require precise control over processing parameters, such as gas flow rate, temperature, and reinforcement concentration, to achieve the desired material properties [14–17].

3.2.2 Solid-State Processing

Solid-state processing methods involve fabricating composites at temperatures below the melting point of the matrix material, which minimizes the issues associated with liquid-state processing, such as porosity, segregation, and interfacial reactions. Solid-state techniques are particularly useful for fabricating magnesium-based composites with nano-scale reinforcements, as they help maintain the integrity of the reinforcements during processing.

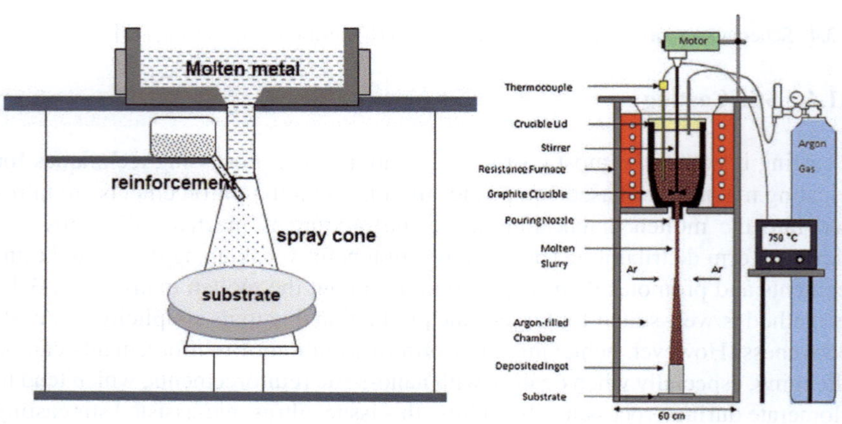

Fig. 3.5 Spray and disintegrated melt deposition [5, 18]

3.2.2.1 Powder Metallurgy (PM)

Powder metallurgy (PM) is one of the most widely used solid-state processing techniques for fabricating metal matrix composites. In PM, magnesium and reinforcement powders are blended together, compacted into a desired shape, and then sintered at elevated temperatures to create a solid composite (Fig. 3.6). PM offers precise control over the reinforcement distribution and avoids many of the issues associated with molten metal processes, such as segregation and agglomeration [19, 20].

The primary advantage of PM is its ability to produce composites with a uniform distribution of reinforcements, even at high reinforcement concentrations. However, the process is typically limited to small, simple geometries due to the need for compaction and sintering. Additionally, the mechanical properties of PM composites are often limited by the porosity and weak bonding between the matrix and reinforcement phases. To address these issues, advanced PM techniques, such as microwave sintering and hot isostatic pressing (HIP), have been developed to improve the density and mechanical properties of the resulting composites [22].

3.2.2.2 Hybrid Microwave Sintering Technique

Microwave sintering has recently gained attention as an energy-efficient method for consolidating powder materials. Initially, this technique was primarily applied to ceramics; however, recent studies have demonstrated its effectiveness in densifying metallic materials as well. It is important to understand that microwave heating operates fundamentally differently from conventional heating methods. In

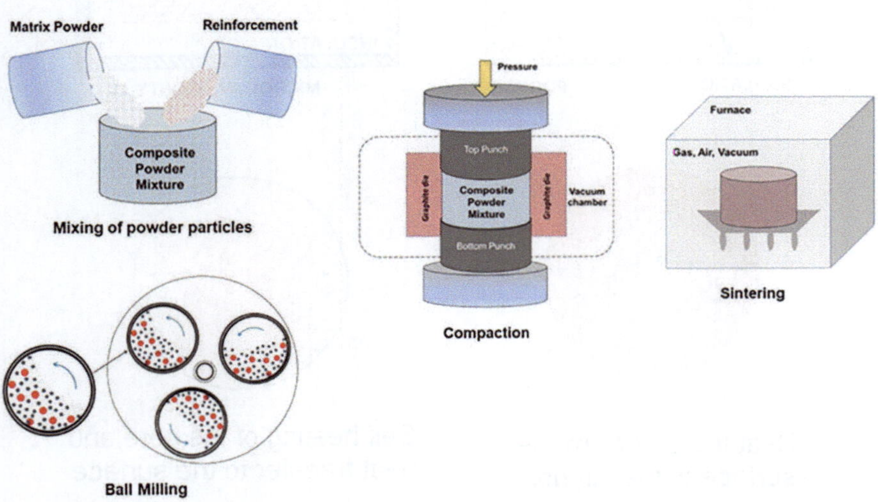

Fig. 3.6 Schematic of powder metallurgy route [21]

traditional processes, heat is transferred from the material's surface to its interior through conduction, convection, and radiation. In contrast, microwave heating causes self-heating of the material's core due to dielectric and magnetic losses, which occur when the material interacts with the electric and magnetic fields. Because microwaves generate an inverse temperature gradient, the material heats rapidly from the inside out, leading to a significant reduction in processing time—often by more than 80% (Fig. 3.7). In recent years, hybrid microwave sintering, which uses susceptors, has been introduced to address temperature and microstructural variations across sample thickness, further improving the consistency and quality of the sintered materials [23–25].

3.2.2.3 Additive Manufacturing (AM)

Additive manufacturing (AM) techniques, such as selective laser melting (SLM) and binder jetting, have revolutionized the fabrication of metal matrix composites, offering unprecedented control over the geometry and composition of the final

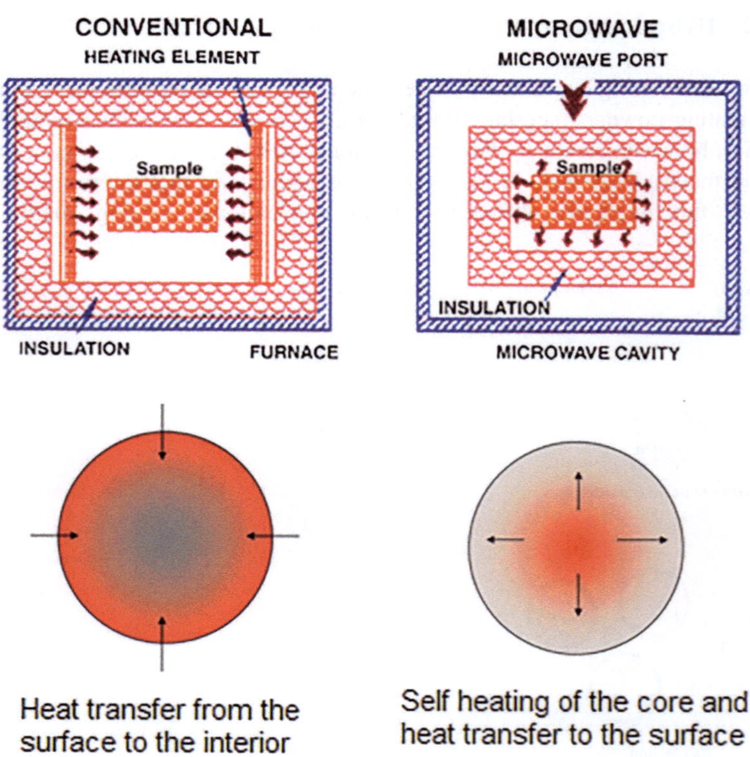

Fig. 3.7 Schematic of microwave sintering and heat transfer principle [23]

material. AM techniques involve building up the composite layer by layer either by selectively melting and fusing powder particles using a high-energy laser or by depositing layers of reinforcement and matrix powders using a binder material. SLM, in particular, has proven effective for fabricating magnesium-based composites with complex geometries and tailored reinforcement distributions (Fig. 3.8). The high-energy laser used in SLM allows for precise control over the melting and solidification of the material, resulting in a fine-grained microstructure with improved mechanical properties. Binder jetting, on the other hand, offers a more cost-effective approach to AM, as it does not require high-energy lasers or expensive equipment (Fig. 3.8) [13, 22, 26–31].

Recently, extrusion-based deposition methods are also being explored for the processing of Mg-based materials. This method involves a syringe to extrude a paste made from powdered material onto a base plate, which moves to shape the desired 3D structure layer by layer (Fig. 3.9). Once the part is deposited, it undergoes drying without any heat treatment. A key advantage of this process is its suitability for producing magnesium implants, as healing drugs can be incorporated directly into the paste. However, compared to fusion-based additive manufacturing methods, the resulting parts often exhibit lower mechanical strength [33].

Friction stir processing (FSP), another solid-state technique was also used to modify the surface of magnesium-based composite materials. In this process, frictional heat is generated at the interface between a rotating tool and the matrix material, facilitating the dispersion of reinforcements (Fig. 3.10). Unlike fusion-based methods, FSP does not involve melting, which helps maintain the integrity of the material while achieving effective reinforcement distribution [34, 35].

While AM techniques are particularly well-suited for fabricating magnesium-based composites for aerospace and biomedical applications, where weight reduction and complex geometries are critical, the high cost of AM equipment and the need for precise control over processing parameters can limit its widespread adoption for large-scale production.

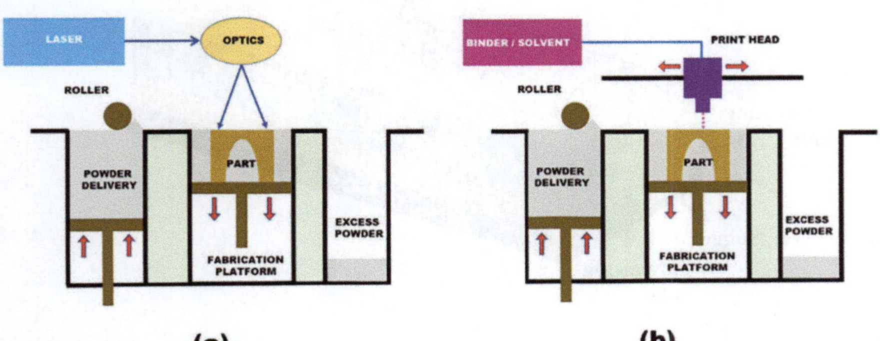

Fig. 3.8 Schematic of (**a**) SLM, (**b**) binder/binder-less jetting, and (**c**) laser deposition methods [32]

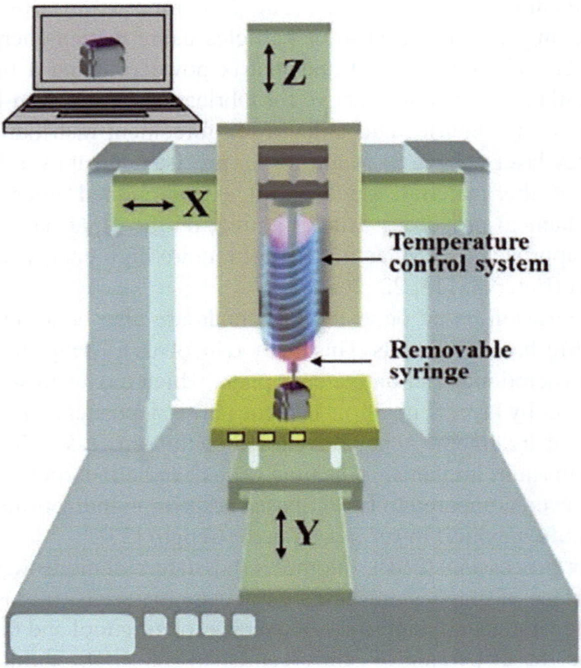

Fig. 3.9 Set-up for paste extrusion deposition. (Adapted from [33])

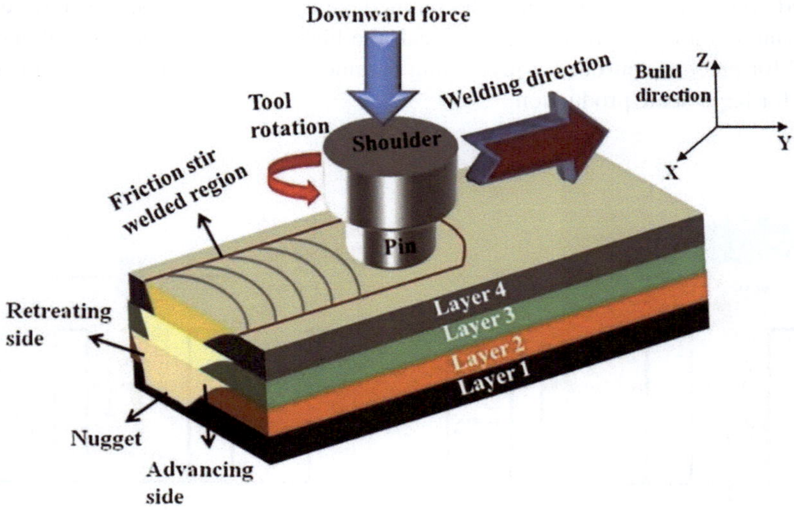

Fig. 3.10 Solid-state welding. (Adapted from [34])

3.2.2.4 Severe Plastic Deformation (SPD)

Severe plastic deformation (SPD) techniques, such as equal channel angular pressing (ECAP) and high-pressure torsion (HPT), are solid-state methods that involve the application of intense plastic strain to the material to refine its microstructure and improve its mechanical properties. SPD techniques have been used to produce ultra-fine-grained magnesium-based composites with improved strength, ductility, and corrosion resistance [36].

In ECAP, the composite is forced through a die with two channels intersecting at an angle, subjecting the material to intense shear deformation. The repeated deformation process refines the grain structure, resulting in a material with superior mechanical properties. HPT, on the other hand, involves subjecting the composite to high compressive and torsional forces, leading to significant grain refinement and improved mechanical performance.

3.2.3 In Situ Synthesis

In situ synthesis involves the formation of the reinforcement phase within the magnesium matrix during processing through chemical reactions and phase transformations [37]. This method offers several advantages, including improved bonding between the matrix and reinforcement phases and the ability to create composites with unique microstructures and properties. For example, titanium carbide (TiC) particles can be formed in situ within a magnesium matrix by introducing titanium and carbon precursors into the molten magnesium, which react to form TiC particles. This approach offers improved bonding between the matrix and reinforcement phases and allows for the creation of composites with tailored properties. Similarly, the in situ reaction of elements during the melting and solidification processes aids in the precipitation of very fine reinforcements, such as AlN and SiC particles, which act as strengthening agents, improving the mechanical properties of the composite.

3.3 Novel Reinforcements

Advanced materials such as high-entropy alloys (HEAs), medium-entropy alloys (MEAs), metallic glasses, and other metastable materials are at the forefront of materials science due to their exceptional properties and potential applications. These materials are engineered to exhibit superior mechanical, thermal, and chemical properties, making them ideal candidates for reinforcing magnesium (Mg) and its alloys. The integration of these advanced materials into Mg composites can significantly enhance their performance, making them suitable for high-performance applications in various industries [38].

HEAs and MEAs, composed of multiple principal elements in near-equiatomic proportions, result in a high configurational entropy that stabilizes the solid solution phase. This unique composition leads to exceptional properties such as high strength, ductility, thermal stability, and resistance to wear and corrosion. These properties make HEAs and MEAs ideal for reinforcing Mg alloys, enhancing their mechanical and thermal performance. Metallic glasses, or amorphous metals, are formed by rapidly cooling a molten alloy to prevent the formation of a crystalline structure. This amorphous structure imparts superior strength-to-weight ratios, high hardness, and excellent corrosion resistance. When used as reinforcements in Mg alloys, metallic glasses can significantly improve the mechanical properties and durability of the composite material. Other metastable materials, such as certain intermetallic compounds and nanocrystalline alloys, also offer promising reinforcement capabilities. These materials can be engineered to exhibit specific properties, such as high strength, thermal stability, and resistance to wear, making them suitable for enhancing the performance of Mg alloys [39–42].

The preparation of these advanced materials involves various sophisticated techniques. HEAs and MEAs can be synthesized using methods such as arc melting, powder metallurgy, and mechanical alloying. Mechanical alloying, for instance, involves repeated fracturing and cold welding of powder particles in a high-energy ball mill, leading to the formation of a homogeneous alloy (Fig. 3.11). This method allows for precise control of the alloy's composition and microstructure. Metallic glasses are typically produced through rapid solidification processes, such as melt spinning or splat quenching. These techniques involve cooling the molten alloy at extremely high rates, preventing the formation of crystalline structures and resulting in an amorphous material. These materials are also increasingly available in various forms, including powders and bulk materials, suitable for composite manufacturing. Metallic glasses are also commercially available, though their production is more specialized due to the need for rapid solidification techniques. The availability of these materials in ready-to-use forms facilitates their integration into composite materials, enabling the development of advanced Mg alloy composites with enhanced performance characteristics [43].

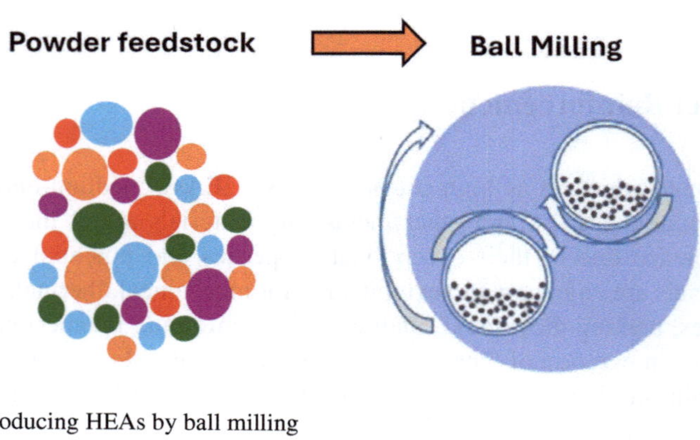

Fig. 3.11 Producing HEAs by ball milling

In this regard, the incorporation of hollow particles, such as glass, ceramic, or polymer microspheres, into Mg alloys yields lightweight syntactic foam composites. These composites have a closed-cell foam-like structure, providing excellent mechanical properties, high specific strength, and low density. The use of hollow particles reduces the overall weight of the composite while maintaining or enhancing its strength and stiffness. This makes syntactic foam composites particularly suitable for applications in the aerospace, automotive, and marine industries, where weight reduction is crucial without compromising structural integrity. One of the primary advantages of syntactic foams is their ability to absorb energy during impact, making them ideal for crash-resistant applications. The hollow microspheres act as energy-absorbing elements, deforming under load and dissipating the impact energy. This property makes syntactic foams highly effective for applications such as automotive crash structures, aerospace components, and protective gear. The hollow structure of the microspheres reduces the overall density of the material, while their spherical shape helps distribute stress evenly throughout the composite, improving its mechanical properties [44–47].

Despite their potential, incorporating hollow particles into the Mg matrix presents several challenges. One major challenge is achieving a uniform distribution of hollow particles within the matrix, as agglomeration can lead to weak points in the composite. Additionally, the bonding between the hollow particles and the Mg matrix must be strong enough to ensure load transfer and prevent particle debonding under stress. Processing techniques such as stir casting, powder metallurgy, and infiltration methods are being explored to overcome these challenges and achieve high-quality syntactic foam composites. The mechanical properties of syntactic foams are highly dependent on the size, shape, and material properties of the microspheres, which must be carefully selected to achieve the desired performance. Recent developments in additive manufacturing and in situ synthesis techniques have opened up new opportunities for fabricating syntactic foams with tailored properties. For example, additive manufacturing techniques such as selective laser melting (SLM) allow for precise control over the distribution of microspheres within the matrix, resulting in composites with improved mechanical performance. In situ synthesis methods, such as in situ precipitation of nano-scale reinforcements, offer the potential to create syntactic foams with enhanced strength and energy absorption capabilities. In general, these hollow particles can be purchased off the shelf for ready use as reinforcement, and in some cases, they are also coated with suitable elements to ensure better wettability and bonding with the Mg matrix.

3.4 Summary

Magnesium-based metal matrix composites have gained significant attention for their lightweight, high-strength properties, making them ideal for weight-sensitive applications such as aerospace, automotive, and biomedical fields. Magnesium's low density (1.74 g/cm^3) compared to materials like aluminum and steel makes it an attractive option for producing materials that combine reduced weight with enhanced mechanical properties. Over time, innovations in these composites have incorporated advanced reinforcements, such as nano-scale reinforcements, metastable phases, and syntactic foams. These reinforcements not only improve the mechanical properties of the materials but also their functional attributes, including thermal stability, wear resistance, and energy absorption. Nano-reinforced magnesium composites stand out for their exceptional strength-to-weight ratio, while syntactic foams, with hollow microspheres, provide lightweight materials ideal for crash-resistant applications, thanks to their energy-absorbing capabilities.

The fabrication of magnesium-based composites involves various methods, broadly categorized into liquid-state processing, solid-state processing, and in-situ synthesis. Liquid-state techniques, such as melt infiltration, ultrasonic infiltration, squeeze casting, and stir casting, are commonly used due to their scalability. Melt infiltration, for example, uses molten magnesium to infiltrate a porous reinforcement preform, though challenges related to poor wettability are addressed through methods like pressure-assisted infiltration. Ultrasonic method enhances the dispersion of nano-scale reinforcements by using high-frequency waves, while squeeze casting applies mechanical pressure to ensure a dense and homogeneous composite. Stir casting, a more cost-effective method, involves mechanically stirring the reinforcements into molten magnesium. Advanced liquid-state techniques, like spray deposition and disintegrated melt deposition, enable fine-grain structures and precise control over reinforcement distribution, though they require relatively complex equipment and processing control.

Solid-state processing methods, such as powder metallurgy, additive manufacturing, and severe plastic deformation (SPD), are used to fabricate composites at temperatures below the melting point of magnesium, which helps avoid issues like porosity and segregation. PM allows for precise reinforcement distribution by blending magnesium and reinforcement powders, though it is often limited to small, simple geometries. Additive manufacturing techniques like selective laser melting and binder jetting enable the fabrication of complex geometries and precise reinforcement placement, with the former providing high-quality, fine-grained microstructures. SPD techniques, such as equal channel angular pressing and high-pressure torsion, refine the composite microstructure through intense plastic deformation, thereby improving strength, ductility, and corrosion resistance.

In situ synthesis methods involve forming reinforcement phases directly within the magnesium matrix through chemical reactions during processing. This approach offers enhanced bonding between the matrix and reinforcement phases and allows for the creation of composites with unique microstructures, improving mechanical properties. Novel reinforcements, including high-entropy alloys, medium-entropy alloys, metallic glasses, and other metastable materials, are increasingly integrated into magnesium composites. HEAs and MEAs exhibit exceptional strength, thermal stability, and corrosion resistance, making them ideal for reinforcing magnesium alloys. Metallic glasses, formed by rapid cooling, offer high hardness and superior strength-to-weight ratios, while other metastable materials, like intermetallic compounds and nanocrystalline alloys, further enhance the composite's performance.

The incorporation of hollow particles, such as glass, ceramic, or polymer microspheres, into magnesium matrices form syntactic foam composites that are lightweight and have excellent mechanical properties. These forms provide energy absorption during impact, making them suitable for applications like automotive crash structures and aerospace components. Despite challenges in achieving uniform distribution and strong bonding between the microspheres and the matrix, methods such as stir casting, powder metallurgy, and advanced additive manufacturing are used to optimize the fabrication process. These developments are crucial for creating magnesium-based composites that meet the growing demands for high-performance, lightweight materials in various industries.

References

1. K.U. Kainer, *Basics of Metal Matrix Composites* (Wiley, 2006). https://doi.org/10.1002/3527608117.ch1
2. A.J. Cook, P.S. Werner, Pressure infiltration casting of metal matrix composites. Mater. Sci. Eng. A **144**, 189–206 (1991). https://doi.org/10.1016/0921-5093(91)90225-C
3. N. Altinkok, A. Demir, I. Ozsert, Processing of Al_2O_3/SiC ceramic cake preforms and their liquid Al metal infiltration. Compos. Part Appl. Sci. Manuf. **34**, 577–582 (2003). https://doi.org/10.1016/S1359-835X(03)00125-8
4. A. Daoud, Wear performance of 2014 Al alloy reinforced with continuous carbon fibers manufactured by gas pressure infiltration. Mater. Lett. **58**, 3206–3213 (2004). https://doi.org/10.1016/j.matlet.2004.06.012
5. P. Garg, A. Jamwal, D. Kumar, K.K. Sadasivuni, C.M. Hussain, P. Gupta, Advance research progresses in aluminium matrixcomposites: manufacturing & applications. J. Mater. Res. Technol. **8**, 4924–4939 (2019). https://doi.org/10.1016/j.jmrt.2019.06.028
6. J. Pan, D.M. Yang, H. Wan, X.F. Yin, Application of ultrasonic infiltration in metal matrix composites. Key Eng. Mater. **104–107**, 275–282 (1995). https://doi.org/10.4028/www.scientific.net/KEM.104-107.275
7. H. Dieringa, Processing of magnesium-based metal matrix nanocomposites by ultrasound-assisted particle dispersion: a review. Metals **8**(6), 431 (2018). https://doi.org/10.3390/met8060431
8. S. Jiang, L.J. Huang, Q. An, L. Geng, X.J. Wang, S. Wang, Study on titanium-magnesium composites with bicontinuous structure fabricated by powder metallurgy and ultrasonic infiltration. J. Mech. Behav. Biomed. Mater. **81**, 10–15 (2018). https://doi.org/10.1016/j.jmbbm.2018.02.017
9. S. Jayalakshmi, S. Kailas, S. Seshan, E. Fleury, Properties of squeeze cast Mg-6Zn-3Cu alloy and its Saffil alumina short fibre reinforced composites. J. Mater. Sci. **41**, 3743–3752 (2006). https://doi.org/10.1007/s10853-005-4484-0
10. H. Uozumi et al., Fabrication process of carbon nanotube/light metal matrix composites by squeeze casting. Mater. Sci. Eng. A **495**, 282–287 (2008). https://doi.org/10.1016/j.msea.2007.11.088
11. S. Tzamtzis, N.S. Barekar, N. Hari Babu, J. Patel, B.K. Dhindaw, Z. Fan, Processing of advanced Al/SiC particulate metal matrix composites under intensive shearing—a novel Rheo-process. Compos. Part Appl. Sci. Manuf. **40**, 144–151 (2009). https://doi.org/10.1016/j.compositesa.2008.10.017
12. L.Y. Chen et al., Processing and properties of magnesium containing a dense uniform dispersion of nanoparticles. Nature **528**(7583), 539–543 (2015). https://doi.org/10.1038/nature16445

13. S. Seetharaman, M. Gupta, Fundamentals of metal matrix composites, in *Reference Module in Materials Science and Materials Engineering*, (Elsevier, 2021). https://doi.org/10.1016/B978-0-12-803581-8.00001-1
14. W.C. Harrigan, Commercial processing of metal matrix composites. Mater. Sci. Eng. A **244**, 75–79 (1998). https://doi.org/10.1016/s0921-5093(97)00828-9
15. J. Wood, Spray atomisation and deposition. Surf. Eng. **13**, 28–30 (1997). https://doi.org/10.1179/sur.1997.13.1.28
16. M. Gupta, M.O. Lai, C.Y. Soo, Processing-microstructure-mechanical properties of an Al-Cu/SiC metal matrix composite synthesized using disintegrated melt deposition techn ique. Mater. Res. Bull. **30**(12), 1525–1534 (1995). https://doi.org/10.1016/0025-5408(95)00141-7
17. M. Gupta, W.L.E. Wong, Magnesium-based nanocomposites: lightweight materials of the future. Mater. Charact. **105**, 30–46 (2015). https://doi.org/10.1016/j.matchar.2015.04.015
18. S. Murugan, Q.B. Nguyen, M. Gupta, Synthesis of magnesium based nano-composites, in *Magnesium-The Wonder Element for Engineering/Biomedical Applications*, (IntechOpen, 2019)
19. J.A. Slotwinski, E.J. Garboczi, P.E. Stutzman, C.F. Ferraris, S.S. Watson, M.A. Peltz, Characterization of metal powders used for additive manufacturing. J. Res. Natl. Inst. Stand. Technol. Ogy. **119**, 460–493 (2014). https://doi.org/10.6028/jres.119.018
20. C. Padmavathi, A. Upadhyaya, D. Agrawal, Effect of microwave and conventional heating on sintering behavior and properties of Al-Mg-Si-Cu alloy. Mater. Chem. Phys. **130**, 449–457 (2011). https://doi.org/10.1016/j.matchemphys.2011.07.008
21. M. Gupta, S. Seetharaman, Magnesium based nanocomposites for cleaner transport, in *Nanotechnology for Energy Sustainability*, (Wiley, 2017). https://doi.org/10.1002/9783527696109.ch33
22. B. AlMangour, D. Grzesiak, J.M. Yang, Selective laser melting of TiB2/H13 steel nanocomposites: influence of hot isostatic pressing post-treatment. J. Mater. Process. Technol. **244**, 344–353 (2017). https://doi.org/10.1016/j.jmatprotec.2017.01.019
23. M. Penchal Reddy, R.A. Shakoor, A.M.A. Mohamed, M. Gupta, Microwave rapid sintering of Al-metal matrix composites: a review on the effect of reinforcements, microstructure and mechanical properties. Metals **6**, 143 (2016). https://doi.org/10.3390/met6070143
24. M. Gupta, W.W.L. Eugene, *Microwaves and Metals* (Wiley, 2011). https://doi.org/10.1002/9780470822746
25. W.L.E. Wong, M. Gupta, Development of Mg/Cu nanocomposites using microwave assisted rapid sintering. Compos. Sci. Technol. **67**(7–8), 1541–1552 (2007). https://doi.org/10.1016/j.compscitech.2006.07.015
26. M.P. Behera, T. Dougherty, S. Singamneni, Conventional and additive manufacturing with metal matrix composites: a perspective. Proc. Manuf. **30**, 159–166 (2019). https://doi.org/10.1016/j.promfg.2019.02.023
27. D. Manfredi et al., Additive manufacturing of Al alloys and aluminium matrix composites (A MCs), in *Light Metal Alloys Applications*, (IntechOpen, 2014). https://doi.org/10.5772/58534
28. E. Herderick, Additive manufacturing of metals: a review. Mater. Sci. Technol. Conf. Exhib. 2011 MS T11 **2**(176252), 1413–1425 (2011)
29. M. Ziaee, N.B. Crane, Binder jetting: a review of process, materials, and methods. Addit. Manuf. **28**, 781–801 (2019). https://doi.org/10.1016/j.addma.2019.05.031
30. B. AlMangour, D. Grzesiak, J.M. Yang, Nanocrystalline TiC-reinforced H13 steel matrix nanocomposites fabrica ted by selective laser melting. Mater. Des. (1980–2015) **96**, 150–161 (2016). https://doi.org/10.1016/j.matdes.2016.02.022
31. O.H. Famodimu, *Additive Manufacturing of Aluminium-Metal Matrix Composite Developed Through Mechanical Alloying* (University of Wolverhampton, 2016)
32. S. Seetharaman et al., Research updates on the additive manufacturing of nickel based superal loys, in *Solid Freeform Fabrication 2016: Proceedings of the 27th Annual International Solid Freeform Fabrication Symposium—An Additive Manufacturing Conference, no. i*, (University of Texas at Austin, 2016), pp. 469–486

33. M.M. Farag, H.S. Yun, Effect of gelatin addition on fabrication of magnesium phosphate-based scaffolds prepared by additive manufacturing system. Mater. Lett. **132**, 111–115 (2014). https://doi.org/10.1016/j.matlet.2014.06.055
34. S. Palanivel, P. Nelaturu, B. Glass, R.S. Mishra, Friction stir additive manufacturing for high structural performance t hrough microstructural control in an Mg based WE43 alloy. Mater. Des. (1980–2015) **64**, 934–952 (2015). https://doi.org/10.1016/j.matdes.2014.09.082
35. R.S. Mishra, Z.Y. Ma, Friction stir welding and processing. Mater. Sci. Eng. R. Rep. **50**(1–2), 1–78 (2005). https://doi.org/10.1016/j.mser.2005.07.001
36. A. Yamashita, Z. Horita, T.G. Langdon, Improving the mechanical properties of magnesium and a magnesium alloy through severe plastic deformation. Mater. Sci. Eng. A **300**(1–2), 142–147 (2001). https://doi.org/10.1016/S0921-5093(00)01660-9
37. S. Tekumalla, R. Shabadi, C. Yang, S. Seetharaman, M. Gupta, Strengthening due to the in-situ evolution of ß1′ Mg-Zn rich phase in a ZnO nanoparticles introduced Mg-Y alloy. Scr. Mater. **133**, 29–32 (2017). https://doi.org/10.1016/j.scriptamat.2017.02.019
38. A. Babenko et al., Comprehensive insights into recent innovations: magnesium-inclusive high-entropy alloys. J. Magnes. Alloys **12**(4), 1311–1345 (2024)
39. S.A. Krishna, N. Noble, N. Radhika, B. Saleh, A comprehensive review on advances in high entropy alloys: fabrication and surface modification methods, properties, applications, and future prospects. J. Manuf. Process. **109**, 583–606 (2024). https://doi.org/10.1016/j.jmapro.2023.12.039
40. D.B. Miracle, O.N. Senkov, A critical review of high entropy alloys and related concepts. Acta Mater. **122**, 448–511 (2017). https://doi.org/10.1016/j.actamat.2016.08.081
41. D. Kumar, R. Seetharam, K. Ponappa, A review on microstructures, mechanical properties and processing of high entropy alloys reinforced composite materials. J. Alloys Compd. **972**, 172732 (2024). https://doi.org/10.1016/j.jallcom.2023.172732
42. Z. Yuan, W. Tian, F. Li, Q. Fu, Y. Hu, X. Wang, Microstructure and properties of high-entropy alloy reinforced aluminum matrix composites by spark plasma sintering. J. Alloys Compd. **806**, 901–908 (2019). https://doi.org/10.1016/j.jallcom.2019.07.185
43. S. Jayalakshmi, R. Singh, M. Gupta, Metallic glasses as potential reinforcements in Al and Mg matrices: a review. Technologies **6**(2), 40 (2018). https://doi.org/10.3390/technologies6020040
44. G. Anbuchezhiyan, B. Mohan, D. Sathianarayanan, T. Muthuramalingam, Synthesis and characterization of hollow glass microspheres reinforced magnesium alloy matrix syntactic foam. J. Alloys Compd. **719**, 125–132 (2017). https://doi.org/10.1016/j.jallcom.2017.05.153
45. J.H. Chen, P.S. Liu, S. Song, Preparation and compression performance of porous magnesium alloy composite with ceramic hollow spheres. J. Alloys Compd. **894**, 162397 (2022). https://doi.org/10.1016/j.jallcom.2021.162397
46. S. Sankaranarayanan, Q.B. Nguyen, R. Shabadi, A.H. Almajid, M. Gupta, Powder metallurgy hollow fly ash cenospheres' particles reinforced magnesium composites. Powder Metall. **59**(3), 188–196 (2016). https://doi.org/10.1080/00325899.2016.1139339
47. S. Prasadh, V. Manakari, G. Parande, R.C.W. Wong, M. Gupta, Hollow silica reinforced magnesium nanocomposites with enhanced mechanical and biological properties with computational modeling analysis for mandibular reconstruction. Int. J. Oral Sci. **12**(1), 31 (2020). https://doi.org/10.1038/s41368-020-00098-x

Chapter 4
Magnesium-Based Composites with Micron-Sized Reinforcements

This chapter explores the advancements in magnesium matrix composites reinforced with either ceramic fibers or particulate materials such as silicon carbide (SiC), boron carbide (B_4C), titanium carbide (TiC), aluminum oxide (Al_2O_3), and boron nitride (BN). Employing diverse fabrication methods, including melt-stir casting, powder metallurgy, spark plasma sintering, and equal channel angular extrusion, researchers have achieved significant improvements in the mechanical, microstructural, and tribological properties of magnesium composites. Reinforcements have been found to enhance grain refinement, tensile strength, hardness, wear resistance, and thermal stability, making these composites suitable for demanding environments. Submicron and nano-sized reinforcements have proven particularly effective, although challenges like particle clustering, reduced ductility at high reinforcement levels, and porosity require advanced techniques such as ultrasonic stirring and sintering optimization. These studies demonstrate the importance of tailoring particulate size, volume fraction, and processing conditions to balance strength, ductility, and structural integrity. Magnesium composites reinforced with optimized ceramic phases show great potential for aerospace, automotive, and other applications requiring lightweight, high-performance materials.

4.1 Introduction

Magnesium-based metal matrix composites (Mg-MMCs) are advanced materials that combine a lightweight magnesium matrix with reinforcing phases such as ceramics, fibers, or particles. These composites are designed to improve the mechanical, thermal, and wear properties of pure magnesium, making them ideal for applications requiring high strength-to-weight ratios. Mg-MMCs exhibit enhanced stiffness, superior fatigue resistance, and improved dimensional stability under high temperatures. The incorporation of reinforcements in the form of fibers or particles in general significantly enhances the load-bearing capacity and wear resistance of

S. Seetharaman et al., *Magnesium-based Syntactic, Metastable and Nano-composites*, https://doi.org/10.1007/978-3-031-96514-2_4

the magnesium matrix. Due to their lightweight and enhanced performance, Mg-MMCs are widely used in aerospace, automotive, and biomedical sectors. Ongoing research focuses on optimizing reinforcement distribution, reducing production costs, and improving corrosion resistance to expand their industrial applications further.

4.2 Magnesium Composites with Ceramic Fibers

AM100 alloy reinforced with saffil alumina short fibers (15–25% volume fraction) was produced by squeeze casting [1]. At room temperature, the composites exhibited hardness values nearly double that of the base alloy, reaching 165 BHN for the 25% fiber fraction compared to 85 BHN for the alloy. While the unreinforced alloy showed significant strength reduction at elevated temperatures (e.g., one-third of room temperature strength at 200 °C), the composites maintained superior tensile strength due to fiber reinforcement (Fig. 4.1). However, elongation was compromised, with composites displaying significantly reduced ductility compared to the base alloy. Fractographic analysis revealed matrix cracking and fiber pull-outs at higher temperatures, emphasizing the composites' temperature-resistance mechanism.

Hu et al. [2] explored the interfacial and fracture behavior of AE44 alloy-based composites reinforced with short Al_2O_3 fibers, fabricated using squeeze casting. The AE44 alloy, composed of 4% Al, 4.1% RE elements, and 0.3% Mn, was chosen for its creep resistance and fracture toughness at elevated temperatures. The incorporation of 15 vol% fibers in a silica binder matrix led to significant improvements in tensile strength and stiffness. The composite's ultimate tensile strength increased to 221.62 MPa, compared to 197.69 MPa for the base alloy, while the elastic modulus rose from 43.6 GPa to 72.3 GPa (Fig. 4.2). However, elongation decreased

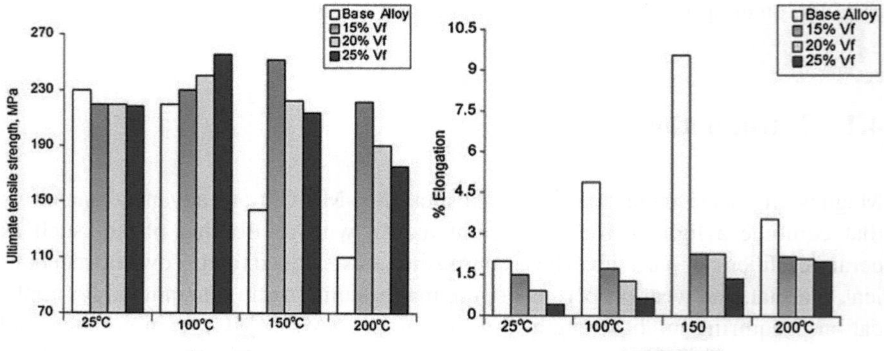

Fig. 4.1 Mechanical properties of saffil alumina short fibers reinforced AM100 alloy composites [1]

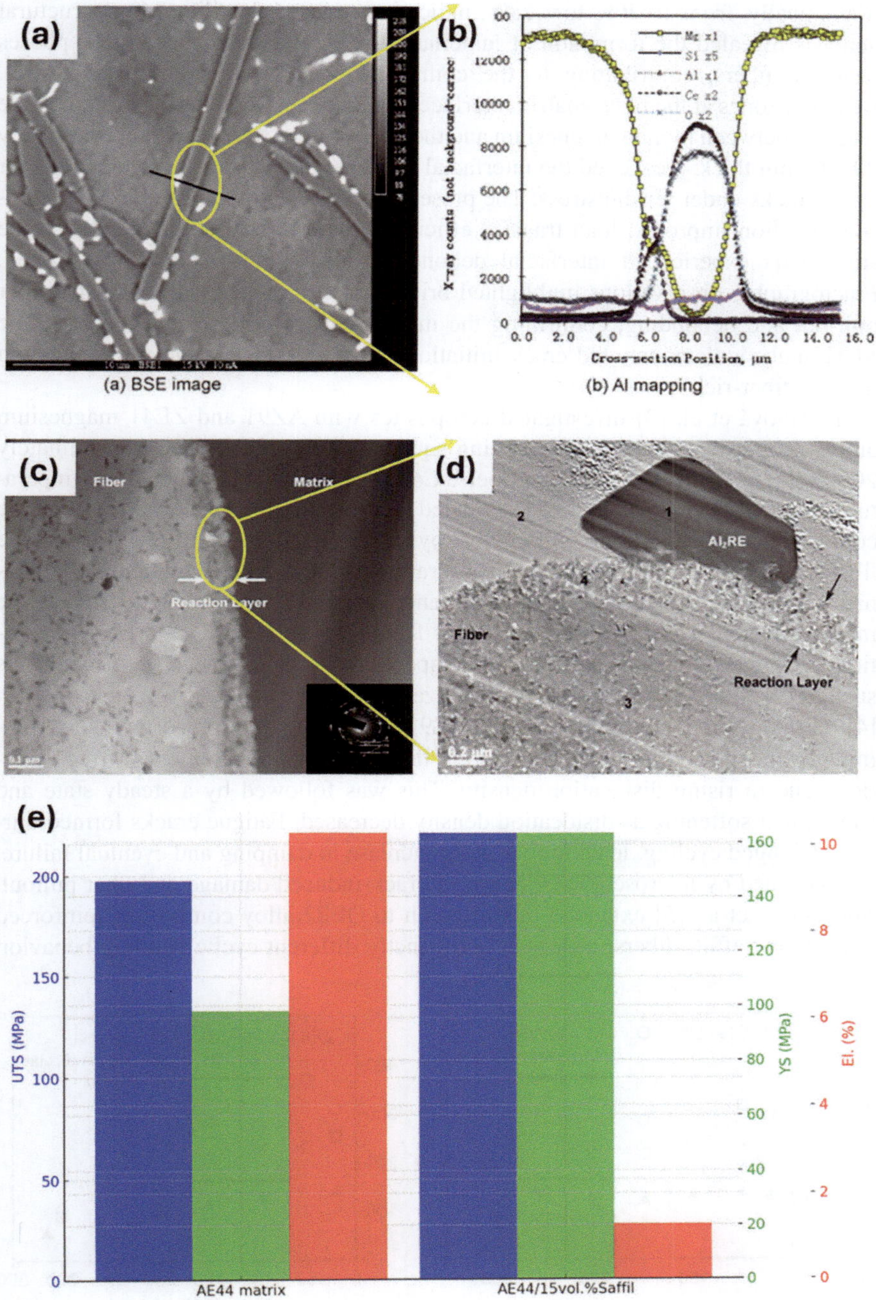

Fig. 4.2 Interfacial characteristics (**a–d**) and mechanical properties (**e**) of short Al_2O_3 fiber reinforced AE44 alloy-based composites (from Hu et al. [2])

dramatically, from 10.26% to 1.28%, indicating reduced ductility. Microstructural analysis revealed the formation of lamellar $Al_{11}RE_3$ and particulate Al_2RE phases near the fibers, contributing to the composite's enhanced strength (Fig. 4.2). Reaction zones at the fiber–matrix interface contained an MgO layer, formed by the reaction between molten magnesium and the silica binder. This layer, approximately 80–100 nm thick, weakened the interfacial bonding and acted as initiation sites for microcracks under tensile stress. The presence of fibers aligned parallel to the tensile direction improved load transfer efficiency, while fibers perpendicular to the stress axis experienced interfacial debonding, leading to fracture propagation. Fractographic observations highlighted brittle fracture modes with extensive fiber pull-out and debonding, confirming the impact of weak interfacial bonding. The SEM analysis demonstrated crack initiation in the interfacial region, propagating through fiber-rich zones.

Trojanová et al. [3] investigated composites with AZ91 and ZE41 magnesium alloys as matrices, manufactured using squeeze casting to achieve approximately 20% volume fraction of alumina fibers. Compression tests conducted at temperatures ranging from 20°C to 300°C revealed significantly higher flow stresses in the composites compared to unreinforced alloys at temperatures up to 250 °C, with the fiber effect diminishing at higher temperatures (Fig. 4.3). Strengthening mechanisms were identified as load transfer and increased dislocation density at the matrix–fiber interface, with contributions from thermal stresses and fiber orientation. However, softening at elevated temperatures was attributed to mechanisms such as cross-slip and dynamic recrystallization. Further studies by Trojanová et al. [4] on the cyclic bending behavior of AZ91 composites reinforced with short alumina fibers showed an initial increase in the logarithmic decrement of free vibrations due to rising dislocation density. This was followed by a steady state and subsequent softening as dislocation density decreased. Fatigue cracks formed during prolonged cycling, leading to a sharp increase in damping and eventual failure, as confirmed by microscopic evidence of crack-induced damage and fiber pullout. Mielczarek et al. [5] extended this research to QE22 alloy composites reinforced with short Saffil® fibers, observing a distinctly different cyclic bending behavior.

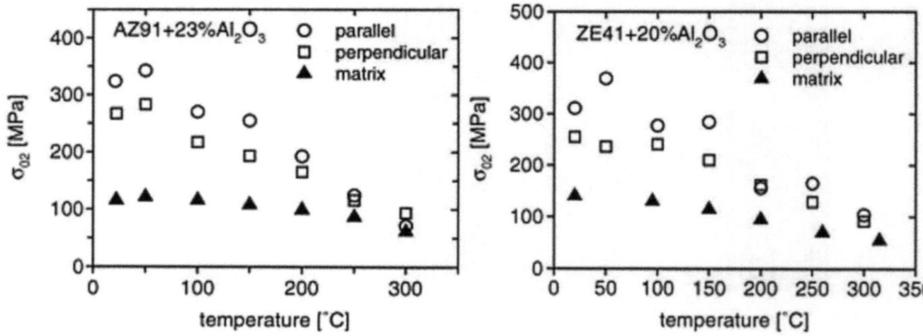

Fig. 4.3 Temperature dependence of yield stress for saffil fiber reinforced AZ91 and ZE41 alloys in two orientations fiber plane [3]

Unlike AZ91 composites, QE22 composites displayed an initial decrease in the logarithmic decrement due to reduced mobile dislocation density, followed by a quasi-steady state with minimal damping fluctuations over numerous cycles. The absence of brittle intermetallic phases in QE22 delayed crack formation, with no significant damage or fiber pullout observed until the final stages of fatigue life. The improved ductility and fatigue resistance of QE22 were attributed to its more homogeneous microstructure and lack of crack-prone intermetallic phases.

Sklenička et al. [6] investigated the creep response of AZ91 alloy composite reinforced with 20 vol.% Al_2O_3 fibers. The results revealed remarkable enhancements, with 2–3 orders of magnitude lower creep rates compared to the unreinforced matrix (Fig. 4.4). This was attributed to the effective load transfer from the matrix to the fibers and the formation of $Mg_{17}(Al,Zn)_{12}$ precipitates at the fiber interface, which strengthened the material by promoting heterogeneous nucleation. These enhancements were particularly evident at elevated temperatures of 423 K and 473 K, where the composite outperformed its unreinforced counterpart in terms of creep resistance and structural integrity under stress.

Bakkar et al. [7] investigated the corrosion behavior of AS41 alloy reinforced with 20 vol% alumina fibers (Al_2O_3/Mg MMC) fabricated using squeeze casting, exploring its performance in various aqueous environments. The composite demonstrated comparable corrosion resistance to the base alloy under neutral and alkaline conditions but showed significantly higher susceptibility in chloride-rich environments, such as 3.5% NaCl. Here, the corrosion current density (Icorr) increased threefold, and the corrosion rate was seven times higher than the matrix alloy after prolonged exposure (Fig. 4.5). High-aluminum zones (HA) near the fibers, containing 9 wt% aluminum compared to the 3 wt% in the matrix, acted as cathodic sites, while low-aluminum zones (LA) were preferentially corroded. Hydrogen evolution tests showed reduced overall corrosion for the composite when adjusted for the inert alumina fraction, but localized pitting was observed, with pits propagating through LA zones while avoiding HA regions. SEM analysis revealed that the alumina fibers remained intact, limiting pit expansion around their sites. The study concluded that while alumina reinforcement enhanced localized corrosion resistance through microstructural modification, the composite's vulnerability in chloride-rich environments necessitates improved processing and protective strategies for application in aggressive conditions.

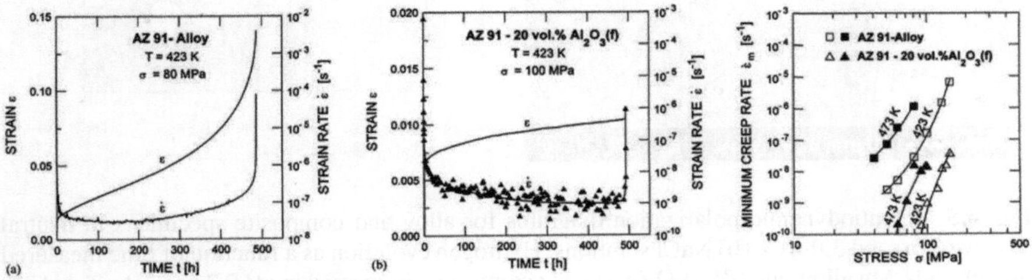

Fig. 4.4 Creep response of AZ91 alloy and its composite containing upto 20 vol.% Al_2O_3 fiber reinforcement [6]

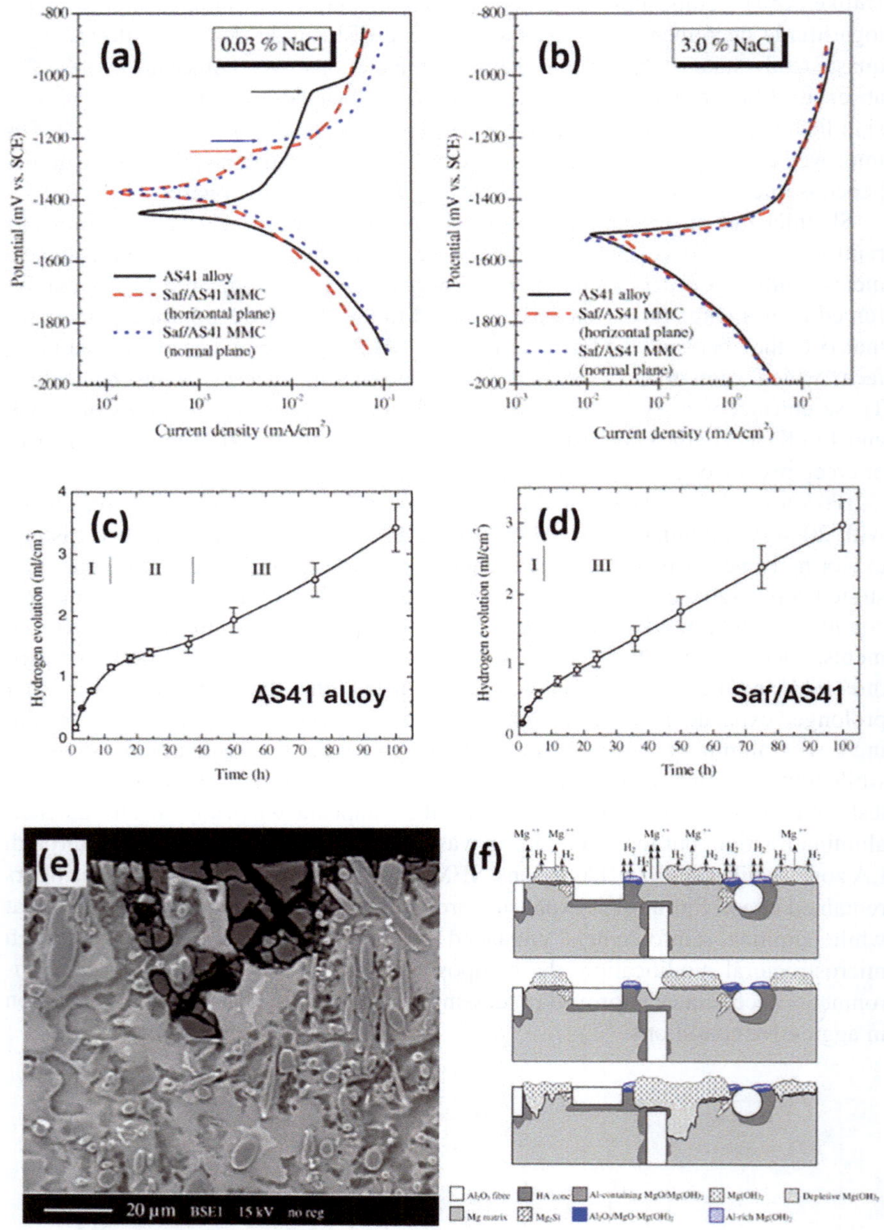

Fig. 4.5 Potentiodynamic polarization diagrams for alloy and composite specimens in neutral 0.03 wt% (**a**) and 3.0 wt% (**b**) NaCl solutions. Hydrogen evolution as a function of time measured on: (**c**) AS41 Mg alloy, and (**d**) Al_2O_3/AS41 Mg composite, immersed in pH 9.3 borax-buffered 3% NaCl solution. (**e**) SEM cross-sectional micrograph of Al_2O_3/AS41 Mg composite following 144 h immersion in pH 9.3 borax-buffered 3% NaCl solution, and (**f**) Schematic representative model showing the corrosion mechanism of Al_2O_3/Mg MMC (from [7])

Rauber et al. [8] investigated the microstructure and mechanical behavior of AZ91 and AJ62 magnesium alloys reinforced with SiC particles using injection molding. The study highlighted the uniform distribution of SiC particles in the matrix without significant agglomeration due to optimized processing temperatures and short contact times, which mitigated reactions between the matrix and the particles. Experimental results revealed notable improvements in hardness, and strength with increasing SiC content (Fig. 4.6). The authors attributed these enhancements to a combination of mechanisms, including load transfer, grain size refinement, and dislocation density increase caused by thermal mismatch. Furthermore, comparisons were made between different particles to optimize lightweight design, concluding that particles with high stiffness and low density, such as B_4C, offered the best potential.

Mark et al. [9] used ultra-high modulus carbon fiber (P100S) as reinforcement to improve the tensile strength of AZ91D alloy. This enhancement was particularly significant at high temperatures, where the failure mode shifted from being driven by the fiber/MgO interface at lower temperatures to becoming matrix-dominated at elevated temperatures. They also investigated the mechanical stability of these composites over 100 thermal cycles between ±100 °C and found that the bending strength remained unaffected while there was a reduction in bending modulus and interlaminar shear strength during the initial cycles [10]. This reduction was attributed to changes in matrix microhardness, interfacial sliding, and microvoid formation. In a follow-up study [11], the same authors examined thermally induced strain responses and developed a model for hysteresis and residual plastic strain caused by matrix yielding and interfacial debonding. Recently, Sinan et al. [12] explored the feasibility of using recycled short carbon fibers to enhance the mechanical properties of AZ91 alloy prepared via a high-shear dispersion technique. In this study, the carbon fiber length ranged from 100 μm to 500 μm, and the fiber content varied between 2 wt.% and 5 wt.%. The resulting composites were evaluated for their microstructural characteristics and performance under indentation, tension, compression, and creep tests at various temperatures (25–200 °C). The addition of

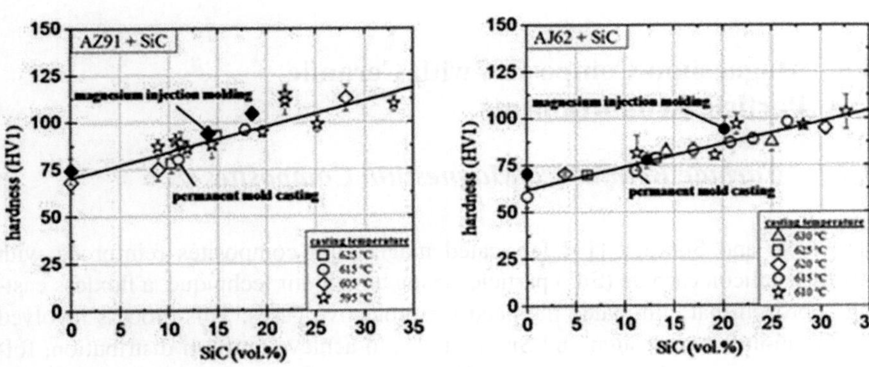

Fig. 4.6 Microstructural and mechanical properties of Sic fiber reinforced AZ91 and AJ62 alloys

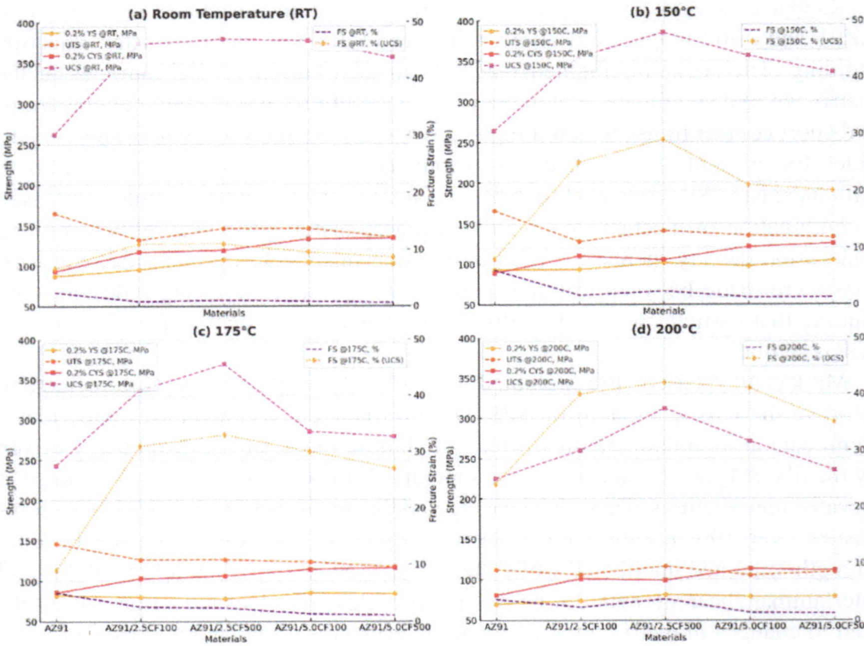

Fig. 4.7 Mechanical properties of AZ91 alloy and its carbon fiber reinforced composites (100 and 500 refer to CF length)

recycled carbon fibers improved the hardness and compressive strength compared to the unreinforced alloy. However, tensile strength and ductility were compromised due to issues related to porosity and fiber-matrix interface quality (Fig. 4.7). Regarding creep performance, the composites showed enhanced potential for creep resistance at elevated temperatures, although the minimum creep rate remained comparable to that of the unreinforced alloy (Fig. 4.8).

4.3 Magnesium Composites with Ceramic Particle Reinforcements

4.3.1 *Carbide Reinforced Magnesium Composites*

Saravanan and Surappa [13] fabricated magnesium composites reinforced with 30 vol.% silicon carbide (SiC) particles using a melt-stir technique, a fluxless casting approach that eliminated the need for protective gases. This process involved stirring molten magnesium and SiC particles to achieve uniform distribution, followed by casting and extrusion at 500°C with a 13:1 extrusion ratio. Microstructural analysis revealed that the composite had a finer grain size of 20 μm compared to

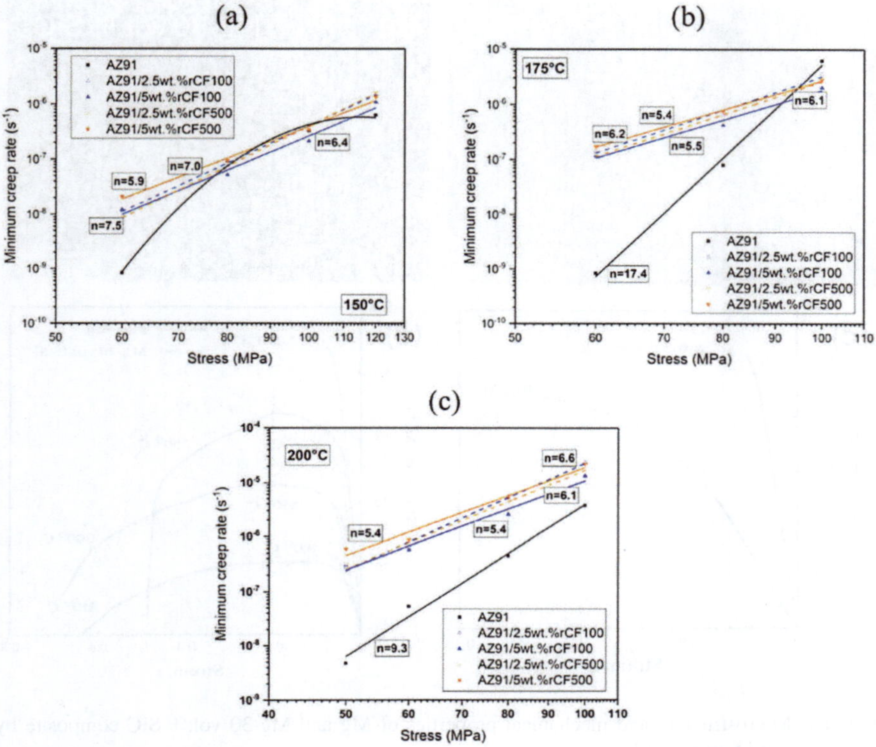

Fig. 4.8 Plots of minimum creep rate corresponding of AZ91 alloy and its carbon fiber reinforced composites over applied creep stress at (**a**) 150 °C, (**b**) 175 °C, and (**c**) 200 °C

50 μm in pure magnesium, with no reaction products observed at the SiC-matrix interface, indicating excellent wetting. Mechanical testing at room temperature showed a significant increase in stiffness (40%) and ultimate tensile strength (UTS, 30%) over unreinforced magnesium (Fig. 4.9). The composite retained superior mechanical properties up to 400 °C, unlike pure magnesium, which softened beyond 350 °C (Table 4.1). Tribological studies indicated a two-order magnitude reduction in wear rate for the composite compared to pure magnesium during pin-on-disc tests, attributed to the reinforcing effect of SiC particles. The study concluded that SiC reinforcement significantly enhances both mechanical and tribological performance without requiring complex processing conditions.

The influence of extrusion temperature on Mg-SiC composites containing 11.5 wt.% SiC, synthesized via disintegrated melt deposition, was investigated [14]. Extrusion temperatures ranged from 350 to 100 °C, and their effects on microstructure and mechanical properties were studied. The composite was produced by superheating magnesium and SiC particles, followed by deposition through argon gas jets and hot extrusion with a 13:1 reduction ratio. Lower extrusion temperatures resulted in better alignment of SiC particles, reduced clustering, and minimal

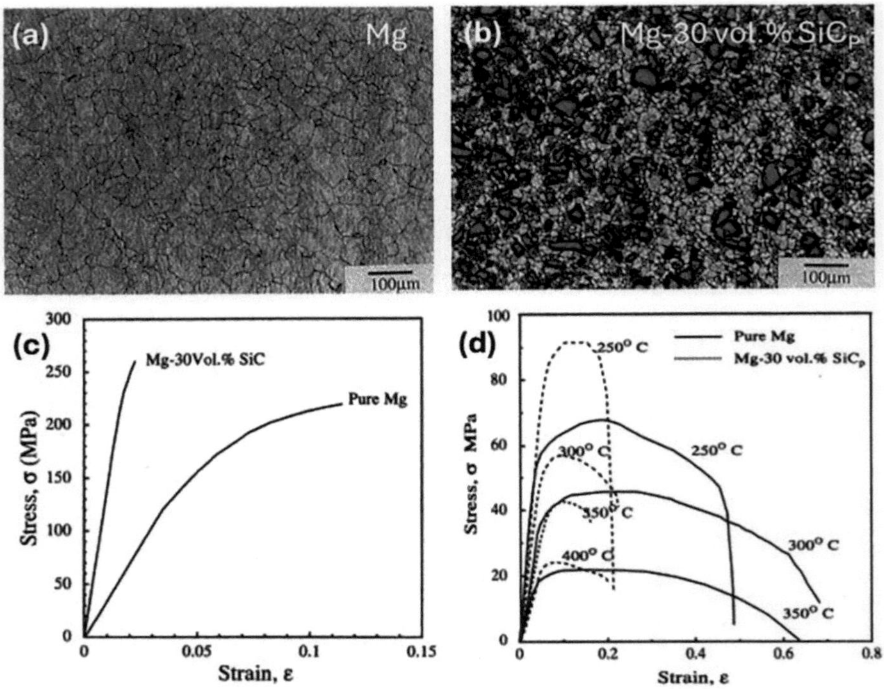

Fig. 4.9 Microstructure and mechanical properties of Mg and Mg-30 vol.% SiC composite by Saravanan and Surappa [13]

porosity. These microstructural improvements enhanced mechanical properties, including significant increases in hardness, elastic modulus, and 0.2% yield strength, while UTS and ductility remained stable. The improvements were attributed to grain refinement and enhanced particle–matrix interactions at lower temperatures.

The role of sub-micron SiC particles in magnesium composites was examined by Reddy et al. [15]. In this study, it was found that the smaller particulates achieved better distribution and stronger interfacial bonding within the matrix. This led to improvements in hardness, wear resistance, and strength. Controlled processing ensured uniform particulate distribution without agglomeration. Microstructural analysis confirmed minimal porosity and uniform dispersion of SiC, which were critical to the enhanced mechanical and tribological properties. Another study used fluxless casting which involved the stirring of molten magnesium and SiC particles under an inert argon atmosphere, followed by casting to synthesize Mg-SiC composites with varying SiC content [16]. The resultant composites were extruded before testing for microstructural and other property characterisation. Microstructural analysis revealed that increasing SiC content resulted in finer grain sizes, better particle distribution, and improved interfacial integrity. X-ray diffraction studies showed no oxide phases or reaction products at the interface, except at very high SiC contents. Mechanical testing demonstrated that hardness and elastic modulus

Table 4.1 Microstructural and mechanical properties of Mg-based metal matrix composites reinforced with micron-sized particle reinforcements

References	Matrix	Reinforcement			Process	Density (g/cc)	Porosity	Average Grain Size (μm)	Hardness	Tensile properties			Compressive properties			Remarks
		Type	Vol.%	Wt.%						Yield Strength (MPa)	Ultimate strength (MPa)	Failure Strain (%)	Yield Strength (MPa)	Ultimate strength (MPa)	Failure Strain (%)	
Kandemir et al. [12]	AZ91	–		0	High shearing dispersion	1.818			55.5 ± 2.9	87 ± 12	165 ± 14	2.4 ± 0.3	93 ± 3	262 ± 7	6.6 ± 1.7	
		CF_{100}		2.5		1.807			58.4 ± 5.6	95 ± 2.8	132 ± 13	0.8 ± 0.2	117 ± 4	373 ± 10	11 ± 0.8	
		CF_{500}				1.817			59.9 ± 2.2	107 ± 7	146 ± 11	1 ± 0.5	119 ± 7	379 ± 5	11 ± 0.2	
		CF_{100}		5		1.818			64.7 ± 3.0	104 ± 4	146 ± 11	0.8 ± 0.2	133 ± 4	377 ± 2	9.5 ± 0.2	
		CF_{500}				1.804			61.1 ± 4.2	102 ± 6	135 ± 17	0.6 ± 0.2	134 ± 5	356 ± 12	8.6 ± 0.8	
Saravanan et al. [13]	Mg	SiC		30	SC					229	258	2				40 μm
S. C. V. Lim and M. Gupta [14]	Mg	SiC350	6.6	11.5	DMD + HE	1.829	0.48	19 ± 6	42.7 ± 0.55	119 ± 3	197 ± 3	3.5 ± 0.19				
		SiC250	6.6	11.5		1.828	0.50	15 ± 4	44.4 ± 0.55	139 ± 7	201 ± 12	3.3 ± 0.65				
		SiC150	6.6	11.5		1.827	0.56	8 ± 2	45.3 ± 0.67	148 ± 3	204 ± 8	3.2 ± 0.89				
		SiC100	6.5	11.4		1.826	0.56	7 ± 3	45.7 ± 0.71	148 ± 7	201 ± 7	3.2 ± 1.07				
Reddy et al. [15]	Mg	SiC		0	DMD + HE					153	207	9.2				0.6 μm
				2.7						182	219	2.1				
				5.8						171	221	1.5				
				9						155	207	1.4				
				2.7						200	233	2.9				HT 5 h
				9						168	213	3.6				HT 5 h
Lim et al. [16]	Mg	SiC		0	DC					110	199	10.2				25 μm
				4.3						112	191	6.8				
				8.7						112	177	4.7				
				16.8						114	154	1.8				

(continued)

Table 4.1 (continued)

		Reinforcement								Tensile properties			Compressive properties			
References	Matrix	Type	Vol.%	Wt.%	Process	Density (g/cc)	Porosity	Average Grain Size (μm)	Hardness	Yield Strength (MPa)	Ultimate strength (MPa)	Failure Strain (%)	Yield Strength (MPa)	Ultimate strength (MPa)	Failure Strain (%)	Remarks
Muhammad et al. [18]	Mg	SiC		0	SPS + HE					115						0.5 –25 μm
				5						130						
				10						140						
				15						120						
	AZ31	SiC		5						160	–	–				
				10						165	–	–				
				15						150	–	–				
Xie et al. [20]	Mg-9Al	SiC		0	Liquid sintering			6.6		276	327.2					
				15				3.3		339	363.5					
				0	Solid sintering			12.0		239	345					
				15				7.3		271	355.0					
Deng et al. [21]	AZ91	SiC	0		Stir casting	1.8099	0.0056	7	82	225	352					0.2 μm
			0.5			1.8164	0.0248	5.9	94	260	300					
			1			1.8228	0.0560	2.8	96	275	330					
			1.5			1.8292	0.0795	2.6	99	280	360					
			2			1.8356	0.1067	3	100	290	365					
			3			1.8489	0.1375	3.2	102	260	350					
			5			1.8739	0.2726	3.5	115	255	330					

References	Matrix	Reinforcement			Process	Density (g/cc)	Porosity	Average Grain Size (μm)	Hardness	Tensile properties			Compressive properties			Remarks
		Type	Vol.%	Wt.%						Yield Strength (MPa)	Ultimate strength (MPa)	Failure Strain (%)	Yield Strength (MPa)	Ultimate strength (MPa)	Failure Strain (%)	
Wang et al. [22]	AZ91	SiC		0	SC + HE250C/ R12					225	350	8				10 μm/ Extr. Temp. (250 C & 350 C), Extr. Ratio (R12 & R5)
				10						225	290	2				
				0	SC + HE350C/ R12					170	290	20				
				10						250	320	5				
				0	SC + HE350C/ R5					170	290	–				
				10						230	300	–				
Nie et al. [23]	AZ91	SiC		0	SC(U)					70	130	2				1 μm
				3						100	155	1.8				
				5						105	140	0.7				
Deng et al. [24]	AZ91	SiC	2		Stir casting			3		280	370	–				0.2 μm SiC
								4		250	320	–				
								3.6		250	310	–				
			5					3.4		260	325	–				5 μm SiC
								2.5		270	330	–				
								2.4		270	315	–				
			10					3.6		–	325	–				10 μm SiC
								1.5		300	385	–				
								1.5		280	350	–				
Gui et al. [25]	AZ91	SiC		15	SC(V) + T4 HT					84	225	7.2				12.8 μm
				15						178	218	1.1				
Soon et al. [26]	Mg	SiC		0	DMD + HE					153	219	3.1				35 μm
				33.6						158	205	12.1				

(continued)

Table 4.1 (continued)

References	Matrix	Reinforcement			Process	Density (g/cc)	Porosity	Average Grain Size (μm)	Hardness	Tensile properties			Compressive properties			Remarks
		Type	Vol.%	Wt.%						Yield Strength (MPa)	Ultimate strength (MPa)	Failure Strain (%)	Yield Strength (MPa)	Ultimate strength (MPa)	Failure Strain (%)	
Nie et al. [27]	AZ91	SiC		0	SC					216	332	15.5				0.2 μm
				2						285	377	4.2				
Kumar et al. [28]	AZ91	SiC	0		Stir casting	1.80	0.53		60 ± 1.8	121 ± 2.8	188 ± 4.5			308 ± 5.3		
			2			1.81	1.32		77 ± 2.4	73 ± 1.6	114 ± 3.4			316 ± 2.8		
			5			1.85	1.51		85 ± 3.5	64 ± 1.2	138 ± 2.6			323 ± 3.5		
			8			1.88	1.92		102 ± 2.5	97 ± 1.4	141 ± 2.7			331 ± 6.0		
			11			1.92	2.11		108 ± 3	126 ± 2.0	196 ± 3.8			340 ± 2.5		
Jiang et al. [35]	AZ91	TiC		0	SHS					–	160	9				
				10						–	214	4				
Zhang et al. [36]	AZ91	TiC		0	BM + DC					95	198	2.5				
				3.1						115	235	1				
Xi et al. [38]	ZK51	Ti alloy		0	PM					229	315	8.5				
				10						295	386	5.6				
	ZK51	SiC		0	PM					229	315	8.5				37 μm
Karuppusamy et al. [39]	AZ91	WC		1	Stir casting					152	200	0.080	106	300	0.175	
				1.5						132	190	0.125	120	325	0.160	
				3						128	187	0.132	128	342	0.142	
Hassan and Gupta [40]	Mg	Al_2O_3	0		DMD	1.7397	0.02	49 ± 8		97 ± 2	173 ± 1					
			0.7			1.7541	0.04	6 ± 2		214 ± 4	261 ± 5					
			1.1			1.7636	0.06	6 ± 1		200 ± 1	256 ± 1					
			2.5			1.7897	0.32	4 ± 1		222 ± 2	281 ± 5					

References	Matrix	Reinforcement			Process	Density (g/cc)	Porosity	Average Grain Size (μm)	Hardness	Tensile properties			Compressive properties			Remarks
		Type	Vol.%	Wt.%						Yield Strength (MPa)	Ultimate strength (MPa)	Failure Strain (%)	Yield Strength (MPa)	Ultimate strength (MPa)	Failure Strain (%)	
Wong et al. [41]	Mg	Al_2O_3		0	MWS—PM + HE					117	168	2.8				0.3 μm
				5						160	215	3.9				
				2.5						130	167	4				
Cay et al. [44]	Mg	Al_2O_3		0		1.74	~0									
				0.05		1.57	10									
				1		1.57	10									
		–		0		1.13	35									Porous
Rahmani et al. [46]	Mg	Al_2O_3	0			1.71	1.6		41.7 ± 2.3							
			10			1.91	2.5		47.4 ± 2.5							
			20			2.08	4.6		58.2 ± 1.7							
			30			2.18	9.2		64.9 ± 9.9							
		MgO	10			1.81	5.8		44.9 ± 0.5							
			20			1.93	8.4		49.3 ± 2.2							
			30			2.02	11.8		52.6 ± 2.5							
Hassan et al. [49]	Mg	Al_2O_3		1.1	PM + HE					172	227	16.8				1 μm
				1.1						182	237	12.1				0.3 μm
	Mg	Al_2O_3		0.7	DMD + HE					214	261	12.5				0.3 μm
				1.1						200	251	8.6				
				2.5						222	281	4.5				

(continued)

Table 4.1 (continued)

References	Matrix	Reinforcement			Process	Density (g/cc)	Porosity	Average Grain Size (μm)	Hardness	Tensile properties			Compressive properties			Remarks
		Type	Vol.%	Wt.%						Yield Strength (MPa)	Ultimate strength (MPa)	Failure Strain (%)	Yield Strength (MPa)	Ultimate strength (MPa)	Failure Strain (%)	
Deng et al. [50]	AZ91	SiC		10	SSM-PDC + HT (T4) + HF					116	139	0.8				Before forging
				10						–	222	0.5				Forg. @ 320 °C
				10						–	239	0.5				Forg. T 370 °C
				10						238	251	0.8				Forg. T 420 °C
				10						226	242	0.8				Forg. T 470 °C
				10						215	230	0.8				Forg. T 520 °C

BM ball milling, DC die casting, *DMD* disintegrated melt deposition, *HE* hot extrusion, *Ext. T* extrusion temperature, *Ext. R* extrusion ratio, *HF* hot forging, *Forg. T* forging temperature, *HT* heat treatment, *LPS* liquid phase sintering, *PM* powder metallurgy, *MWS-PM* microwave sintering assisted powder metallurgy, *SC* stir casting, *SC(U)* ultrasonic stir casting, *SC(V)* vacuum stir casting, *SPS* spark plasma sintering, *SSS-PDC* semi-solid stirring and pressure die casting, *SHS* self-propagating high-temperature synthesis increased linearly with SiC content, but higher reinforcement levels reduced ductility and UTS due to brittleness introduced by the particulate phase.

Kumar et al. [17] utilized powder metallurgy to fabricate Mg-SiC composites with SiC reinforcement levels of 0%, 4%, 8%, and 12% by weight. Magnesium and SiC powders were blended, compacted under a pressure of 650 MPa, and sintered at temperatures ranging from 500 to 670 °C. The resulting composites were evaluated for microstructural, density, and mechanical properties. The 12 wt.% SiC composite exhibited the highest mechanical performance, including significant improvements in hardness, compressive strength, and impact strength compared to lower reinforcement levels and pure magnesium (Fig. 4.10). Microstructural analysis showed uniform particle distribution and strong bonding at the particle-matrix interface. Density measurements confirmed low porosity, contributing to the superior mechanical properties.

Using spark plasma sintering technology, Muhammad et al. [18] fabricated AZ31 alloy composites reinforced with SiC particles. This study identified the optimal sintering temperatures for pure Mg and AZ31/SiC composites as 585 and 552 °C, respectively. Uniform dispersion of SiC particles was achieved at SiC contents up to 10 wt%; beyond this, agglomeration reduced the quality of interfacial bonding, leading to a decline in mechanical properties. Mechanical testing revealed that hardness increased with higher SiC content, while tensile strength peaked at 10 wt% SiC, achieving 140 MPa for Mg-SiC and 170 MPa for AZ31-SiC composites. At higher SiC content, tensile strength declined due to particle clustering. Microstructural analysis indicated the absence of reaction products at the SiC-Mg interface, maintaining the composite's integrity. In a similar research [19], Mg-3 wt% Al alloy composites reinforced with varying amounts of SiC (3%, 6%, and 9% by weight) were fabricated using powder metallurgy and process included ball milling, compaction at 5 tons, and sintering at 450 °C under an argon atmosphere. SEM and XRD analyses confirmed uniform distribution of SiC in the matrix and the absence of intermetallic compounds (Fig. 4.11). Mechanical testing demonstrated that density, hardness, and compressive strength improved with increasing SiC content (Fig. 4.11). The composite with 9 wt% SiC showed the highest compressive strength (169 MPa) and hardness (66.9 Hv). Corrosion resistance also improved significantly, with the lowest corrosion rate recorded for the 9 wt% SiC composite (0.000142 mm/year).

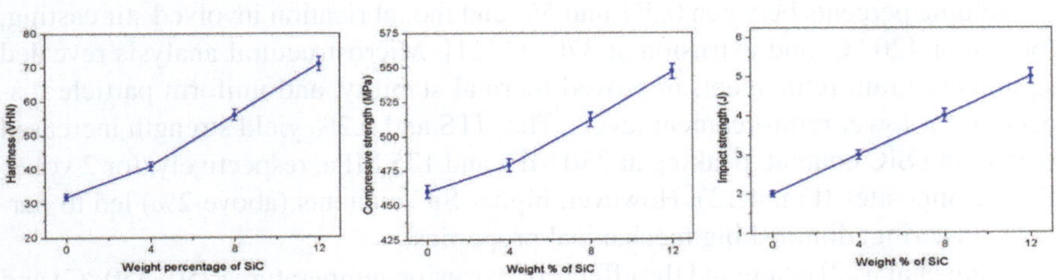

Fig. 4.10 Mechanical properties of Mg/SiC composites fabricated by Kumar et al. [17]

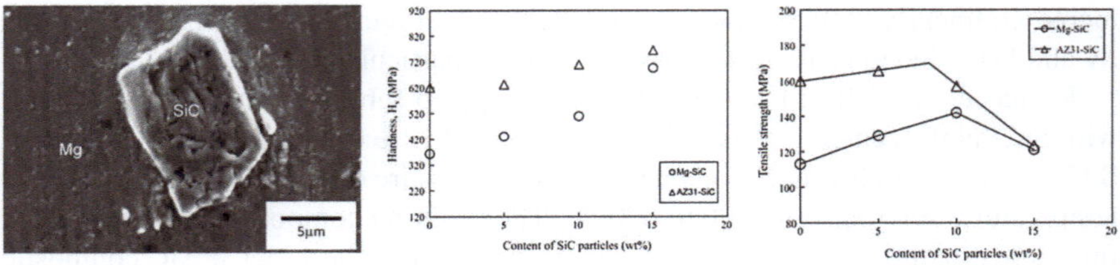

Fig. 4.11 Microstructure and mechanical properties of SiC reinforced AZ31 alloy composite [18]

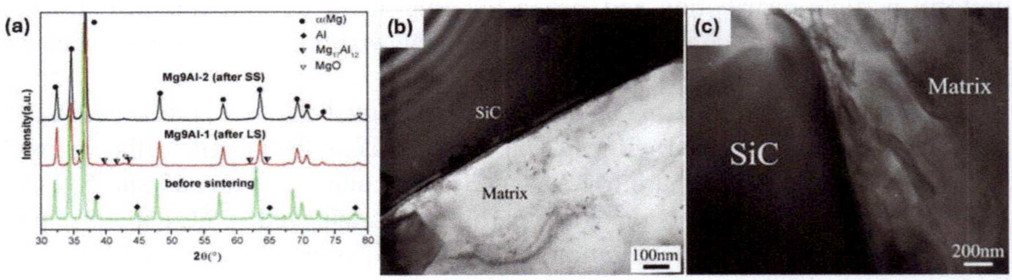

Fig. 4.12 (**a**) XRD pattern corresponding to developed materials and (**b, c**) TEM images showing the interfacial properties of solid phase sintered 15 vol.% SiCp/Mg-9Al composites from Wen Xie et al. [20]

Wen Xie et al. [20] explored the effects of liquid-phase sintering and solid-phase sintering on Mg-9Al alloy and 15 vol.% SiCp/Mg-9Al composites produced via powder metallurgy. Liquid phase sintering was performed at 470 °C for 1 h, while solid phase sintering was conducted at 427 °C for 10 h. Mechanical testing revealed that composites processed via liquid phase sintering exhibited significantly higher ultimate tensile strength of 363.5 MPa and yield strength of 339 MPa, marking 15% and 22.8% improvements, respectively, compared to the matrix alloy. These enhancements were attributed to grain refinement and the presence of $Mg_{17}Al_{12}$ phases as observed in the XRD results (Fig. 4.12a). In contrast, composites produced by solid phase sintering displayed lower strength but higher ductility due to the absence of these phases as seen in Fig. 4.12b, c. Liquid phase sintering also yielded finer grain sizes of 3.3 μm compared to 6.6 μm in the matrix alloy.

The influence of submicron SiC particles (0.2 μm) on AZ91 alloy was examined for volume percents between 0.5% and 5% and the fabrication involved stir casting, forging at 420 °C, and extrusion at 370 °C [21]. Microstructural analysis revealed significant grain refinement, improved thermal stability, and uniform particle dispersion for lower reinforcement levels. The UTS and 0.2% yield strength increased with rising SiC content, peaking at 230 MPa and 175 MPa, respectively, for 2 vol.% SiCp composites (Fig. 4.13). However, higher SiC contents (above 2%) led to particle clustering, diminishing mechanical properties.

Wang et al. [22] examined the effects of extrusion temperature (250–350 °C) and extrusion ratios (5:1 and 12:1) on AZ91 magnesium composites reinforced with

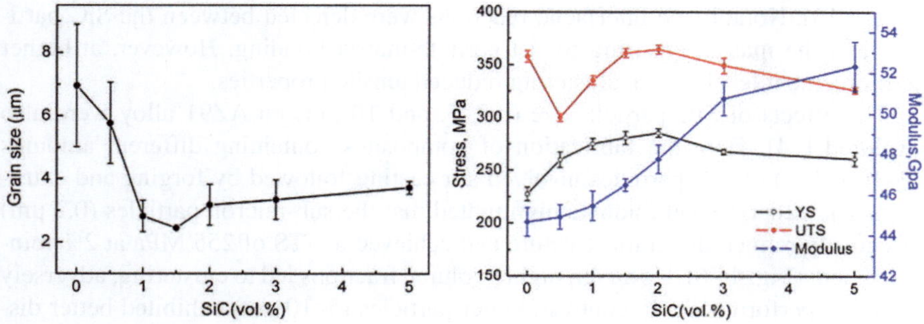

Fig. 4.13 Microstructural and mechanical properties of SiC particles reinforced AZ91 alloy composites from [21]

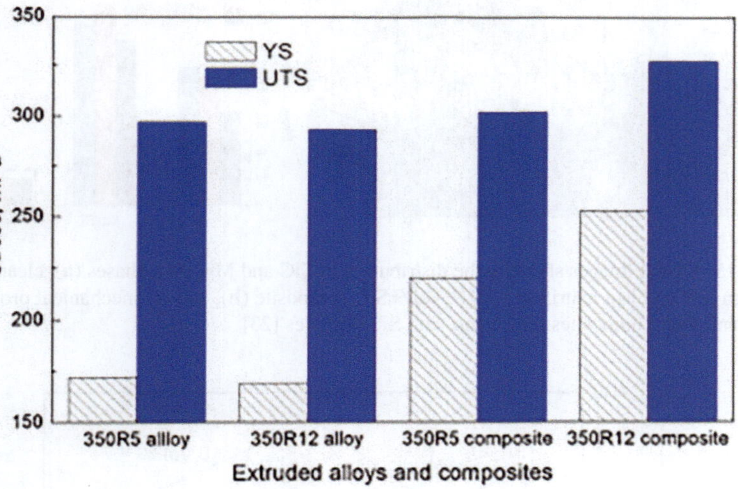

Fig. 4.14 Mechanical properties of SiC reinforced AZ91 alloy composites from Wang et al. [22]

10 vol.% SiC particles. Stir casting and subsequent extrusion reduced particle clustering and refined the microstructure. Higher extrusion temperatures (up to 350 °C) and ratios (12:1) improved particle alignment and mechanical properties. The optimal combination yielded a UTS of 275 MPa and a yield strength of 215 MPa (Fig. 4.14). However, excessive extrusion led to particle cracking, emphasizing the need for parameter optimization to balance strength and ductility.

The research work by Nie et al. [23] employed a novel semisolid stirring and ultrasonic vibration technique to fabricate composites reinforced with 1 μm SiC particles at 3% and 5% volume fractions. This method ensured excellent particle dispersion, leading to significant grain refinement. The UTS and 0.2% yield strength were 245 MPa and 200 MPa, respectively, for the 3 vol.% SiCp/AZ91 composite. Lamellar $Mg_{17}Al_{12}$ phases formed, contributing to enhanced strength and ductility

(Fig. 4.15). Notably, no interfacial reactions were detected between the SiC particles and the matrix, ensuring robust particle-matrix bonding. However, at higher reinforcement levels (5%), clustering reduced tensile properties.

The effects of SiC particle size (0.2, 5, and 10 μm) on AZ91 alloy were also explored [24]. Here the fabrication of composites containing different amounts (2–10 vol.%) of SiC particles involved stir casting followed by forging and extrusion. The experimental findings highlighted that the sub-micron particles (0.2 μm) effectively refined the grain structure and achieved a UTS of 255 MPa at 2% reinforcement (Fig. 4.16). However, higher volume fractions led to clustering, adversely affecting performance. In contrast, larger particles (5–10 μm) exhibited better distribution and less clustering. However, the grain refinement was found as minimal and hence the study concluded that the submicron sized SiC particles are more

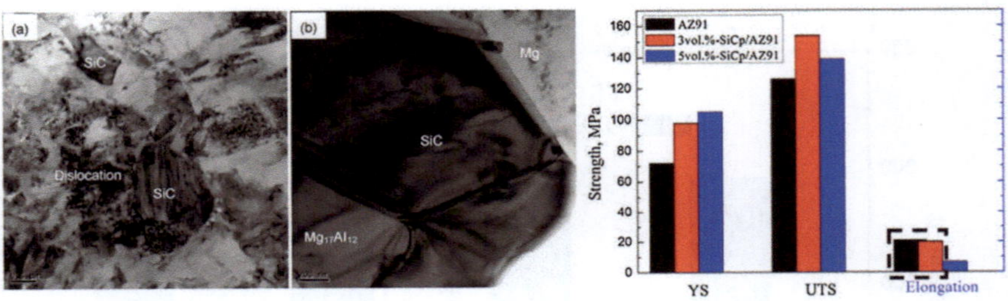

Fig. 4.15 STEM images showing the distribution of SiC and $Mg_{17}Al_{12}$ phases (**a**), clear interface between SiC and Mg matrix in AZ91/5vol.%SiC composite (**b**), and (**c**) mechanical properties of AZ91 and their composites reinforced with SiC particles [23]

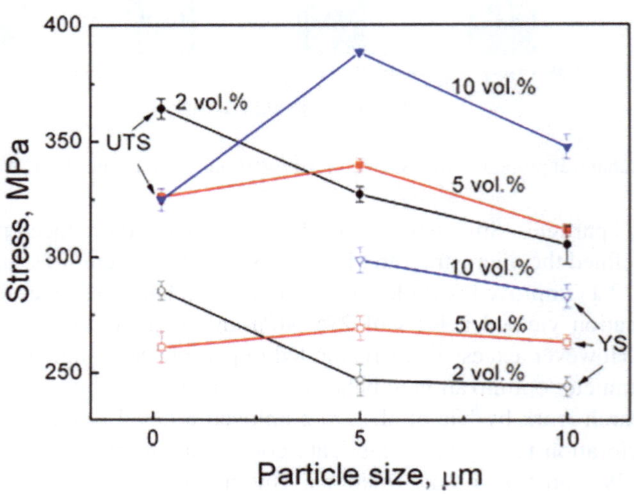

Fig. 4.16 Effect of particle size on the mechanical strength of SiC reinforced Mg composite (from [24])

effective for grain refinement and strengthening although a careful control in reinforcement distribution is required to achieve the intended benefits.

Gui et al. [25] studied the Mg–Al9Zn/15SiCp composite produced using vacuum stir casting, demonstrating a uniform distribution of SiC particles with minimal porosity (~1%) and reduced grain size compared to the unreinforced alloy. Mechanical testing revealed a 33% increase in elastic stiffness (from 42.7 to 57.0 MPa) and a 112% enhancement in yield strength (from 84 to 178 MPa), although elongation dropped from 7.2% to 1.1%. At higher temperatures (750 °C), interfacial reactions between SiC and Mg led to Mg_2Si phase formation, which compromised melt fluidity and mechanical integrity. The composite exhibited a mixed ductile-brittle fracture behavior, indicating improved strength but reduced ductility due to the reinforcement. In another study, Soon and Gupta [26] investigated the recyclability of Mg/SiC composites using the DMD technique. The characterization of extruded composites revealed marginal reductions in SiC content (from 12.7 wt.% to 11 wt.%) and porosity (from 1.11% to 0.21%) over two recycling cycles. Mechanical tests showed improved properties for the recycled composite, with yield strength and ultimate tensile strength reaching 285 MPa and 377 MPa, respectively. Microstructural studies indicated no adverse effects on grain morphology or interfacial integrity, while fracture analysis showed minimal debonding and predominant SiC particle breakage. Magnesium composites reinforced with submicron SiC particles (0.2 μm) were fabricated using room-temperature deformation [27]. The reinforcement facilitated dislocation multiplication and hindered motion, resulting in enhanced mechanical properties. The composite achieved a yield strength of 285 MPa and an ultimate tensile strength of 377 MPa, outperforming the unreinforced AZ91 alloy by about 40%. Submicron SiC particles promoted dislocation pileups and increased twin density, contributing to significant strengthening via load transfer, grain refinement, and dislocation pinning mechanisms (Fig. 4.17).

The impact of varying SiC reinforcement levels (2%, 5%, 8%, and 11% by weight) on AZ91 alloy was investigated by Kumar et al. [28]. Fabricated through vacuum-assisted stir casting, the composites demonstrated an 80% increase in microhardness at 11% reinforcement. Yield strength and UTS showed slight increases, while density and porosity rose with higher reinforcement levels, reaching 2.11% at 11% SiC. X-ray diffraction confirmed the presence of Mg, SiC, and $Mg_{17}Al_{12}$ phases without oxidation, due to the inert processing atmosphere. Fractography revealed mixed ductile-brittle failure, with microcracks increasing at higher reinforcement levels. Compressive testing showed improved resistance to shear failure, highlighting the balance between enhanced strength and increased porosity (Fig. 4.18).

Jiang et al. [29] explored Mg composites (MMCs) reinforced with boron carbide (B_4C) at 10%, 15%, and 20% volume fractions using a PM technique. The composites exhibited a unique necklace-like microstructure characterized by B_4C particulates distributed along the magnesium matrix, with minimal microporosity observed in all samples. X-ray diffraction (XRD) revealed the formation of MgB_2 and MgO phases, attributed to the reactions between Mg and B_4C during processing. These

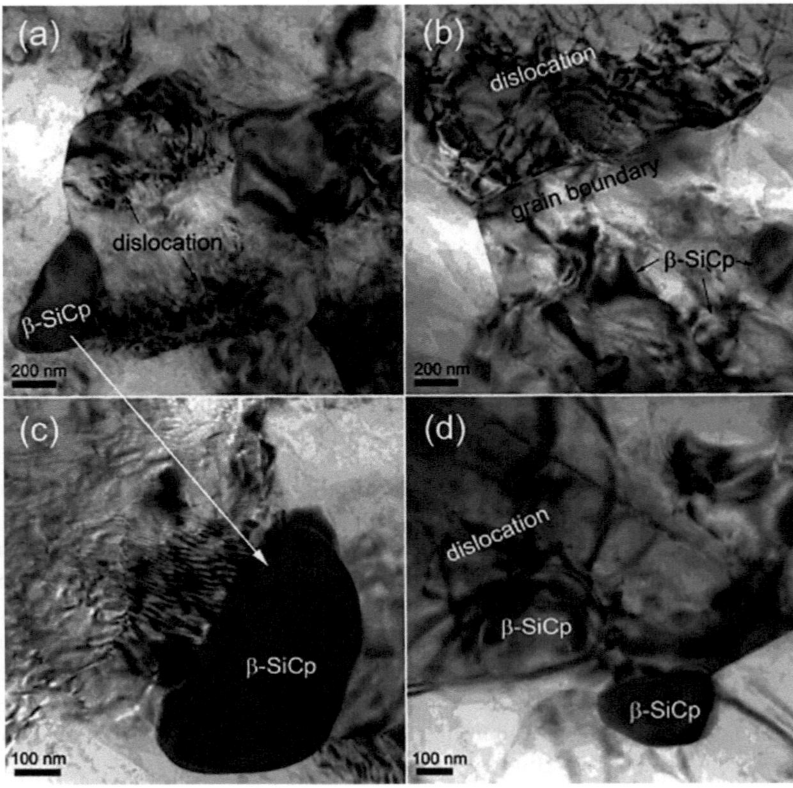

Fig. 4.17 TEM micrographs of submicron SiCp/AZ91 composite at the strain of 0.9%. (**a**) Dislocations near submicron SiC particle; (**b**) dislocations arriving at grain boundary; (**c**) high magnification of (**a**) shows stress fringes near submicron SiC particle; (**d**) submicron SiC particles pinning grain boundary [27]

phases were responsible for a significant improvement in hardness and wear resistance compared to as-cast magnesium, and their effectiveness increased with higher B_4C content. However, attempts to fabricate composites with 30% B_4C failed due to excessive porosity, highlighting a limitation in achieving higher reinforcement volumes using this method.

In a similar study, Jutanaiman and Syahrial [30] produced magnesium composites reinforced with varying amounts of boron carbide (B_4C) particles (2%, 4%, 6%, and 8% by volume) by stir casting. The best mechanical properties were observed with 8% B_4C, achieving a hardness of 73 HRH, ultimate tensile strength (UTS) of 64 MPa, and an impact strength of 0.11 J/mm^2 (Fig. 4.19). XRD analysis confirmed the presence of MgB_2, MgO, B_4C, and $Mg_2B_2O_5$ phases (Fig. 4.19). The formation of these compounds improved mechanical properties by impeding dislocation movement and acting as load transfer elements. SEM analysis demonstrated

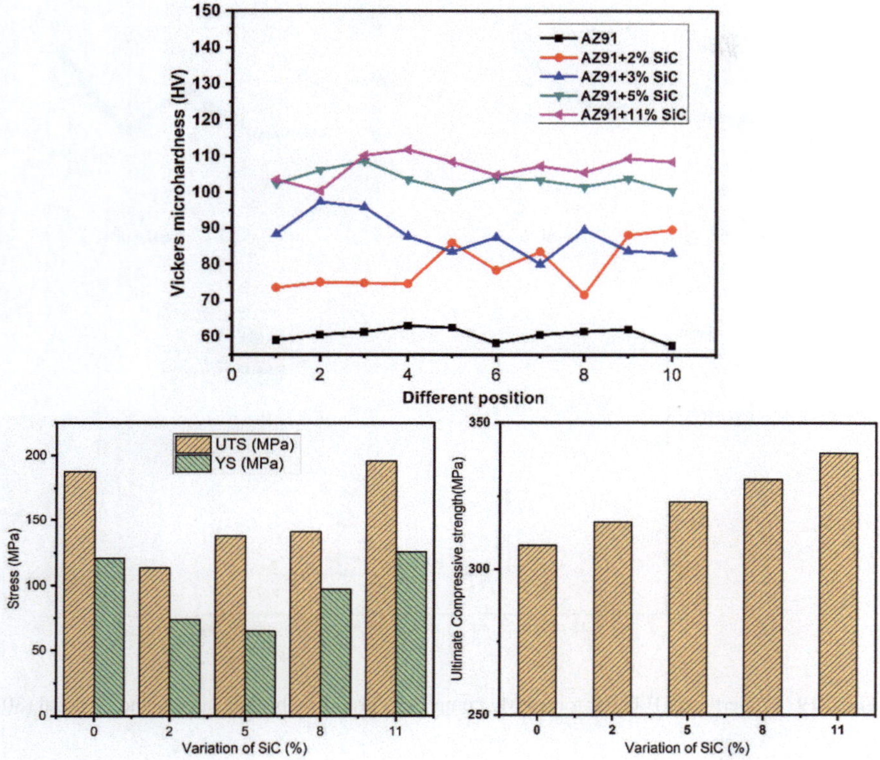

Fig. 4.18 Mechanical properties of AZ91 alloy and composites with different percentages of SiC [28]

uniform dispersion of B_4C particles, which also contributed to reduced wear rates (0.0023 mm^3/m).

The mechanical and tribological performance of AZ91 magnesium composites reinforced with B_4C (5%, 10%, 15%, and 20% by volume) were evaluated [31]. The composites displayed significant improvements in hardness, tensile strength, and compressive strength with increasing B_4C content (Fig. 4.20). Notably, the bending strength of the composite reached 497 MPa at 20% B_4C reinforcement, while compressive strength and tensile strength also showed consistent enhancement (Fig. 4.20). B_4C particles acted as barriers to dislocation motion and promoted load transfer, thereby improving overall mechanical performance. The study underscored the role of ceramic phases in optimizing the strength and durability of magnesium alloys, with applications in industries demanding lightweight yet strong materials.

In another study, Cafri et al. [32] produced magnesium composites containing 80% B_4C using infiltration techniques. The composites had a specific weight of 2.44 g/cm^3, with a microstructure comprising MgB_2 and MgB_2C_2 phases (Fig. 4.21).

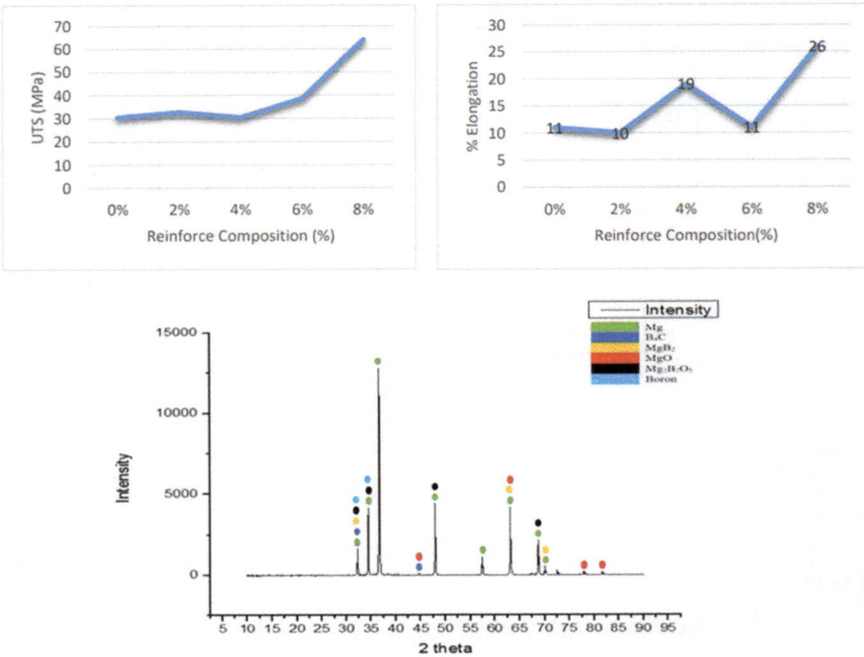

Fig. 4.19 Properties of B_4C reinforced Mg composites produced by Jutanaiman and Syahrial [30]

These newly formed phases bonded the original B_4C particles and residual Mg, significantly enhancing the mechanical properties. The thermodynamic analysis confirmed the feasibility of reactions between B_4C and Mg, which were critical for achieving strong interfacial bonding. Despite the high ceramic content, the composites retained excellent structural integrity and lightweight characteristics, making them suitable for applications requiring high strength-to-weight ratios.

Ghasali et al. [33] compared microwave and spark plasma sintering techniques for fabricating Mg-B_4C composites containing 5 wt% B_4C. The spark plasma sintered composites exhibited superior mechanical properties, including a microhardness of 92 Vickers and a bending strength of 191 MPa, compared to 87 Vickers and 171 MPa in microwave-sintered samples. XRD analysis confirmed the formation of MgB_2 and Mg_2C_3 phases in both sintering methods, which improved interfacial bonding and mechanical strength. The spark plasma sintering technique ensured better homogeneity and fewer defects, highlighting its advantages for producing high-performance magnesium composites.

The effects of B_4C reinforcement (5% and 10% by weight) and their size were investigated on the formability and strain hardening index of AZ91 matrix composites [34]. In this study, the composite reinforced with 10% nano B_4C exhibited the highest relative density, formability stress index, and strain hardening index

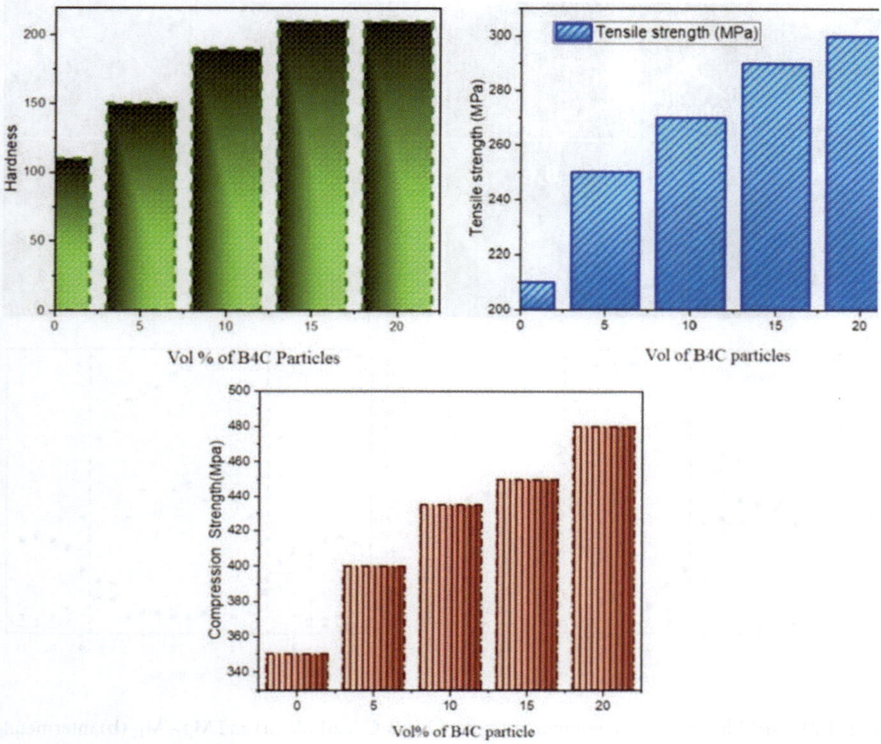

Fig. 4.20 Mechanical properties of B_4C reinforced Mg composites from [31]

(Fig. 4.22). The addition of nano-sized particles improved workability and plastic deformation under triaxial stress conditions. SEM images revealed uniform dispersion of nano B_4C particles, which enhanced the overall mechanical performance of the composite. This research emphasized the critical role of reinforcement size and distribution in achieving optimal mechanical properties.

Jiang et al. [35] explored the use of a TiC–Al master alloy produced via self-propagating high-temperature synthesis to reinforce AZ91 composites by semi-solid slurry stirring. Here, the incorporation of fine, spherical TiC particles (5 μm) into the Mg matrix enhanced the ultimate tensile strength (214 MPa), hardness (83 HB), and wear resistance compared to the unreinforced AZ91 alloy. These enhancements were attributed to the clean and uncontaminated particle-matrix interfaces and improved wetting between the in-situ reinforcement and Mg matrix (Fig. 4.23).

Magnesium composites reinforced with 8 wt.% TiC particles were produced using the remelting and dilution (RD) method [36]. The addition of TiC refined grain size to 12 μm, leading to enhanced yield strength (115 MPa) and ultimate tensile strength (235 MPa). However, elongation decreased significantly to 1.0%. Damping capacity tests revealed superior performance compared to unreinforced

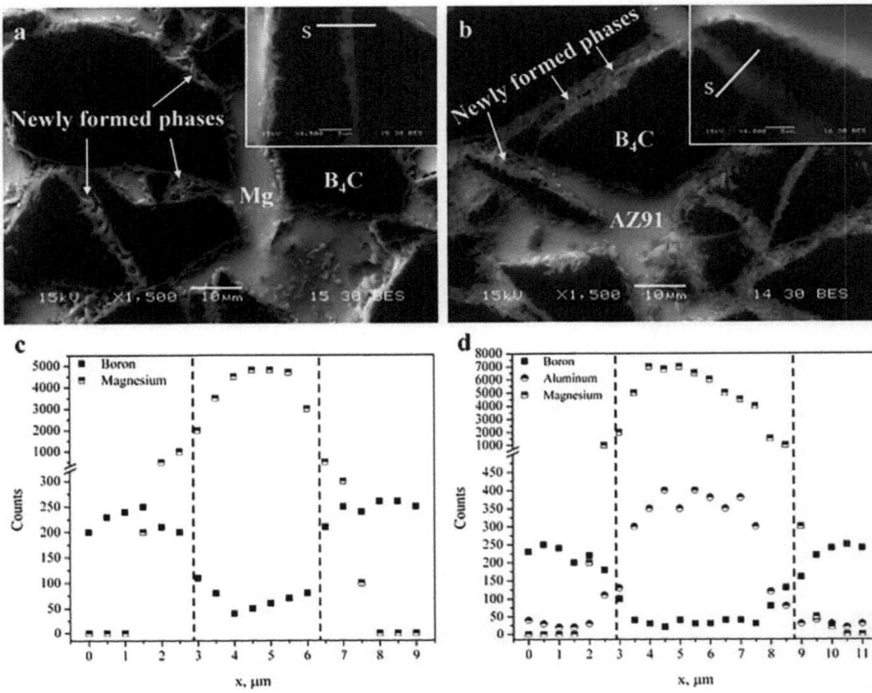

Fig. 4.21 SEM images a high volume fraction of MgB_2C_2 carbide (**a**) and $Mg_{17}Al_{12}$ (**b**) intermetallics in Mg/B4C composite produced by Cafri et al. [32]. Figures (**c**) and (**d**) present the corresponding EDS spectra, confirming the elemental distribution

AZ91, with damping peaks observed at lower strain amplitudes and temperatures (Fig. 4.24). These enhancements were attributed to increased dislocation density, grain boundary slip, and the formation of compound twins, which facilitated energy dissipation under cyclic loading.

Balakrishnan et al. [37] employed the friction stir processing method to fabricate AZ31/TiC composites with varying TiC content (0%, 6%, 12%, and 18% by volume). SEM analysis confirmed uniform dispersion of TiC particles without clustering or interfacial reactions (Fig. 4.25). The process produced a defect-free microstructure with enhanced mechanical integrity. The absence of voids alongside the better interfacial bonding between TiC particles and the Mg matrix contributed to significant improvements in mechanical properties, highlighting the potential of friction stir processing method for producing high-performance magnesium composites.

In another study, 10 vol.% Ti–6Al–4 V particulates (TAp) were added to ZK51 alloy by powder metallurgy method and their effectiveness were compared with SiC particulates (SiCp) [38]. The processing involved mixing of powder particles followed by cold pressing at 300 MPa, sintering at 530 °C, hot pressing at 380 °C

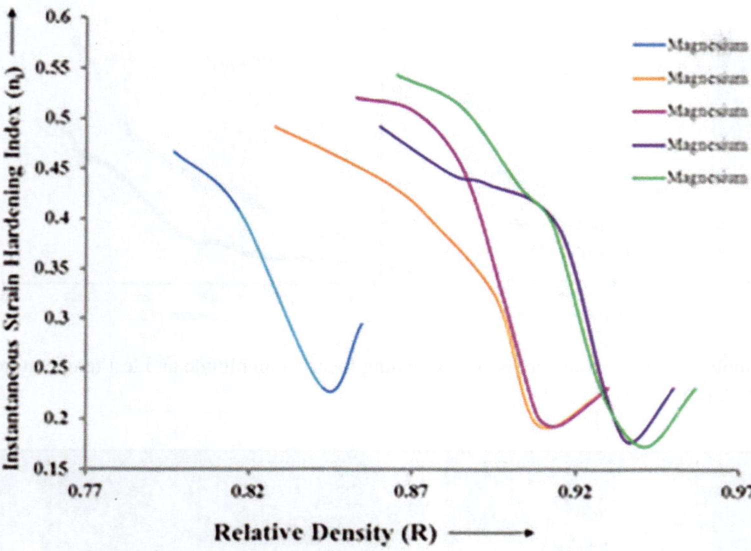

Fig. 4.22 Influence of instantaneous strain hardening index (ni) pertaining to relative density (R) for Mg alloy, Mg-5% micro-B_4C, Mg-10% micro-B_4C, Mg-5% nano-B_4C and Mg-10% nano-B_4C composites

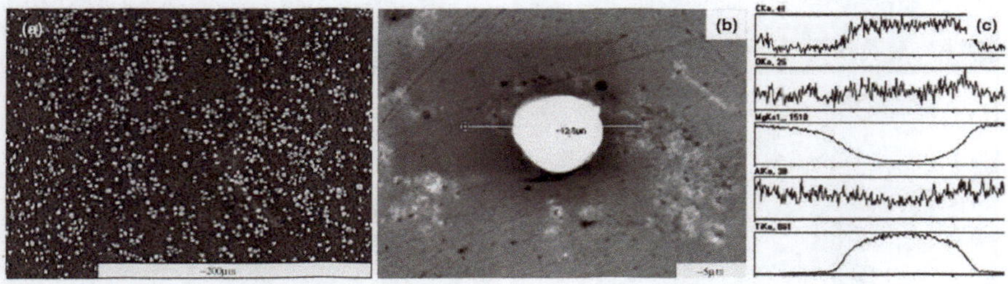

Fig. 4.23 SEM images showing the (**a**) distribution of TiC particles in AZ91 alloy composites and (**b, c**) their interfacial characteristics from Jiang et al. [35]

under 250 MPa, and hot extrusion at 350 °C. A solution heat treatment and aging at 145 °C were also carried out. XRD analyses confirmed the presence of $MgZn_2$ precipitates and MgO, and aging tests revealed double-hardening peaks, with TA_p/ZK51 composites reaching peak hardness faster (6 and 17 h) than ZK51 alloy (8 and 21 h). SEM and optical micrographs demonstrated uniform TA_p dispersion, strong interfacial bonding, and reduced grain size (8 μm compared to 13 μm in ZK51), attributed to TA_p's role in restricting grain boundary motion and enhancing recrystallization. Mechanical testing revealed significant improvements in ultimate tensile strength (UTS) from 315 MPa (ZK51) to 386 MPa (TA_p/ZK51), yield strength from 229 MPa to 295 MPa, and elastic modulus from 46.2 to 52.8 GPa, while elongation

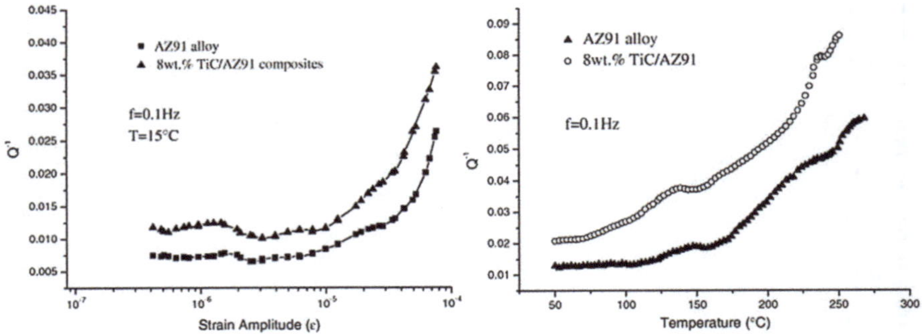

Fig. 4.24 Damping capacity of materials corresponding to stress amplitude and test temperature from Zhang et al. [36]

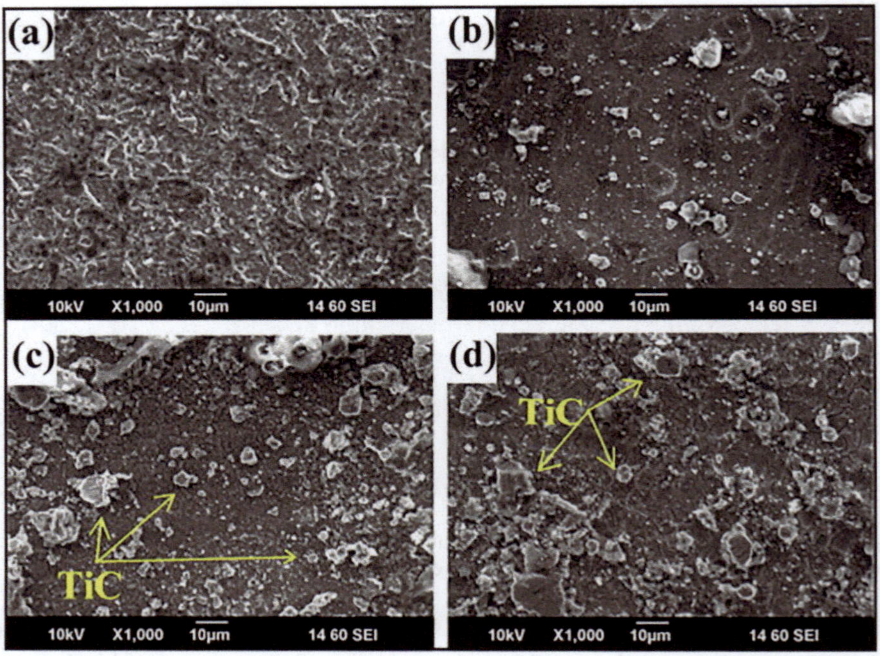

Fig. 4.25 SEM Micrograph of AZ31/TiC MMC containing TiC; (**a**) 0 vol.%, (**b**) 6 vol.%, (**c**) 12 vol.% and (**d**) 18 vol.% (from Balakrishnan et al. [37])

decreased from 8.5% to 5.6% (Fig. 4.26). The improved properties were ascribed to grain refinement, dislocation restriction, and load transfer facilitated by TA_p. Compared to SiC_p/ZK51 composites, TA_p/ZK51 exhibited better ductility and UTS despite a slightly lower elastic modulus, as the softer and more ductile TA_p resulted in fewer defects and better interfacial bonding than the brittle SiC_p.

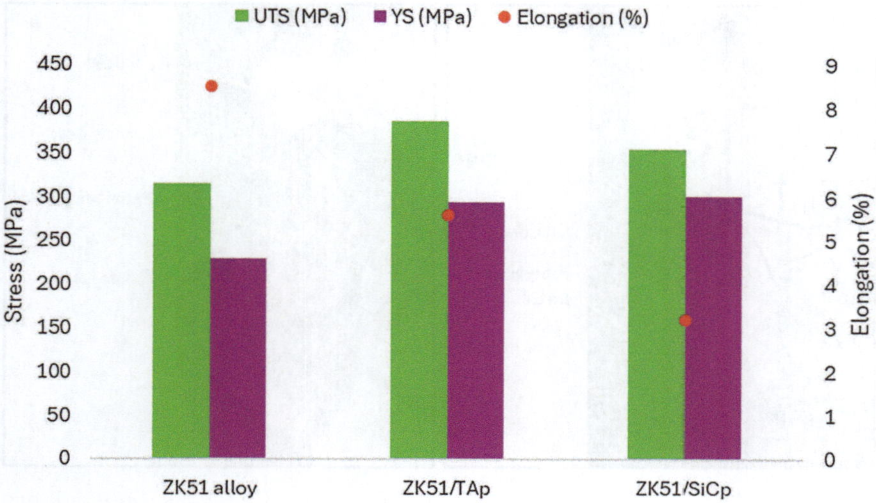

Fig. 4.26 Mg composites reinforced with Ti–6Al–4V particulates from [38]

Karuppusamy et al. [39] investigated AZ91 magnesium alloy composites reinforced with 1.5% and 3.0% tungsten carbide (WC) particulates fabricated via vacuum stir casting. Post-fabrication, the composites underwent T6 heat treatment, involving solutionizing at 420 °C and artificial aging at 175 °C to refine microstructure and enhance mechanical properties. EDS and XRD analyses confirmed the incorporation of WC and the precipitation of secondary $Mg_{17}Al_{12}$ phases, which contributed to increased hardness and strength. Tensile tests showed a reduction in yield (152 MPa for AZ91 to 132 MPa for AZ91/1.5WC) and ultimate tensile strength (200 MPa for AZ91 to 187 MPa for AZ91/3.0WC), but a notable 65% improvement in ductility due to uniform particle dispersion and grain refinement. Conversely, compressive properties improved significantly, with ultimate compressive strength increasing from 300 MPa for AZ91 to 342 MPa for AZ91/3.0WC. Quasi-static load tests on pistons fabricated from these composites revealed enhanced load-bearing capacity, with WC reinforced pistons exhibiting 30% (AZ91/1.5WC) and 70% (AZ91/3.0WC) higher resistance compared to unreinforced AZ91 pistons (Fig. 4.27). The pistons also demonstrated a weight reduction of up to 35% compared to traditional materials, making them suitable for lightweight applications.

4.3.1.1 Oxide Particulate Reinforced Magnesium Composites

The effects of Al_2O_3 reinforcement on magnesium composites have been explored extensively across various studies, revealing significant improvements in mechanical, thermal, and microstructural properties, as well as process-specific benefits. Hassan and Gupta [40] studied the effects of submicron-scale Al_2O_3 particle

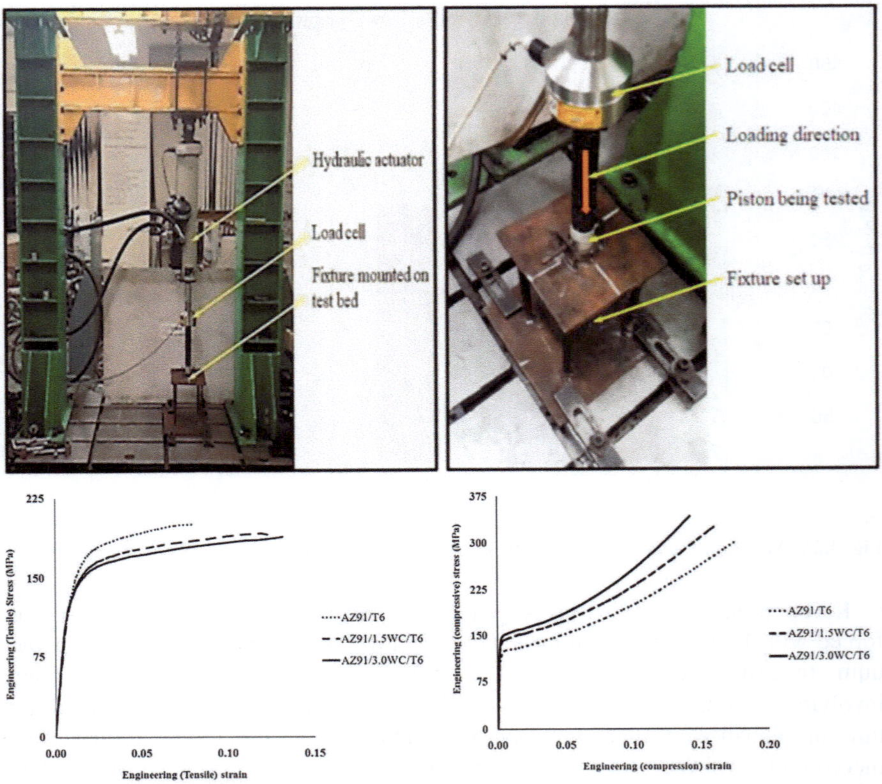

Fig. 4.27 Quasi-static load testing of piston in Instron PL25R structural system, and respective stress–strain curves under tensile and compressive loading [39]

reinforcements on the properties of magnesium, using the disintegrated melt deposition (DMD) method followed by hot extrusion. Mechanical testing revealed significant improvements in yield strength, tensile strength, and ductility (Fig. 4.28). Among the composites, the one containing 1.1 vol.% of 0.3 μm Al_2O_3 particles exhibited superior tensile properties compared to those reinforced with either micron-scale or nanoscale particles. The incorporation of submicron Al_2O_3 particles also enhanced ductility and work of fracture, while achieving substantial grain refinement and reducing porosity, demonstrating the effectiveness of submicron reinforcements in optimizing composite performance.

Microwave-assisted rapid sintering was also utilized to fabricate Mg composites with submicron Al_2O_3 particles [41]. Here, the sintered composites exhibited improved hardness, elastic modulus, and tensile properties, with increasing alumina content correlating with better performance (Fig. 4.29). The coefficient of thermal expansion decreased with higher reinforcement levels, demonstrating enhanced thermal stability. This study highlighted the benefits of microwave sintering for achieving uniform particle dispersion and optimizing composite properties.

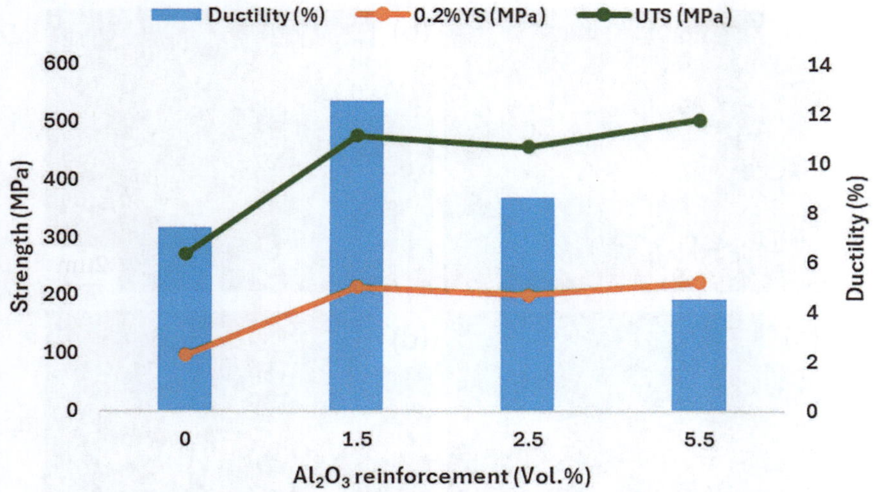

Fig. 4.28 Mechanical properties of sub-micron size Al_2O_3 particles reinforced Mg comoposites from Hassan and Gupta [40]

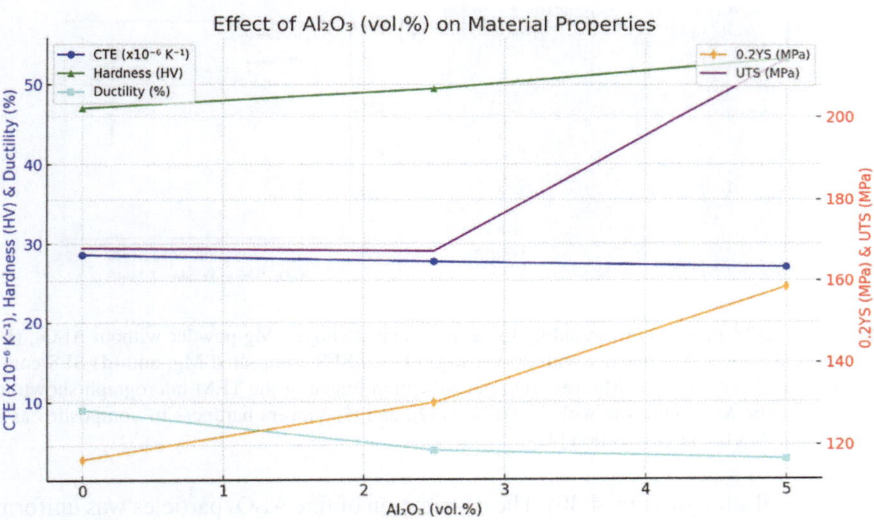

Fig. 4.29 Thermal and mechanical properties of sub-micron-sized Al_2O_3 reinforced Mg composites fabricated by PM assisted microwave sintering (from [41])

Fujiwara et al. [42] explored the mechanical milling (MM) and spark plasma sintering (SPS) processes for producing Mg composites with 10–20 vol% Al_2O_3. The results revealed that the formation of $Mg_{17}Al_{12}$ and MgO phases during the SPS process significantly contributed to the composites' enhanced hardness and

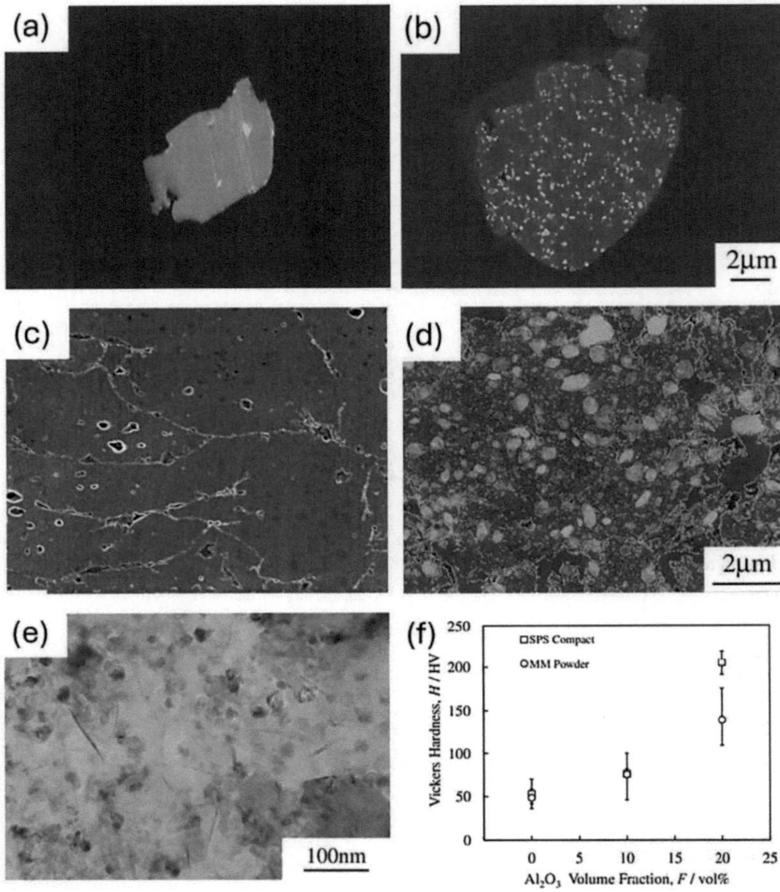

Fig. 4.30 SEM images corresponding to (**a**) mechanical alloyed Mg powder without Al_2O_3, (**b**) mechanical alloyed Mg powder with 20 vol% Al_2O_3, (**c**) SPS compacted Mg, and (**d**) SPS compacted Mg SPS compacted Mg. (**e**) High magnification image of the TEM micrograph shown in Fig. 4.8 of the Mg composite with 20 vol% Al_2O_3, and (**f**) Vickers hardness of composites as a function of Al_2O_3 volume fraction [42]

mechanical strength (Fig. 4.30). The distribution of fine Al_2O_3 particles was uniform (Fig. 4.30), and the microstructure showed notable grain refinement. These improvements were particularly pronounced for composites with 20 vol% Al_2O_3, highlighting the synergistic effects of microstructural changes and phase formation.

Stir casting method was employed by Kumar et al. [43] to incorporate 0–12 vol% Al_2O_3 into magnesium. The study revealed that increasing Al_2O_3 content consistently enhanced hardness while tensile strength peaked at 9% (Fig. 4.31). The microstructural analysis confirmed strong interfacial bonding, contributing to the superior mechanical performance of the composites. This work demonstrated that

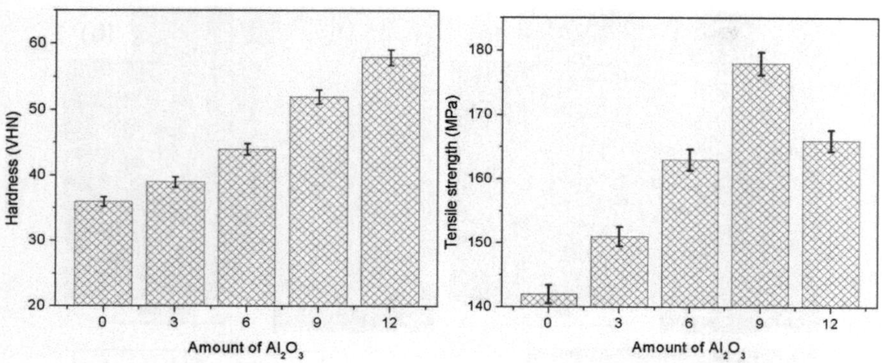

Fig. 4.31 Mechanical properties of Mg-Al_2O_3 composites (from [43])

stir casting is a viable method for large-scale production of Mg-Al_2O_3 composites with optimized properties.

Cay et al. [44] synthesized magnesium/alumina composites using powder metallurgy and evaluated their mechanical properties. The composites had porosities of 10%, 28%, and 38%, with average pore sizes of 25 μm, 70 μm, and 100 μm, respectively. The stress–strain curves exhibited three distinct regimes, with higher porosities correlating to reduced yield strength and Young's modulus. For example, samples with 10% porosity showed higher yield strength than those with 38% porosity. The anisotropy in mechanical properties was attributed to the directional alignment of magnesium walls during the compacting process. Samples tested along the in-plane orientation showed superior strength and modulus compared to the normal orientation, highlighting their potential for optimized structural designs.

The properties of the AZ91 alloy were significantly enhanced through in situ-formed oxides (MgO, Al_2O_3, and $MgAl_2O_4$) generated via ultrasonic treatment [45]. This process facilitated the uniform dispersion of oxide particles, reduced porosity, and improved mechanical properties (Fig. 4.32). Composites containing 6.5 vol.% oxides demonstrated a 43% increase in yield strength and a 64% improvement in hardness compared to the base alloy (Fig. 4.32). Furthermore, wear resistance improved markedly; the AZ91–6.5-UST composite exhibited a 27% reduction in wear rate under a 14.7 N load compared to unprocessed AZ91 alloy. The study emphasized UST's critical role in achieving fine particle dispersion and robust reinforcement-matrix interfaces, which contributed to enhanced wear behavior and superior mechanical performance.

Using equal channel angular extrusion (ECAE) method, Rahmani et al. [46] fabricated Mg-Al_2O_3 composites with reinforcement levels up to 30 vol%. These composites exhibited significant improvements in hardness and compressive strength, achieving a 55% and 74% increase, respectively, over pure magnesium (Fig. 4.33). Additionally, ECAE produced finer grain sizes and enhanced the material's corrosion resistance, though higher reinforcement levels led to reduced relative density. This study highlighted ECAE as a promising technique for improving the

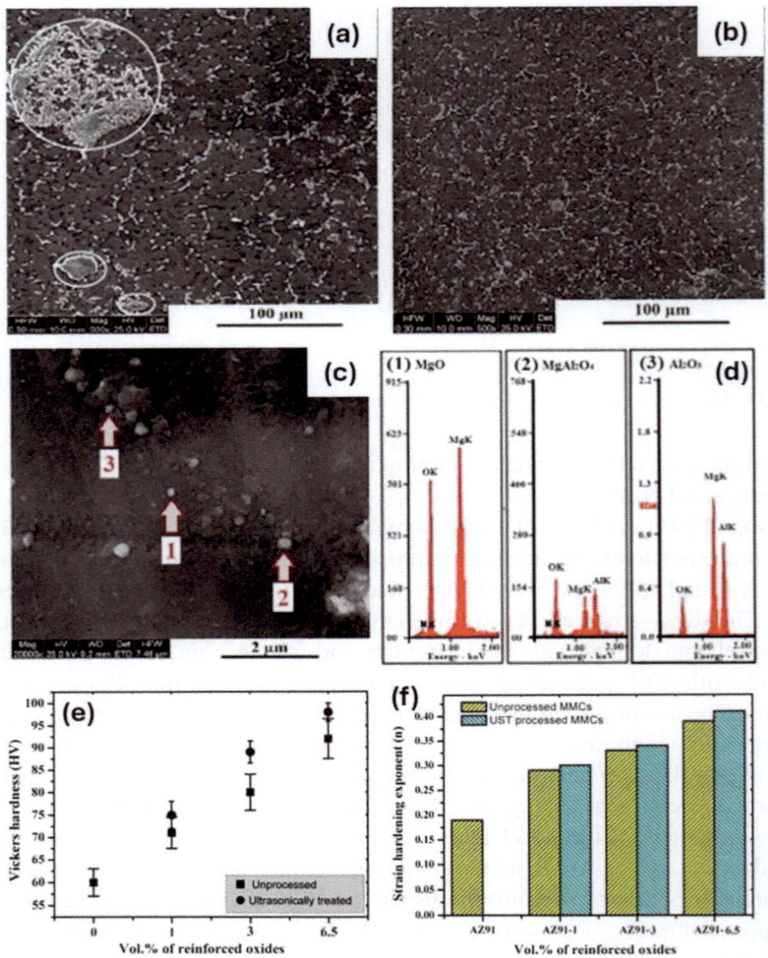

Fig. 4.32 SEM micrographs of 6.5% oxide reinforced AZ91 composite before (**a**) and after ultrasonic treatment (**b, c**). EDS patterns of the particles of different morphology identified in (**d**). Vickers microhardness (**e**) and strain hardening exposure (**f**) of the developed AZ91 composites [45]

mechanical and corrosion properties of magnesium composites. The surface modification of AZ61 alloys through friction stir processing and Al_2O_3 reinforcement was investigated by Ashokkumar et al. [47]. The addition of up to 15 vol% Al_2O_3 significantly increased the ultimate tensile strength (UTS) to 630 MPa and hardness to 300 HV. These enhancements were attributed to the severe plastic deformation during FSP, which refined the grains and ensured uniform particle dispersion.

The reinforcement of Mg alloys with boron nitride (BN) particles has been shown to significantly enhance their mechanical properties and microstructural

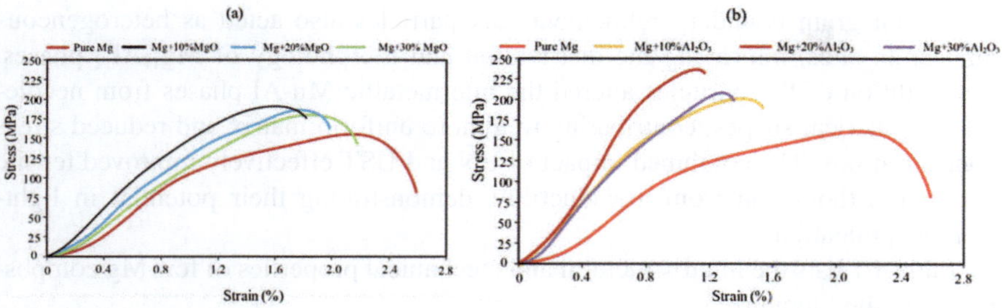

Fig. 4.33 Compressive stress–strain graphs of (**a**) Mg–Al_2O_3 composite samples (**b**) Mg–MgO composite samples fabricated by ECAE (from [46])

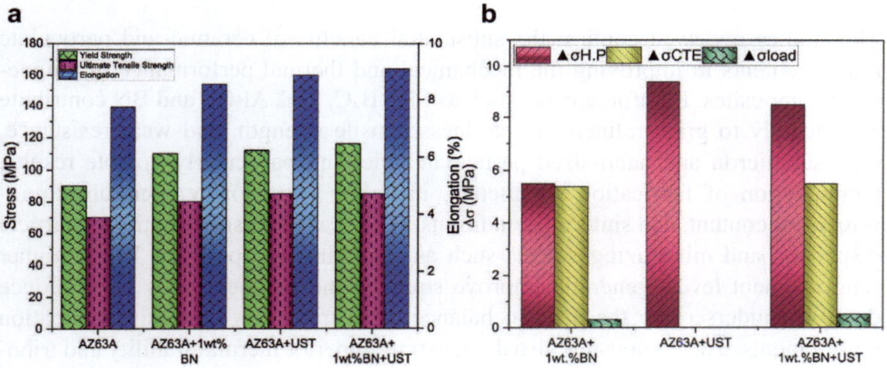

Fig. 4.34 (**a**) Mechanical properties of AZ63A and its composites reinforced with BN particles, and (**b**) strengthening contribution from Hall-petch, CTE mismatch and load bearing mechanisms [48]

characteristics. Vairamuthu et al. [48] used the hybrid mechanical and ultrasonic-assisted stir casting method to produce AZ63A alloy-based composite reinforced with BN particles. In this study, the introduction of BN particles led to a remarkable grain size reduction from 202 μm in the base alloy to 79 μm with 1 wt% BN addition. This 61% reduction was further enhanced by ultrasonic treatment (UST), which reduced the grain size to 50 μm. The hybrid method combining BN and UST yielded a grain size of 54 μm, demonstrating the synergistic effects of BN and ultrasonic cavitation. The refined grains improved the tensile properties, with the ultimate tensile strength increasing from 138 MPa in the base alloy to 163 MPa in the hybrid alloy. The yield strength rose from 96 MPa to 114 MPa, and elongation increased from 1.4% to 2.1%, marking improvements of 18%, 19%, and 50%, respectively (Fig. 4.34). Mechanically, BN refinement influenced the AZ63A alloy through Hall-Petch strengthening, load transfer, and mismatch in thermal expansion coefficients. The Hall-Petch effect dominated the strength enhancement due to significant grain boundary refinement. BN particles also acted as heterogeneous nucleation sites, improving the distribution and morphology of $Mg_{17}Al_{12}$ phases. The addition of BN particles altered the intermetallic Mn-Al phases from needle-like to spherical shapes, contributing to a more uniform matrix and reduced stress concentrations. The combined impact of BN and UST effectively improved tensile strength without compromising ductility, demonstrating their potential in lightweight applications.

Table 4.1 lists the microstructural and mechanical properties of few Mg composites from the literature.

4.4 Summary

The studies reviewed confirm the substantial benefits of ceramic and particulate reinforcements in improving the mechanical and thermal performance of magnesium composites. Reinforcements such as SiC, B_4C, TiC, Al_2O_3, and BN contribute significantly to grain refinement, hardness, tensile strength, and wear resistance, with submicron and nano-sized particulates yielding particularly notable results. Optimization of fabrication parameters, including extrusion temperatures, reinforcement content, and sintering conditions, is critical for ensuring uniform particle dispersion and minimizing defects such as clustering and porosity. While higher reinforcement levels generally improve strength and hardness, they often reduce ductility, underscoring the need to balance these trade-offs based on application requirements. The composites also demonstrate superior thermal stability and tribological properties, maintaining performance at elevated temperatures. Emerging techniques, such as hybrid reinforcements and ultrasonic stirring, are paving the way for further optimization, enhancing both microstructural and mechanical characteristics. Despite these advancements, challenges like interfacial reactions, scalability, and porosity persist, requiring future research to integrate computational and experimental approaches for industrial-scale applications. Overall, magnesium composites reinforced with ceramic phases hold immense promise for lightweight, high-performance applications in aerospace, automotive, and structural engineering.

References

1. S. Jayalakshmi, S.V. Kailas, S. Seshan, Tensile behaviour of squeeze cast AM100 magnesium alloy and its Al_2O_3 fibre reinforced composites. Compos. Part Appl. Sci. Manuf. **33**(8), 1135–1140 (2002). https://doi.org/10.1016/S1359-835X(02)00049-0
2. B. Hu, L. Peng, B.R. Powell, M.P. Balough, R.C. Kubic, A.K. Sachdev, Interfacial and fracture behavior of short-fibers reinforced AE44 based magnesium matrix composites. J. Alloys Compd. **504**(2), 527–534 (2010). https://doi.org/10.1016/j.jallcom.2010.05.155
3. Z. Trojanová, V. Gärtnerová, P. Lukáč, Z. Drozd, Mechanical properties of Mg alloys composites reinforced with short Saffil® fibres. Proc. 9th Int. Symp. Phys. Mater. ISPMA 9 **378**(1), 19–26 (Sep. 2004). https://doi.org/10.1016/j.jallcom.2003.12.047

4. Z. Trojanová, A. Mielczarek, W. Riehemann, P. Lukáč, Cyclic bending and the damping behaviour of short fibre-reinforced magnesium alloy AZ91. Compos. Sci. Technol. **66**(3), 585–590 (2006). https://doi.org/10.1016/j.compscitech.2005.05.033
5. A. Mielczarek, Z. Trojanová, W. Riehemann, P. Lukáč, Influence of mechanical cycling on damping behaviour of short fibre-reinforced magnesium alloy QE22. Proc. 14th Int. Conf. Intern. Frict. Mech. Spectrosc. **442**(1), 484–487 (2006). https://doi.org/10.1016/j.msea.2006.02.228
6. V. Sklenička, M. Svoboda, M. Pahutová, K. Kuchařová, T.G. Langdon, Microstructural processes in creep of an AZ91 magnesium-based composite and its matrix alloy. Mater. Sci. Eng. A **319–321**, 741–745 (2001). https://doi.org/10.1016/S0921-5093(01)01023-1
7. A. Bakkar, V. Neubert, Corrosion characterisation of alumina–magnesium metal matrix composites. Corros. Sci. **49**(3), 1110–1130 (2007). https://doi.org/10.1016/j.corsci.2006.07.002
8. C. Rauber, A. Lohmüller, S. Opel, R.F. Singer, Microstructure and mechanical properties of SiC particle reinforced magnesium composites processed by injection molding. Mater. Sci. Eng. A **528**(19), 6313–6323 (2011). https://doi.org/10.1016/j.msea.2011.04.088
9. M. Russell-Stevens, D.C. Plane, J. Summerscales, P. Schulz, M. Papakyriacou, Effect of elevated temperature on ultimate tensile strength and failure modes of short carbon fibre reinforced magnesium composite. Mater. Sci. Technol. **18**(5), 501–506 (2002). https://doi.org/10.1179/026708302225002083
10. M. Russell-Stevens, R. Todd, M. Papakyriacou, The effect of thermal cycling on the properties of a carbon fibre reinforced magnesium composite. Mater. Sci. Eng. A **397**(1), 249–256 (2005). https://doi.org/10.1016/j.msea.2005.02.025
11. M. Russell-Stevens, R.I. Todd, M. Papakyriacou, Thermal expansion behaviour of ultra-high modulus carbon fibre reinforced magnesium composite during thermal cycling. J. Mater. Sci. **41**(19), 6228–6236 (2006). https://doi.org/10.1007/s10853-006-0318-y
12. S. Kandemir, S. Gavras, H. Dieringa, High temperature tensile, compression and creep behavior of recycled short carbon fibre reinforced AZ91 magnesium alloy fabricated by a high shearing dispersion technique. J. Magnes. Alloys **9**(5), 1753–1767 (2021). https://doi.org/10.1016/j.jma.2021.03.029
13. R.A. Saravanan, M.K. Surappa, Fabrication and characterisation of pure magnesium-30 vol.% SiCP particle composite. Mater. Sci. Eng. A **276**(1), 108–116 (2000). https://doi.org/10.1016/S0921-5093(99)00498-0
14. S.C.V. Lim, M. Gupta, Enhancing the microstructural and mechanical response of a Mg/SiC formulation by the method of reducing extrusion temperature. Mater. Res. Bull. **36**(15), 2627–2636 (2001). https://doi.org/10.1016/S0025-5408(01)00750-4
15. S.U. Reddy, N. Srikanth, M. Gupta, S.K. Sinha, Enhancing the properties of magnesium using SiC particulates in sub-micron length scale. Adv. Eng. Mater. **6**(12), 957–964 (2004). https://doi.org/10.1002/adem.200400105
16. S.C.V. Lim, M. Gupta, L. Lu, Processing, microstructure, and properties of Mg–SiC composites synthesised using fluxless casting process. Mater. Sci. Technol. **17**(7), 823–832 (2001). https://doi.org/10.1179/026708301101510591
17. S. Dinesh Kumar, M. Ravichandran, S. Sakthivelu, M. Meignanamoorthy, C. Chanakyan, S.V. Alagarsamy, Mechanical properties of magnesium-silicon carbide composite fabricated through powder metallurgy route. First Int. Conf. Adv. Lightweight Mater. Struct. **27**, 1137–1141 (2020). https://doi.org/10.1016/j.matpr.2020.01.592
18. W.N.A.W. Muhammad, Z. Sajuri, Y. Mutoh, Y. Miyashita, Microstructure and mechanical properties of magnesium composites prepared by spark plasma sintering technology. J. Alloys Compd. **509**(20), 6021–6029 (2011). https://doi.org/10.1016/j.jallcom.2011.02.153
19. S. Jayasathyakawin, M. Ravichandran, Experimental investigations on effect of silicon carbide on microstructure and mechanical properties in Mg-3 wt% Al alloy matrix using powder metallurgy. SILICON **14**(14), 9163–9173 (2022). https://doi.org/10.1007/s12633-022-01834-0
20. W. Xie, Y. Liu, D.S. Li, J. Zhang, Z.W. Zhang, J. Bi, Influence of sintering routes to the mechanical properties of magnesium alloy and its composites produced by PM technique. J. Alloys Compd. **431**(1), 162–166 (2007). https://doi.org/10.1016/j.jallcom.2006.05.076

21. K.K. Deng, K. Wu, Y.W. Wu, K.B. Nie, M.Y. Zheng, Effect of submicron size SiC particulates on microstructure and mechanical properties of AZ91 magnesium matrix composites. J. Alloys Compd. **504**(2), 542–547 (2010). https://doi.org/10.1016/j.jallcom.2010.05.159
22. X.J. Wang et al., Influences of extrusion parameters on microstructure and mechanical properties of particulate reinforced magnesium matrix composites. Mater. Sci. Eng. A **528**(21), 6387–6392 (2011). https://doi.org/10.1016/j.msea.2011.04.064
23. K.B. Nie, X.J. Wang, K. Wu, X.S. Hu, M.Y. Zheng, L. Xu, Microstructure and tensile properties of micro-SiC particles reinforced magnesium matrix composites produced by semisolid stirring assisted ultrasonic vibration. Mater. Sci. Eng. A **528**(29), 8709–8714 (2011). https://doi.org/10.1016/j.msea.2011.08.035
24. K.K. Deng, X.J. Wang, Y.W. Wu, X.S. Hu, K. Wu, W.M. Gan, Effect of particle size on microstructure and mechanical properties of SiCp/AZ91 magnesium matrix composite. Mater. Sci. Eng. A **543**, 158–163 (2012). https://doi.org/10.1016/j.msea.2012.02.064
25. M.C. Gui, J.M. Han, P.Y. Li, Microstructure and mechanical properties of Mg—Al9Zn/SiCp composite produced by vacuum stir casting process. Mater. Sci. Technol. **20**(6), 765–771 (2004). https://doi.org/10.1179/026708304225017319
26. L.P. Soon, M. Gupta, Synthesis and recyclability of a magnesium based composite using an innovative disintegrated melt deposition technique. Mater. Sci. Technol. **18**(1), 92–98 (2002). https://doi.org/10.1179/026708301125000276
27. K.B. Nie, K.K. Deng, F.J. Xu, X.J. Wang, K. Wu, Development of microstructure in submicron particles reinforced magnesium matrix composite processed by room temperature deformation. Mater. Chem. Phys. **149–150**, 21–26 (2015). https://doi.org/10.1016/j.matchemphys.2014.11.011
28. A. Kumar, S. Kumar, N.K. Mukhopadhyay, A. Yadav, J. Winczek, Effect of SiC reinforcement and its variation on the mechanical characteristics of AZ91 composites. Materials **13**(21), 4913 (2020). https://doi.org/10.3390/ma13214913
29. Q.C. Jiang, H.Y. Wang, B.X. Ma, Y. Wang, F. Zhao, Fabrication of B4C particulate reinforced magnesium matrix composite by powder metallurgy. J. Alloys Compd. **386**(1), 177–181 (2005). https://doi.org/10.1016/j.jallcom.2004.06.015
30. S. Paulo Hang Jutanaiman, A. Zulfia Syahrial, Characteristics of magnesium/B4C reinforced composite fabricated by stir casting method. IOP Conf. Ser. Mater. Sci. Eng. **924**(1), 012020 (2020). https://doi.org/10.1088/1757-899X/924/1/012020
31. G. Pitchayyapillai, M. Jinnah Sheik Mohamed, G. Dhanraj, R. Malkiya Rasalin Prince, M. Rajeshwaran, A. Mangrulkar, Influence of B4C on mechanical properties of AZ91 magnesium matrix composites. Int. Conf. Virtual Conf. Technol. Adv. Mech. Eng. **59**, 1438–1441 (2022). https://doi.org/10.1016/j.matpr.2021.12.135
32. M. Cafri, H. Dilman, M.P. Dariel, N. Frage, Boron carbide/magnesium composites: processing, microstructure and properties. J. Eur. Ceram. Soc. **32**(12), 3477–3483 (2012). https://doi.org/10.1016/j.jeurceramsoc.2012.04.007
33. E. Ghasali, M. Alizadeh, M. Niazmand, T. Ebadzadeh, Fabrication of magnesium-boron carbide metal matrix composite by powder metallurgy route: comparison between microwave and spark plasma sintering. J. Alloys Compd. **697**, 200–207 (2017). https://doi.org/10.1016/j.jallcom.2016.12.146
34. P. Rajkumar, C. Kailasanathan, A. Senthilkumar, N. Selvakumar, A. JohnRajan, Study on formability and strain hardening index: influence of particle size of boron carbide (B4C) in magnesium matrix composites fabricated by powder metallurgy technique. Mater. Res. Express **7**(1), 016597 (2020)
35. Q.C. Jiang, X.L. Li, H.Y. Wang, Fabrication of TiC particulate reinforced magnesium matrix composites. Scr. Mater. **48**(6), 713–717 (2003). https://doi.org/10.1016/S1359-6462(02)00551-1
36. X. Zhang, L. Liao, N. Ma, H. Wang, Mechanical properties and damping capacity of magnesium matrix composites. Compos. Part Appl. Sci. Manuf. **37**(11), 2011–2016 (2006). https://doi.org/10.1016/j.compositesa.2005.12.007

37. M. Balakrishnan, I. Dinaharan, R. Palanivel, R. Sivaprakasam, Synthesize of AZ31/TiC magnesium matrix composites using friction stir processing. J. Magnes. Alloys **3**(1), 76–78 (2015). https://doi.org/10.1016/j.jma.2014.12.007
38. Y.L. Xi, D.L. Chai, W.X. Zhang, J.E. Zhou, Titanium alloy reinforced magnesium matrix composite with improved mechanical properties. Scr. Mater. **54**(1), 19–23 (2006). https://doi.org/10.1016/j.scriptamat.2005.09.020
39. P. Karuppusamy, K. Lingadurai, V. Sivananth, S. Arulkumar, A study on mechanical properties of tungsten carbide reinforced magnesium metal matrix composites for the application of piston. Int. J. Lightweight Mater. Manuf. **4**(4), 449–459 (2021). https://doi.org/10.1016/j.ijlmm.2021.06.007
40. S.F. Hassan, M. Gupta, Effect of submicron size Al_2O_3 particulates on microstructural and tensile properties of elemental Mg. J. Alloys Compd. **457**(1), 244–250 (2008). https://doi.org/10.1016/j.jallcom.2007.03.058
41. W.L.E. Wong, S. Karthik, M. Gupta, Development of high performance Mg–Al_2O_3 composites containing Al_2O_3 in submicron length scale using microwave assisted rapid sintering. Mater. Sci. Technol. **21**(9), 1063–1070 (2005). https://doi.org/10.1179/174328405X51758
42. H. Fujiwara, S. Hamanaka, S. Kawamori, H. Miyamoto, Effect of microstructure on the mechanical properties of magnesium composites containing dispersed alumina particles prepared using an MM/SPS process. Mater. Trans. **55**(3), 543–548 (2014). https://doi.org/10.2320/matertrans.Y-M2013845
43. S.D. Kumar, M. Ravichandran, M. Meignanamoorthy, S. Sakthivelu, S.V. Alagarsamy, C. Chanakyan, Investigations on properties of Mg-Al_2O_3 composites fabricated via stir casting route. First Int. Conf. Adv. Lightweight Mater. Struct. **27**, 1132–1136 (2020). https://doi.org/10.1016/j.matpr.2020.01.586
44. H. Cay, H. Xu, Q. Li, Mechanical behavior of porous magnesium/alumina composites with high strength and low density. Mater. Sci. Eng. A **574**, 137–142 (2013). https://doi.org/10.1016/j.msea.2013.03.012
45. P.P. Bhingole, G.P. Chaudhari, S.K. Nath, Processing, microstructure and properties of ultrasonically processed in situ MgO–Al_2O_3–$MgAl_2O_4$ dispersed magnesium alloy composites. Compos. Part Appl. Sci. Manuf. **66**, 209–217 (2014). https://doi.org/10.1016/j.compositesa.2014.08.001
46. K. Rahmani et al., Mechanical and corrosion properties of Mg–MgO and Mg–Al_2O_3 composites fabricated by equal channel angular extrusion method. Smart Mater. Manuf. **1**, 100010 (2023). https://doi.org/10.1016/j.smmf.2022.100010
47. M. Ashokkumar, T. Sonar, A. Thirugnanasambandam, P. Vignesh, M. Ivanov, Microstructure and mechanical properties of Al_2O_3/AZ61 Mg alloy surface composite developed using friction stir processing and groove reinforcement filling processes. Mater. Testing **66**(11), 1920–1930 (2024). https://doi.org/10.1515/mt-2024-0132
48. J. Vairamuthu, A. Senthil Kumar, G.D. Sivakumar, S.K. Rajeshwaran, R. Subbiah, Preparation and characterization of AZ63A/boron nitride composites using hybrid mechanical and ultrasonic assisted stir casting. Int. Conf. Virtual Conf. Technol. Adv. Mech. Eng. **59**, 1513–1519 (2022). https://doi.org/10.1016/j.matpr.2022.01.403
49. S.F. Hassan, *Creation of New Magnesium-Based Material Using Different Types of Rein Forcements* (National University of Singapore, 2006)
50. K.K. Deng et al., Microstructure evolution and mechanical properties of a particulate re inforced magnesium matrix composites forged at elevated temperatures. Mater. Sci. Eng. A **527**(6), 1630–1635 (2010). https://doi.org/10.1016/j.msea.2009.10.053

Chapter 5
Magnesium Matrix Composites with Nano-Sized Reinforcements

This chapter explores the role of nanoscale reinforcements in enhancing the properties of magnesium-based composites for a wide range of applications, including structural, aerospace, automotive, and biomedical fields. It highlights the effects of ceramic particles (oxides, carbides, nitrides, and borides), carbon-based reinforcements (carbon nanotubes and graphene), calcium phosphate derivatives (for biodegradable implants), and metallic nanoparticles (e.g., Cu, Al, Ti, and Ni). Key mechanisms such as grain refinement, precipitation hardening, and bioactivity improvements are discussed in detail, along with challenges such as achieving uniform particle dispersion and strong interfacial bonding. Readers will gain insights into advancements in fabrication techniques, the unique benefits of each reinforcement type, and the future outlook for multifunctional magnesium nanocomposites.

5.1 Introduction

Magnesium-based nanocomposites represent an advanced category of materials incorporating nanoscale reinforcements within a magnesium matrix to achieve enhanced mechanical, thermal, and functional properties. The high surface area and superior intrinsic qualities of nano-sized reinforcements significantly enhance the matrix's strength, stiffness, and wear resistance. These materials exhibit exceptional strength-to-weight ratios, improved creep resistance, and enhanced thermal stability, making them well-suited for lightweight, high-performance applications in aerospace, automotive, and biomedical sectors. In comparison to micro-composites, magnesium nanocomposites often demonstrate superior grain refinement, improved ductility, and reduced thermal expansion. The following subsections provide a detailed discussion of the specific properties of magnesium-based nanocomposites, emphasizing their unique benefits, advancements in fabrication techniques, and their promising potential for engineering applications.

S. Seetharaman et al., *Magnesium-based Syntactic, Metastable and Nano-composites*, https://doi.org/10.1007/978-3-031-96514-2_5

5.2 Traditional Hard Ceramic Nanoparticles Reinforced Mg Composites

Hassan et al. [1] explored nano-Al_2O_3 addition to pure magnesium and studied the properties of resulting composites under tensile loads. The study involved processing Mg/Al_2O_3 nanocomposites using two methods: (i) disintegrated melt deposition (DMD) and (ii) powder metallurgy (PM), followed by hot extrusion. The DMD method included stirring the molten composite slurry and disintegrating it using argon gas before depositing it onto a steel mold, while the PM method encompassed blending, compaction, and sintering. It is important to note that sintering was conducted using a conventional heat treatment furnace. The microstructural and mechanical properties of Mg/Al_2O_3 nanocomposites produced by both methods were examined in the hot extruded condition. The results demonstrated enhanced strength and ductility in Mg, attributed to grain refinement and effective dispersion of nano-Al_2O_3 particles (Fig. 5.1). Further research on Mg/Al_2O_3 nanocomposites using microwave sintering, an energy-efficient powder metallurgy method, also showed improvements [2]. The study observed increased yield and tensile strength values due to the incorporation of nano-Al_2O_3 particles, though the failure strain improved only up to a 0.6 vol.% addition.

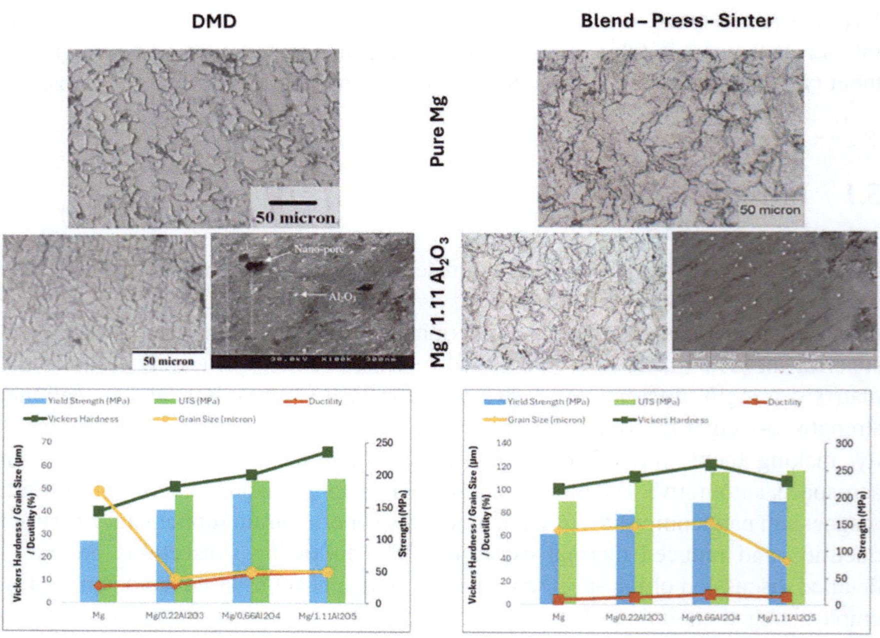

Fig. 5.1 Microstructure and mechanical properties of Al_2O_3 nanoparticles added Mg composites prepared by DMD and powder metallurgy methods from Hassan et al. [1]

In another study, Zaid et al. [3] used the powder metallurgy approach to fabricate nano-Al_2O_3 reinforced Mg composites. Here, the powder raw materials containing up to 5 vol.% nano-Al_2O_3 particles were first mixed for 6 h before compaction and sintering. The densified composites were then tested for a range of properties and the results revealed a substantial improvement in compressive strength and hardness for 3 vol.% nano-Al_2O_3 addition (Fig. 5.2).

In a related study, Shanthi et al. [4] highlighted that a stirring time of 5–7.5 min is critical for achieving effective dispersion of nanoscale reinforcements in magnesium composites, resulting in a balanced combination of strength, ductility, and toughness (Fig. 5.3). However, stirring beyond 7.5 min led to a significant reduction in ductility, accompanied by a shift in fracture behavior from ductile to brittle.

Han et al. [5] also used the DMD method to synthesize AZ41 alloy-based composites with 1 vol.% nano-Al_2O_3 reinforcement. Here, the microstructural analyses revealed a reasonably uniform distribution of nano-sized reinforcement particulates and intermetallics in the matrix which assisted in improving the overall mechanical properties including microhardness, tensile strength and ductility when compared to AZ31 magnesium. In a similar study, Habibnejad-Korayem et al. [6] also studied the impact of nano-sized Al_2O_3 particles on the mechanical and physical properties of pure magnesium and AZ31 magnesium alloy composites in both as-cast and wrought conditions. The addition of nano-Al_2O_3 particles significantly enhanced hardness and tensile strength but resulted in a notable reduction in ductility compared to the monolithic materials (Fig. 5.4). The increase in strength, accompanied by a loss in ductility, was attributed to grain refinement achieved through the relatively uniform distribution of Al_2O_3 nanoparticles within the magnesium matrix. The optimal strength was observed with the addition of 2 wt.% Al_2O_3 nanoparticles.

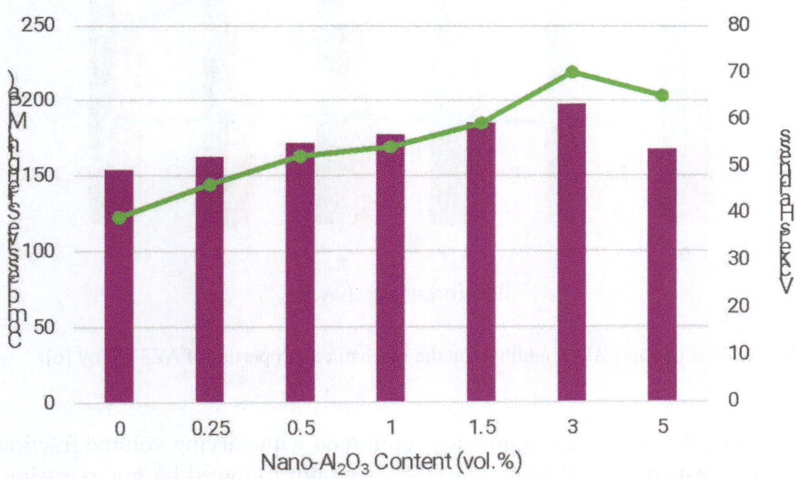

Fig. 5.2 Compressive strength and VHN variation due to nano-Al_2O_3 addition to pure Mg, based on the data from Zaid et al. [3]

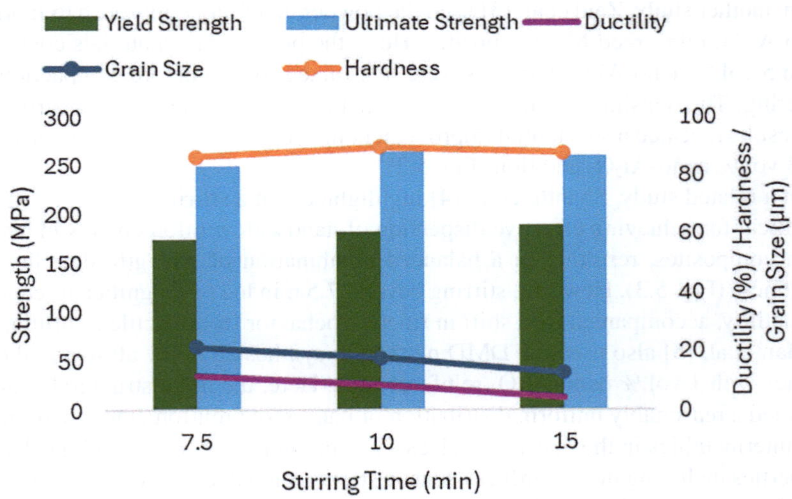

Fig. 5.3 Effect of stirring duration on the microstructure and mechanical properties of $Mg/1.11Al_2O_3$ nanocomposite from Shanthi et al. [4]

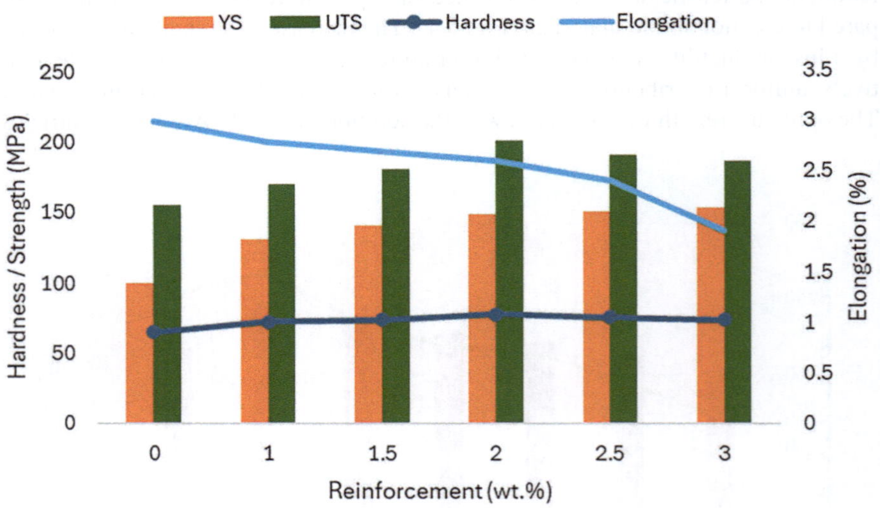

Fig. 5.4 Effects of nano-Al_2O_3 addition on the mechanical properties of AZ31 alloy [6]

Similarly, AZ31B magnesium alloy reinforced with varying volume fractions of nano-Al_2O_3 was produced using the DMD method, followed by hot extrusion [7]. While microstructural analysis showed the fairly uniform distribution of Al_2O_3 nanoparticulates with good interfacial integrity, the results of mechanical property measurements revealed superior ductility for the nanocomposites (i.e., 29.5% in the case of AZ31B/1.11Al_2O_3 when compared to 5.6% for AZ31B) at the expense of tensile yield and ultimate tensile strengths (by 25% and 20% respectively) (Fig. 5.5). On the other hand, the compressive yield and ultimate strengths were reported as improved by 30% and 10%, respectively, without a significant impact on the compressive fracture strain. Authors attributed these changes to the activation of non-basal slip systems by the Al_2O_3 nanoparticles.

In a related study, Paramsothy et al. [8] examined how the type of raw materials and the method of reinforcement addition influence the mechanical properties of DMD-synthesized AZ31/1.11Al_2O_3 nanocomposites. The study found that both nanocomposites exhibited increased microhardness compared to the monolithic AZ31. However, the differences in raw material types and reinforcement integration methods led to distinct material behaviors. The first nanocomposite, prepared using AZ31 rods and nano-Al_2O_3 packed in Al foil, achieved uniform nanoparticle dispersion, which resulted in superior tensile properties, including a 21% increase in ultimate tensile strength (UTS) and a 113% improvement in elongation compared to monolithic AZ31. In contrast, the second nanocomposite, made with AZ31 blocks and nanoparticles packed inside machined holes, exhibited enhanced compressive properties, with a 22% increase in compressive yield strength and a 23% improvement in compressive ductility (Fig. 5.6). The authors attributed these differences in material behavior to variations in crystallographic texture and grain refinement, which were influenced by the nanoparticle distribution.

The microstructure and mechanical properties of Al_2O_3 reinforced Mg-2Zn composites fabricated using powder metallurgy method were studied by Ercetin et al.

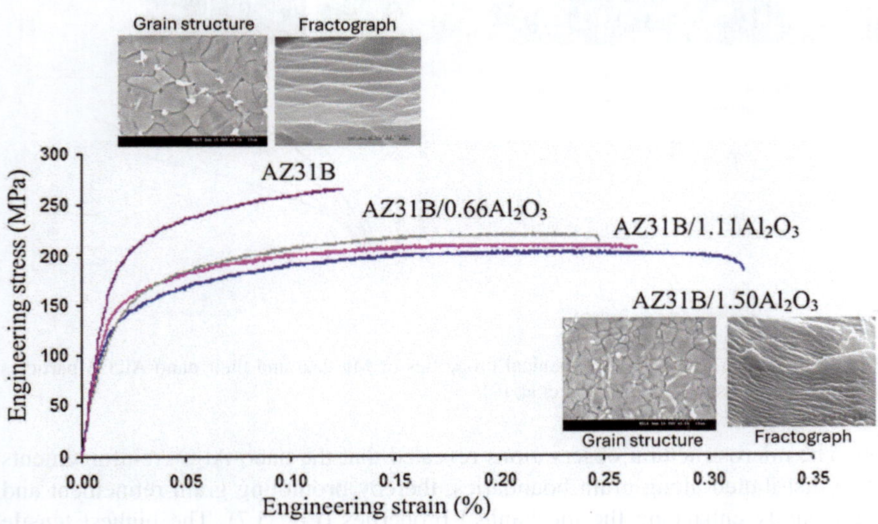

Fig. 5.5 Engineering stress–strain curve of AZ31B and their nano-Al_2O_3 reinforced composites. The representative microstructure and fractographs are shown as insets. From [7]

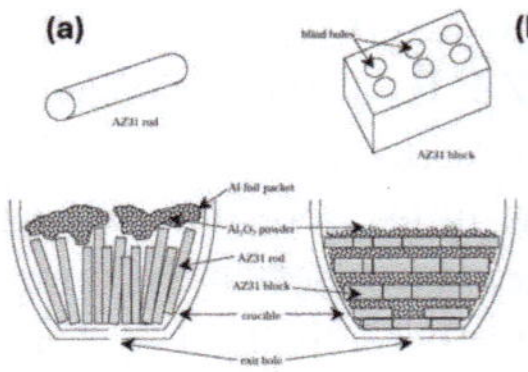

(b)

Material	Grain Size	Hardness	0.2TYS	UTS	TFS	0.2CYS	UCS	CFS
AZ31	4.0 ± 0.9	64 ± 4	172 ± 15	263 ± 12	10.4 ± 3.9	93 ± 9	486 ± 4	19.7 ± 7.2
AZ31/1.11Al_2O_3 (rod)	2.3 ± 0.7	83 ± 5 (+30)	204 ± 8 (+19)	317 ± 5 (+21)	22.2 ± 2.4 (+113)	98 ± 2 (+5)	509 ± 12 (+5)	19.0 ± 2.7 (-4)
AZ31/1.11Al_2O_3 (ingot)	3.1 ± 0.0	86 ± 3 (+34)	144 ± 9 (-16)	214 ± 16 (-19)	29.5 ± 1.9 (+184)	113 ± 8 (+22)	531 ± 14 (+9)	24.2 ± 1.0 (+23)

Fig. 5.6 (**a**) Schematic representation illustrating the alloy raw material and the method of reinforcement incorporation. (**b**) Comparative properties of the AZ31 alloy and its nanocomposite counterparts. From [8]

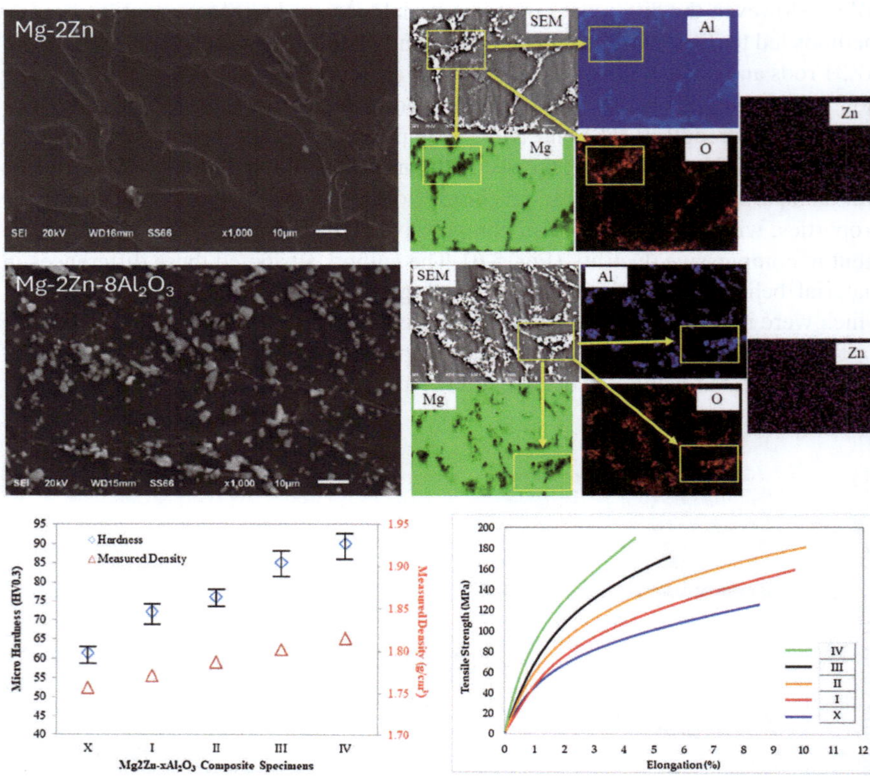

Fig. 5.7 Microstructural and mechanical properties of Mg-2Zn and their nano-Al2O3 particles reinforced composites from Ercetin et al. [9]

[9]. The microstructural observations revealed that the nano-Al_2O_3 reinforcements were distributed along grain boundaries, thereby promoting grain refinement and significantly enhancing the mechanical properties (Fig. 5.7). The highest tensile strength (~191 MPa) and hardness were achieved with 8 wt.% Al_2O_3 reinforcement.

Radi et al. [10] investigated the microstructural stability and mechanical behavior of AZ31 magnesium alloy composites reinforced with nano-Al_2O_3 particles (ranging from 0.5 to 2 wt.%) processed via equal channel angular pressing (ECAP). Their study found that the addition of 2 wt% nano-Al_2O_3 particles, combined with ECAP processing, significantly refined the microstructure, reducing the average grain size from 19.9 μm in the extruded state to 4.3 μm in the ECAPed state through grain boundary pinning (Fig. 5.8). Mechanical testing demonstrated improved hardness and enhanced grain boundary strengthening in the composites compared to the monolithic AZ31 alloy.

Paramsothy et al. [11] investigated the effects of Al_2O_3 nanoparticle reinforcement on AZ31 and ZK60A magnesium alloys, synthesized via disintegrated melt deposition (DMD) and hot extrusion. The addition of 1.5 vol% Al_2O_3 nanoparticles significantly enhanced tensile and compressive properties. For AZ31, tensile yield strength, ultimate tensile strength, and failure strain increased by 19%, 21%, and 113%, respectively, while compressive yield strength and ultimate compressive strength showed modest improvements. In ZK60A, ultimate tensile and compressive strengths increased by 13% and 7%, respectively, with substantial

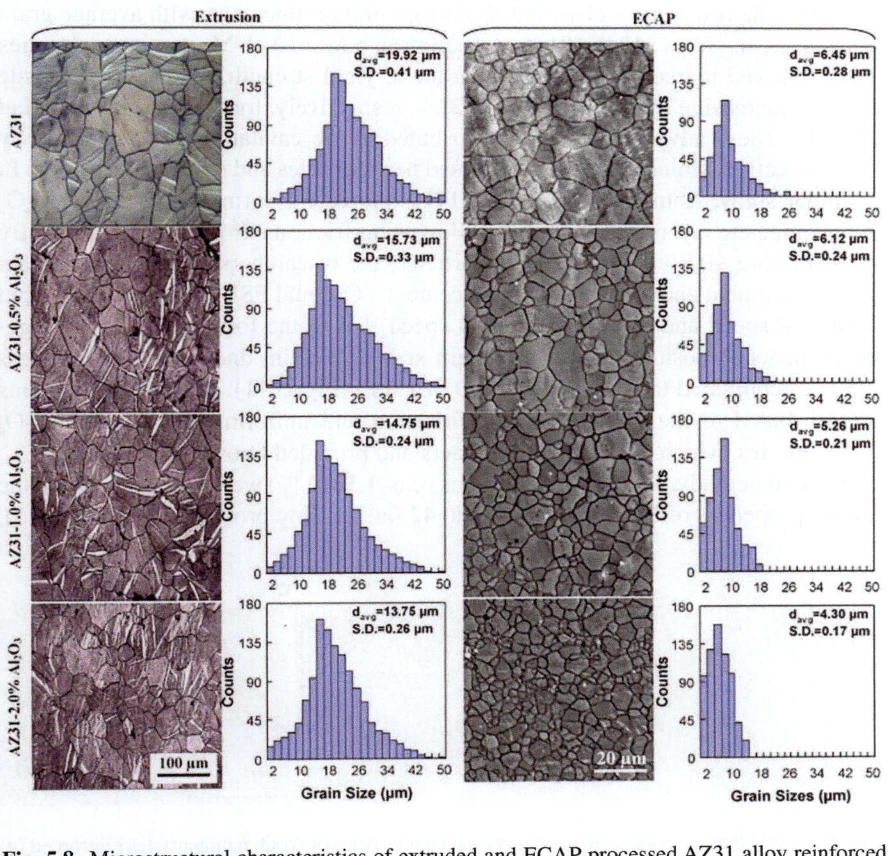

Fig. 5.8 Microstructural characteristics of extruded and ECAP processed AZ31 alloy reinforced with different amounts of nano-Al_2O_3 particles from Radi et al. [10]

improvements in failure strain. These enhancements were attributed to mechanisms such as Orowan strengthening, load transfer, and dislocation generation due to mismatches in modulus and thermal expansion. Furthermore, the nanoparticles improved mechanical anisotropy and promoted a uniform grain structure without inducing microporosity (Fig. 5.9). Similar enhancements were also observed in Al_2O_3-reinforced AZ81 alloy composites, demonstrating the consistent benefits of Al_2O_3 nanoparticle reinforcement across magnesium alloy systems.

AZ91E alloy composites reinforced with 1–3 wt.% nano-Al_2O_3 particles were fabricated using semi-solid stir casting by Sameer Kumar et al. [12]. The study showed that the addition of up to 2 wt.% nano-Al_2O_3, with a uniform distribution, significantly enhanced the mechanical properties due to grain refinement. Specifically, hardness increased by up to 22.5%, and tensile strength improved by 26.54% in the case of AZ91E/2wt.%Al_2O_3 composite. However, when the reinforcement level exceeded 2 wt.%, particle clustering and poor bonding between the matrix and particles led to a decline in mechanical performance (Fig. 5.10).

The study by Gnanavelbabu et al. [13] examined the influence of ultrasonication power on AZ91D/nano-Al_2O_3 composites prepared via ultrasonication-assisted squeeze casting. By varying ultrasonication power levels (0, 1500, 2000, and 2500 W), the researchers observed significant grain refinement, with average grain size decreasing from 312 to 68 μm at the highest power level. Mechanical properties also improved markedly, with microhardness, yield strength, and ultimate tensile strength increasing by 18%, 48%, and 28%, respectively, for composites treated at 2500 W. These advancements were attributed to the cavitation effect induced by ultrasonication, which efficiently dispersed nanoparticles and reduced clustering. In a similar study, Ahmadkhaniha et al. [14] explored the formation of AZ91/Al_2O_3 nanocomposite layers on magnesium alloy using friction stir processing (FSP). By incorporating ~0.8 vol% Al_2O_3 nanoparticles, the researchers achieved significant grain refinement and mechanical enhancements. Optimal FSP parameters (800 rpm rotational speed and 40 mm/min travel speed) led to the formation of a homogeneous nanocomposite layer with a grain size of 3.6 μm and improved hardness (95 HV) compared to untreated AZ91D (70 HV) (Fig. 5.11). These improvements are attributed to the dynamic recrystallization and uniform dispersion of Al_2O_3 nanoparticles, which acted as grain refiners and provided Orowan strengthening.

In another study, nano-Al_2O_3 additions (0.5–1.5 vol.%) were used to improve the tensile properties of MgY1.06Zn0.76Al0.42 (at.%) alloy processed via DMD [15].

Fig. 5.9 SEM micrographs showing the grain characteristics of Al_2O_3 nanoparticles reinforced (**a**) AZ31 and (**b**) ZK60A alloys. (**c**) Representative TEM micrograph showing the presence of individual nanoparticles and fine intermetallic particles in the nanocomposite [11]

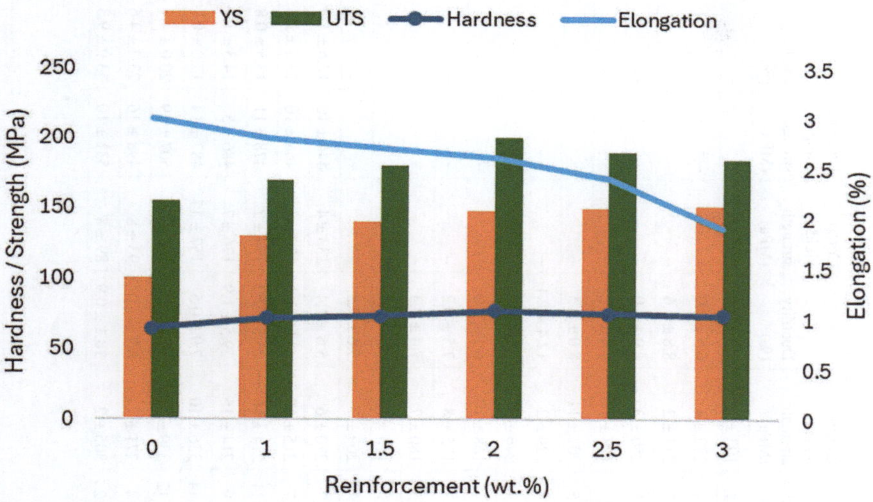

Fig. 5.10 Mechanical properties of nano-Al_2O_3 reinforced AZ91E alloy. (From Sameer Kumar et al. [12])

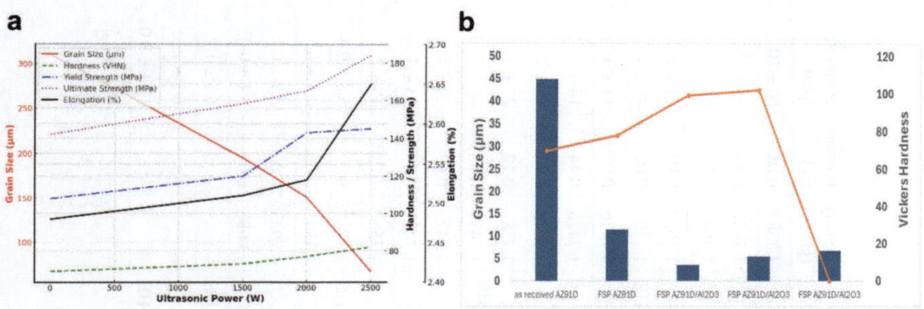

Fig. 5.11 Microstructural and mechanical properties of nano-Al_2O_3 reinforced AZ91 alloy composites. (From (**a**) Gnanavelbabu et al. [13] and (**b**) Ahmadkhaniha et al. [14] respectively)

The alloy's microstructure, after homogenization at 450 °C for 2 h, featured fine Mg-Y-Zn-Al precipitates and nano-Al_2O_3 particles, contributing to grain refinement through particle-stimulated nucleation and dynamic recrystallization. The study also highlighted that nano-Al_2O_3 addition suppressed the formation of 14H long-period stacking ordered (LPSO) phases during solidification. Mechanical testing results under tensile loading confirmed that the grain refinement and weakened basal texture enhanced both the strength and ductility of the MgY1.06Zn0.76Al0.42 alloy nanocomposites. The strength reached its peak at 0.5 vol.% nano-Al_2O_3, while ductility improved with increasing amounts of nano-Al_2O_3 particles.

The summary of grain size and mechanical properties of alumina reinforced magnesium-based composites is presented in Table 5.1.

Table 5.1 Properties of nano-Al_2O_3 reinforced Mg nanocomposites

References	Matrix	Reinforcement type	Wt.%	Vol.%	Process	Density (g/cc)	Porosity (%)	Average grain size (μm)	Hardness (Hv)	Tensile yield strength (MPa)	Ultimate tensile strength (MPa)	Ductility (%)	Comp yield strength (MPa)	Ultimate comp strength (MPa)	Ductility (%)
Hassan and Gupta [1]	Mg	Al_2O_3	0	0	PM-MWS-HE	1.7387	0.08	60 ± 10	37	132 ± 7	193 ± 2	4.2 ± 0.1	–	–	–
			0.5	0.22		1.7421	0.16	61 ± 18	44	169 ± 4	232 ± 4	6.5 ± 2.0	–	–	–
			1.5	0.66		1.7501	0.27	63 ± 16	50	191 ± 2	247 ± 2	8.8 ± 1.6	–	–	–
			2.5	1.11		1.7632	0.09	31 ± 13	70	194 ± 5	250 ± 3	6.9 ± 1.0	–	–	–
			0	0	DMD-HE	1.7397	0.02	49 ± 8	40	97 ± 2	173 ± 1	7.4 ± 0.2	–	–	–
			0.5	0.22		1.7436	0.07	11 ± 3	51	146 ± 5	207 ± 11	8.0 ± 2.3	–	–	–
			1.5	0.66		1.7476	0.44	14 ± 4	56	170 ± 4	229 ± 2	12.4 ± 2.1	–	–	–
			2.5	1.11		1.7623	0.14	14 ± 2	66	175 ± 3	246 ± 3	14.0 ± 2.4	–	–	–
Wong and Gupta [2]	Mg	Al_2O_3		0	PM-MWS-HE	1.738	0.12	27 ± 7	40	116 ± 11	168 ± 10	6.1 ± 2.0	–	–	–
				0.3		1.741	0.32	–	48	119 ± 7	175 ± 8	7.5 ± 0.2	–	–	–
				0.6		1.742	0.67	24 ± 4	54	130 ± 5	180 ± 7	7.4 ± 0.3	–	–	–
				1.0		1.748	0.83	15 ± 3	60	154 ± 5	213 ± 12	6.3 ± 0.4	–	–	–
				1.5		1.756	1.0	18 ± 3	68	148 ± 10	209 ± 7	5.6 ± 0.3	–	–	–
Nguyen & Gupta [7]	AZ31B	Al_2O_3		0	DMD-HE	1.77	0	5.8 ± 0.4	63	201 ± 7	270 ± 6	5.6 ± 1.4	133 ± 4	444 ± 10	12.6 ± 1.0
				0.66		1.7841	0.06	4.5 ± 0.3	79	149 ± 7	215 ± 15	14.6 ± 1.1	172 ± 17	468 ± 10	13.4 ± 1.8
				1.11		1.7933	0.06	3.9 ± 0.1	82	148 ± 11	214 ± 16	25.5 ± 2.2	174 ± 2	478 ± 11	13.5 ± 0.8
				1.5		1.8086	0.07	3.1 ± 0.03	86	144 ± 9	214 ± 16	29.5 ± 1.9	176 ± 7	486 ± 5	14.3 ± 1.2
Paramsothy et al. [16]	AZ81	Al_2O_3		0	DMD-HE	–	–	6.3 ± 1.2	119	225 ± 14	336 ± 10	7.9 ± 0.5	157 ± 17	487 ± 14	17.0 ± 0.1
				1.5		–	–	6.1 ± 1.5	117	212 ± 12	339 ± 11	13.1 ± 1.0	146 ± 11	507 ± 19	20.0 ± 4.1
Paramsothy et al. [17]	ZK60A	Al_2O_3		0	DMD-HE	–	–	11.1 ± 3.0	117	182 ± 4	271 ± 1	6.7 ± 1.0	93 ± 8	498 ± 16	23.2 ± 4.6
				1.5		–	–	10.7 ± 2.5	135	175 ± 2	305 ± 2	18.1 ± 0.9	89 ± 8	531 ± 10	33.9 ± 9.3

Paramsothy et al. [18]	AZ31–AZ91	Al_2O_3		0	DMD-HE	–	–	5.1 ± 0.7	111	207 ± 4	316 ± 6	8.0 ± 0.1	117 ± 15	495 ± 13	19.6 ± 1.9
				1.5		–	–	4.8 ± 1.0	139	232 ± 13	339 ± 10	15.9 ± 0.5	123 ± 16	512 ± 3	18.2 ± 0.7
Tan et al. [15]	$MgY_{1.06}Zn_{0.76}Al_{0.42}$	Al_2O_3		0	DMD-HE	–	–	2.43 ± 1.10	–	376 ± 17	416 ± 5	10.8 ± 0.6	–	–	–
				0.5		–	–	5.95 ± 3.16	–	310 ± 11	355 ± 11	15.6 ± 1.8	–	–	–
				1.0		–	–	4.88 ± 2.35	–	260 ± 7	293 ± 1	16.3 ± 0.7	–	–	–
				1.5		–	–	3.74 ± 1.89	–	239 ± 7	292 ± 2	17.4 ± 1.3	–	–	–
Kumar et al. [12]	AZ91E	Al_2O_3	0		Semi-solid stir casting	1.806	0.22	–	64.95 ± 3	–	–	–	–	–	–
			1			1.820	0.58	–	72.12 ± 1	–	–	–	–	–	–
			1.5			1.829	0.66	–	73.61 ± 2	–	–	–	–	–	–
			2			1.832	1.05	–	77.62 ± 3	–	–	–	–	–	–
			2.5			1.838	1.28	–	75.45 ± 3	–	–	–	–	–	–
			3			1.844	1.51	–	73.77 ± 2	–	–	–	–	–	–

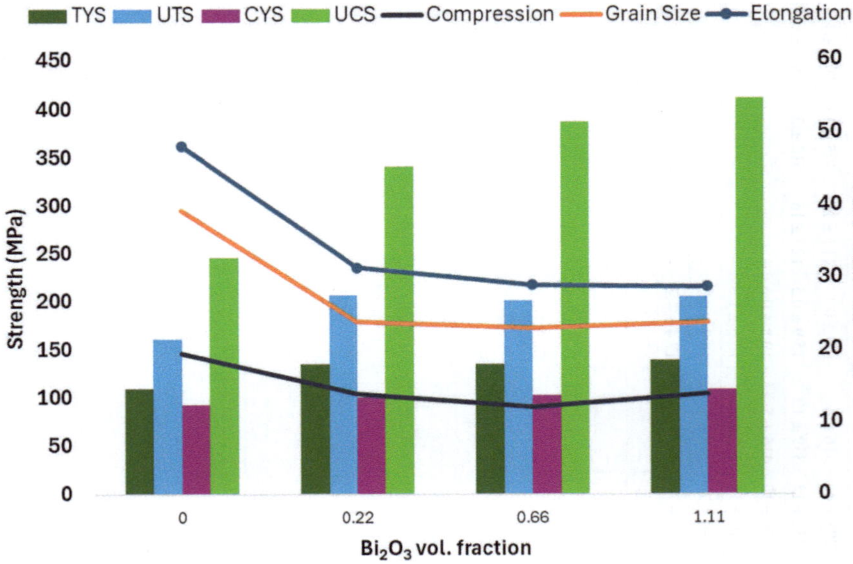

Fig. 5.12 Properties of nano-Bi_2O_3 reinforced Mg composites [19]

Habibi et al. [19] synthesized novel Mg-based nanocomposites reinforced with x = 0.22, 0.66, 1.11 vol.%) Bi_2O_3 nanoparticles via powder metallurgy and microwave-assisted rapid sintering followed by hot extrusion. The developed Mg/xBi_2O_3 nanocomposites exhibited significantly improved mechanical properties compared to pure Mg, with the Mg/0.22Bi_2O_3 formulation showing the best overall performance (Fig. 5.12). Enhancements included tensile yield strength (24%), compressive yield strength (9%), ultimate tensile strength (28%), and ultimate compressive strength (39%) without significant compromise in failure strain. The authors attributed these improvements to key mechanisms including grain size strengthening, geometrically necessary dislocations from thermo-elasto-plastic mismatch, Orowan strengthening, and the formation of Bi_2Mg_3 phase. Despite these enhancements, the textural evolution was found to be minimal.

Mg nanocomposites reinforced with calcium oxide (CaO) nanoparticles with enhanced mechanical properties were developed by DMD method followed by hot extrusion [20]. The microstructural analysis revealed significant grain refinement and the formation of Mg_2Ca phases, contributing to improved strength. The addition of CaO increased both tensile and compressive yield strengths, with the Mg/1CaO composition showing optimal toughness and ductility, as the excess amounts of CaO nanoparticles resulted in agglomeration affecting the load bearing attributes and ductility (Fig. 5.13). In this study, the authors also investigated the ignition resistance of developed nanocomposites and found that the Mg/1.5CaO composite achieved an ignition temperature of 831 °C, much higher than pure magnesium and commercial alloys like AZ31 and AZ91. These improvements are attributed to the

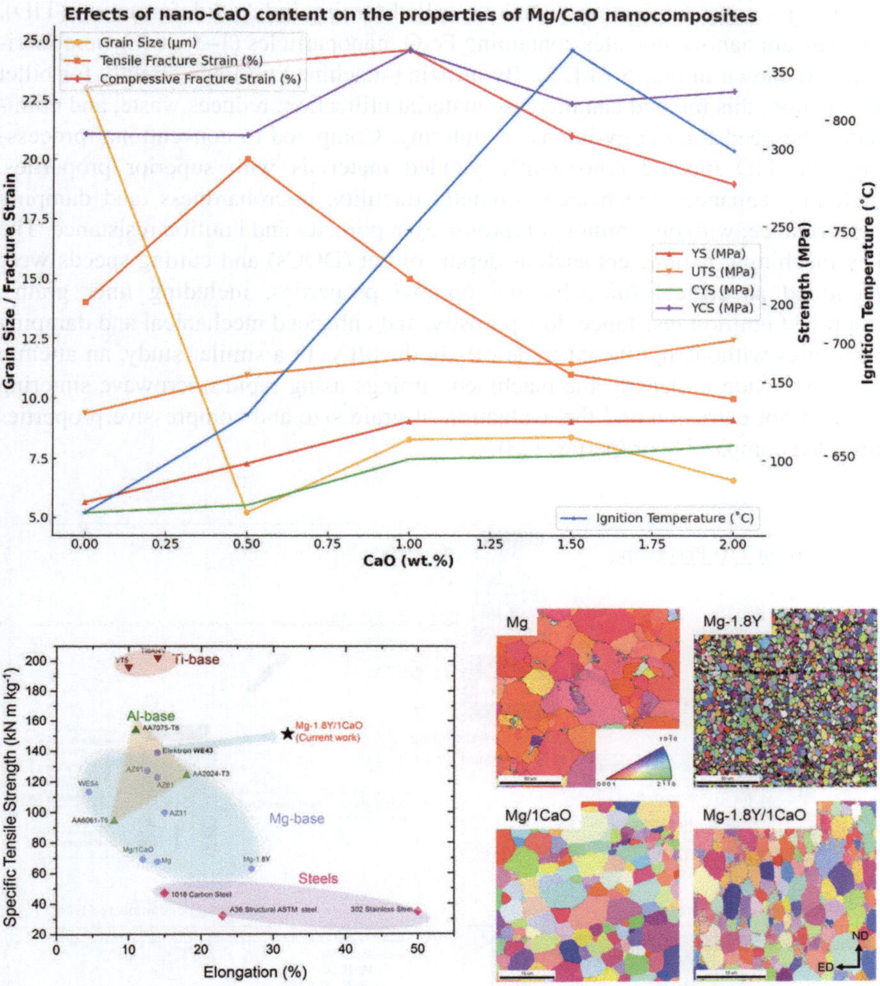

Fig. 5.13 Mechanical and microstructural properties of nano-CaO added Mg and Mg-1.8Y composites. (Adapted from [20, 21])

formation of stable composite oxide layers and the presence of hard Mg_2Ca and CaO phases. In a similar study, a magnesium nanocomposite (Mg-1.8Y/1CaO) with exceptional mechanical properties and ignition resistance, addressing the critical limitations of traditional magnesium alloys was also prepared [21]. It has a tensile strength of approximately 343 MPa, ductility of 30%, and a delayed ignition temperature of 1045 °C, closer to the boiling point of Mg. These enhancements stemmed from the incorporation of yttrium and calcium oxide into the magnesium matrix, which forms secondary phases like Mg_2Ca and in situ nanoparticles of Y_2O_3.

Using a novel fabrication technique called turning induced deformation (TID), magnesium nanocomposites containing Fe_3O_4 nanoparticles (1–3 wt.%) were fabricated as shown in Fig. 5.14 [22]. By utilizing machined turnings directly for billet production, this method can increase material utilization, reduces waste, and eliminated the need for energy-intensive sintering. Compared to conventional processing, the TID method consistently yielded materials with superior properties, including enhanced mechanical strength, ductility, microhardness, and damping performance, with only minor compromises in porosity and ignition resistance. The key machining parameters such as depths of cut (DOCs) and cutting speeds were identified as critical for achieving optimal properties, including finer grains, improved ignition resistance, low porosity, and enhanced mechanical and damping capacities without significant trade-offs in ductility. In a similar study, an attempt was also made to densify the machined turnings using rapid microwave sintering without hot extrusion and the evaluation of grain size and compressive properties revealed comparable properties [23].

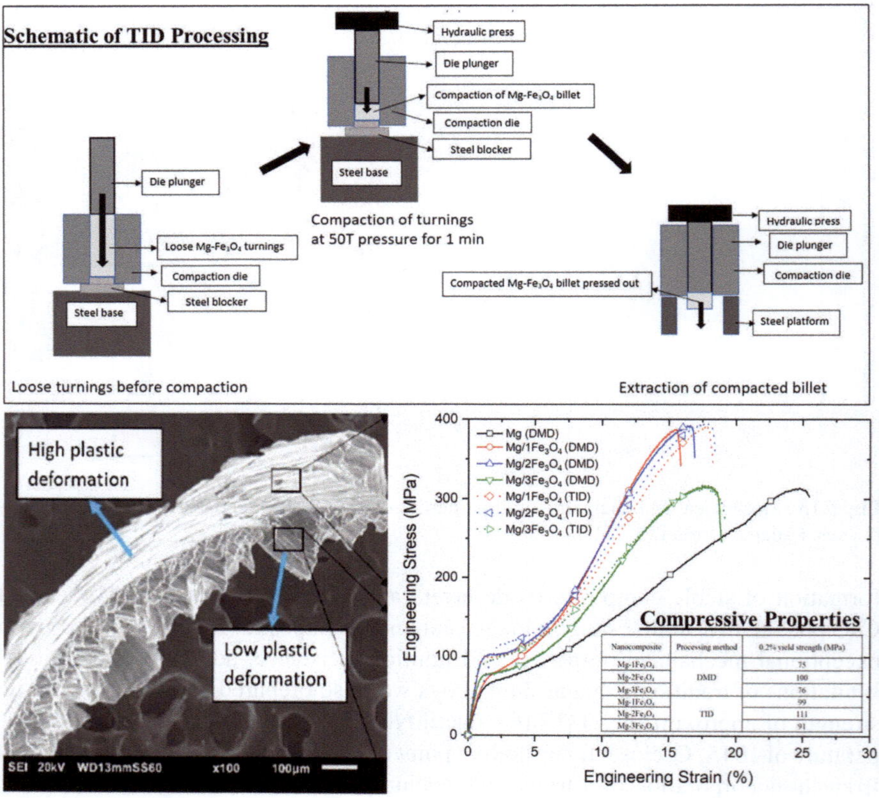

Fig. 5.14 Turning induced deformation processing of Mg/Fe_3O_4 nanocomposites. (Adapted from [22])

Mg/SiO_2 nanocomposites were also synthesized using microwave-assisted PM method and their mechanical and damping properties were investigated [24]. Here, the addition of up to 2 vol.% SiO_2 nanoparticles reduced grain size by 32%, increased hardness by 50%, and improved compressive yield strength by 23% compared to pure magnesium. Damping capacity was also significantly enhanced, with the $Mg/2SiO_2$ composite exhibiting a 239% increase (Fig. 5.15). These improvements aligned the composite's mechanical properties with those of human bone, maintaining an elastic modulus that minimized stress shielding, demonstrating its suitability for orthopedic implants and structural applications.

In another study, Lee et al. [25] used the friction stir processing (FSP) technique to integrate and disperse 5–10 vol.% SiO_2 nanoparticles into an AZ61 Mg alloy matrix. This process significantly refined the grain size of the base alloy from 75 to 0.8 μm upon 10 vol% SiO_2 addition. This also results in the formation of Mg_2Si and MgO phases which together with grain refinement has contributed to the improvement in mechanical properties, including a near doubling of hardness compared to the base material and enhanced yield strength (225 MPa) and tensile strength (251 MPa) (Fig. 5.15). The 10% SiO_2 composite also exhibited high strain rate superplasticity, achieving elongation of over 400% at elevated temperatures.

The study by Kujur et al. [26] explored the development and potential of magnesium (Mg)-based nanocomposites reinforced with Sm_2O_3 nanoparticles for use as bioresorbable orthopedic implants by microwave-assisted powder metallurgy method. Leveraging the benefits of Mg's biocompatibility and mechanical

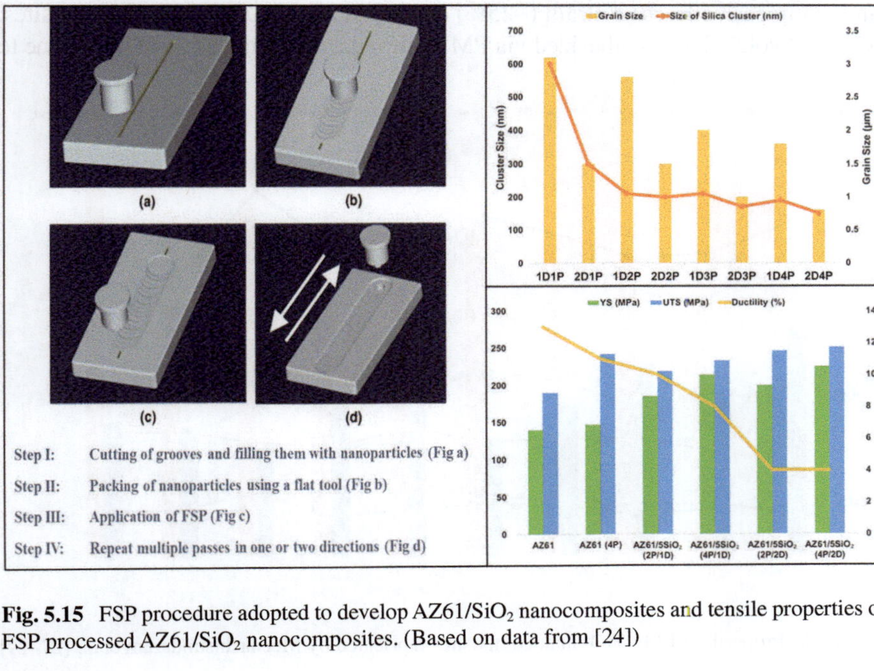

Fig. 5.15 FSP procedure adopted to develop $AZ61/SiO_2$ nanocomposites and tensile properties of FSP processed $AZ61/SiO_2$ nanocomposites. (Based on data from [24])

properties, the incorporation of rare earth oxides via powder metallurgy and microwave sintering significantly enhanced the material's corrosion resistance and cytocompatibility. Mg/1.5 vol.% Sm_2O_3 exhibited a 25% lower corrosion rate (1.028 mm/year) compared to pure Mg, attributed to a refined microstructure and protective $Mg(OH)_2$ layer formation. Cytotoxicity assessments revealed all tested nanocomposites displayed level 0 cytotoxicity, with cell viability exceeding 70% for Sm_2O_3 additions up to 1 vol.%, ensuring in vitro biocompatibility. These findings demonstrate that rare earth oxide reinforcement, rather than alloying, can mitigate toxicity concerns while improving properties of magnesium, making it a promising candidate for temporary implants. Ongoing research aims to evaluate long-term corrosion behavior, osteogenic potential, and cellular interactions to further support clinical translation of these nanocomposites.

Meenashisundaram et al. [27] synthesized Mg–TiO_2 nanocomposites using the disintegrated melt deposition (DMD) technique followed by hot extrusion. The addition of TiO_2 nanoparticulates significantly refined the grain size of Mg, achieving a maximum reduction of ~47% with 2.5 vol.% TiO_2. Mechanical properties improved under tensile loading, with proof stress, ultimate tensile strength (UTS), and tensile fracture strain increasing by ~37%, ~9%, and ~31%, respectively. Under compression, the 0.2% compressive yield strength (CYS) and fracture strain improved by ~85% and ~33%, though the ultimate compressive strength (UCS) decreased by ~7.5%. The inclusion of TiO_2 also reduced tensile-compression asymmetry (TCA), with a minimum value of 1 (Fig. 5.16). The authors further investigated these composites using the microwave-assisted powder metallurgy (PM) method [28]. PM-processed composites exhibited superior tensile ductility (~14.5%) and compressive fracture strain (~25%) compared to DMD-processed composites. Mg-1.98 vol.% TiO_2 synthesized via PM achieved a minimum TCA of ~1.06 due to

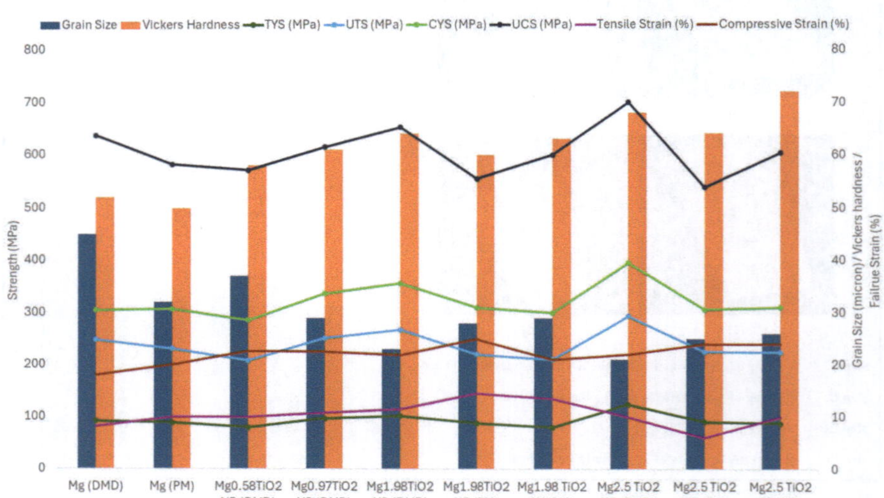

Fig. 5.16 Properties of Mg/TiO_2 nanocomposites fabricated by Meenashisundaram et al. [27–29]

the activation of cross-slip. However, DMD-processed composites outperformed in grain refinement (~47% reduction), hardness (~30% increase) and compressive yield strength for 2.5 vol.% TiO_2.

In another study, the authors compared TiO_2 nanoparticulates (NPs) and nanofibers (NFs) in Mg composites prepared via PM and hot extrusion [29]. The grain size magnesium, with NPs achieved greater reductions (~22% for 2.5 vol.% TiO_2). NFs provided higher hardness (~44% improvement) and superior UCS (~300 MPa), while NPs enhanced tensile ductility (~14.5% fracture strain, 45% higher than pure Mg). The TCA was significantly reduced with NFs and the authors concluded that while TiO_2 NPs are more effective for grain refinement and ductility, TiO_2 NFs excel in hardness and compressive performance.

The summary of processing type, microstructural and mechanical response of Mg nanocomposites reinforced with Bi_2O_3, SiO_2, Sm_2O_3, and TiO_2 nanoparticles is listed in Table 5.2.

Magnesium nanocomposites containing nano-sized Y_2O_3 particles were prepared using powder metallurgy (PM) and disintegrated melt deposition (DMD) methods [1]. Tensile testing revealed significant mechanical property enhancements in the Mg/Y_2O_3 nanocomposites due to grain refinement and efficient distribution of the nanoscale reinforcements. Among the formulations, PM-processed Mg/0.22Y_2O_3 exhibited superior ductility (12%) compared to pure Mg (4%), while DMD-processed Mg/0.66Y_2O_3 demonstrated improved strength (YS: 151 ± 2 MPa, UTS: 202 ± 2 MPa) relative to pure Mg (YS: 132 ± 7 MPa, UTS: 193 ± 2 MPa). Tun and Gupta [32] employed microwave sintering-assisted PM to produce magnesium composites with 0.2 and 0.7 vol.% Y_2O_3 (30–50 nm). The Mg/0.7vol.%Y_2O_3 composite achieved a 22% hardness increase (45 HV), a 17% rise in ultimate tensile strength (244 MPa), and improved ductility (8.6%) compared to pure Mg, with grain refinement reducing grain size to 18 μm. Additionally, the study reported a reduced coefficient of thermal expansion (20.8×10^{-6}/°C) due to yttria's low CTE, enhancing thermal stability. Further research by Tun and Gupta explored the effects of microwave heating rates [33] and extrusion ratios [34] on Mg/0.7Y_2O_3 nanocomposites. A higher heating rate (49 °C/min) produced superior properties, with Mg/0.7Y_2O_3 exhibiting a microhardness of 45 HV, ultimate tensile strength of 244 MPa, and reduced porosity (0.35%) due to uniform yttria dispersion, while lower heating rates (20 °C/min) led to grain coarsening, yttria agglomeration, and diminished properties, including microhardness (38 HV) and tensile strength (175 MPa). Regarding extrusion ratios, a 25:1 ratio yielded optimal mechanical properties, with microhardness reaching 55 HV and ultimate tensile strength improving to 300 MPa (a 55% increase compared to 12:1), along with reduced grain size (18 μm) and enhanced yttria dispersion, though ductility peaked at a 19:1 ratio and decreased at higher ratios due to localized plastic deformation (Fig. 5.17).

Hassan et al. [35] studied Mg/Y_2O_3 nanocomposites using conventional sintering (CS) and microwave-assisted sintering (MS), finding that the processing route significantly affected the microstructure and mechanical properties. CS-processed Mg/0.7Y_2O_3 exhibited superior ductility (15.8% elongation vs. 8.0% for MS) due to better grain refinement (13 μm) and reduced porosity, whereas MS-processed

Table 5.2 Microstructural and mechanical response of Mg nanocomposites reinforced with Bi_2O_3, SiO_2, Sm_2O_3, and TiO_2 nanoparticles

References	Matrix	Reinforcement type	Wt.%	Vol.%	Process	Density (g/cc)	Porosity (%)	Average grain size (μm)	Hardness (Hv)	Tensile yield strength (MPa)	Ultimate tensile strength (MPa)	Ductility (%)	Comp yield strength (MPa)	Ultimate comp strength (MPa)	Ductility (%)	Remarks
Habibi et al. [19]	Mg	Bi_2O_3		0	PM-MWS-HE	–	–	20 ± 3	–	109 ± 09	161 ± 16	8.9 ± 1.1	93 ± 02	246 ± 23	19.4 ± 0.8	–
				0.22		–	–	10 ± 2	–	135 ± 12	207 ± 13	7.5 ± 0.3	101 ± 09	341 ± 08	13.9 ± 0.7	–
				0.66		–	–	11 ± 3	–	135 ± 13	201 ± 15	6.0 ± 0.5	102 ± 11	387 ± 17	12.0 ± 1.5	–
				1.11		–	–	10 ± 2	–	139 ± 13	205 ± 18	4.9 ± 0.6	108 ± 03	411 ± 06	13.6 ± 1.4	–
Kujur et al. [30]	Mg	Sm_2O_3		0	PM-MWS	–	–	25.7 ± 2.6	54	74 ± 3	249 ± 6	17.4 ± 0.3	–	–	–	–
				0.5		–	–	21.4 ± 2.3	64	87 ± 1	285 ± 6	19.8 ± 0.6	–	–	–	–
				1.0		–	–	17.6 ± 1.8	71	118 ± 2	331 ± 7	20.1 ± 0.7	–	–	–	–
				1.5		–	–	12.9 ± 2.2	73	128 ± 5	395 ± 7	17.2 ± 0.5	–	–	–	–
Parande et al. [31]	Mg	SiO_2	0		PM-MWS	–	–	–	–	87 ± 4	128 ± 6	8.7 ± 3.0	–	–	–	–
			0.5			–	–	–	–	85 ± 2	136 ± 1	10.1 ± 0.7	–	–	–	–
			1.0			–	–	–	–	94 ± 3	150 ± 2	14.3 ± 2.1	–	–	–	–
			2.0			–	–	–	–	114 ± 5	171 ± 8	9.5 ± 3.0	–	–	–	–
Meenashisundaram et al. [27–29]	Mg	TiO_2		0	DMD	1.738	0.11	45 ± 2.4	52	92 ± 5	156 ± 6	8.2 ± 0.2	57 ± 3	332 ± 10	18 ± 0	–
				0	PM	1.7377	0.13	32 ± 1.5	50	89 ± 4.5	142 ± 6	10 ± 0.3	76 ± 2	275 ± 4	20 ± 1.5	–
				1.98	DMD	1.7895	0.15	23 ± 5.5	64	102 ± 3	165.5 ± 3	11.5 ± 1	88.3 ± 1	297 ± 1	21.9 ± 1	–
					PM	1.7880	0.24	28 ± 1.5	60	88 ± 10	132 ± 8	14.5 ± 1	89.9 ± 2	245 ± 8	25 ± 2	Particle reinforcement
						1.7845	0.26	29 ± 2	63	80 ± 6	130 ± 9	13.5 ± 2	89.9 ± 7.4	300 ± 9	21 ± 0.5	Nanofiber reinforcement
				2.5	DMD	1.7990	0.18	21 ± 4	68	124 ± 8.8	170 ± 6	10 ± 1	101 ± 9	305.5 ± 11	22 ± 2	–
					PM	1.7965	0.32	25 ± 2.5	64	91.1 ± 5	134 ± 7	6 ± 1	81 ± 0.6	233 ± 6	24 ± 2	Particle reinforcement
						1.7958	0.36	26 ± 3	72	88 ± 5	136 ± 2	10 ± 1	86 ± 1	294 ± 0.3	24 ± 1	Nanofiber reinforcement

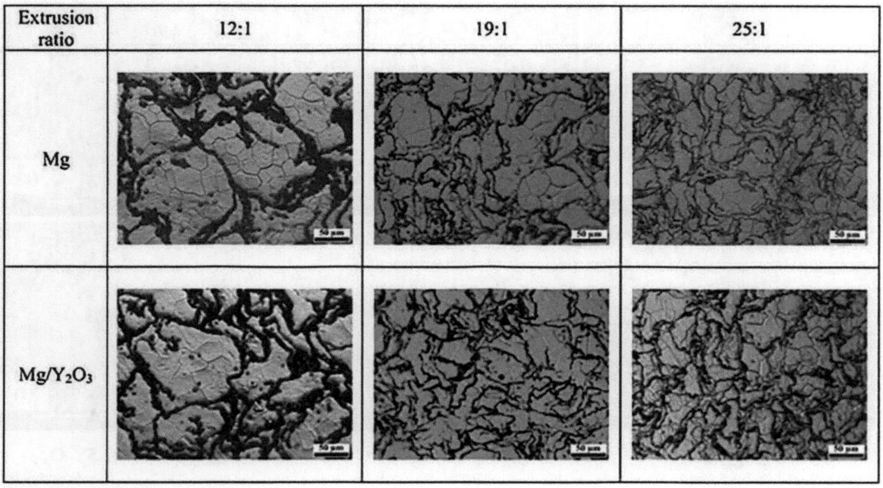

Fig. 5.17 Grain morphology of pure magnesium and Mg/Y_2O_3 nanocomposites extruded at different extrusion ratios. (Adapted from Tun et al. [34])

$Mg/0.7Y_2O_3$ showed higher ultimate tensile strength (244 MPa vs. 202 MPa) owing to enhanced matrix-reinforcement bonding and effective load transfer. Both methods led to substantial hardness and fracture energy improvements, with the CS $Mg/0.2Y_2O_3$ composite achieving a 313% fracture energy increase over pure Mg. Strengthening mechanisms included Hall-Petch strengthening, Orowan looping, and dislocation generation from thermal mismatch, highlighting the advantages of MS for strength and CS for ductility and fracture resistance, demonstrating the versatility of Mg/Y_2O_3 nanocomposites for tailored mechanical properties.

In a similar study, Kumar et al. [36] investigated the effects of nano-Y_2O_3 on the microstructure and mechanical properties of Mg-3Al-2.5La alloy compared to Mg-3Al. The materials were synthesized using DMD technique followed by hot extrusion. With respect to microstructure, the addition of Y_2O_3 significantly refined the grain size (~3.6 μm, a 43% reduction from Mg–3Al–2.5La grain size) and precipitates ($Al_{11}La_3$, Al_2La, and $Al_{2.12}La_{0.88}$) (Fig. 5.18). While the coefficient of thermal expansion (CTE), damping capacity and the compressive strengths improved, the fracture strain of Mg–3Al–2.5La remained unaffected by Y_2O_3 addition. In terms of compression properties, the compressive yield and ultimate strengths improved from ~141 to ~156 MPa and ~456 to ~520 MPa, surpassing those of Mg–3Al respectively.

The summary of processing type, grain size, and mechanical response of Y_2O_3 reinforced magnesium matrices is summarized in Table 5.3.

Tun et al. [37] investigated the mechanical properties of magnesium composites reinforced with nano-ZnO particles. The addition of ZnO (0.5–1.5 vol.%) enhanced microhardness while maintaining similar tensile and compressive yield strengths compared to monolithic Mg, attributed to heterogeneous grain size distribution and

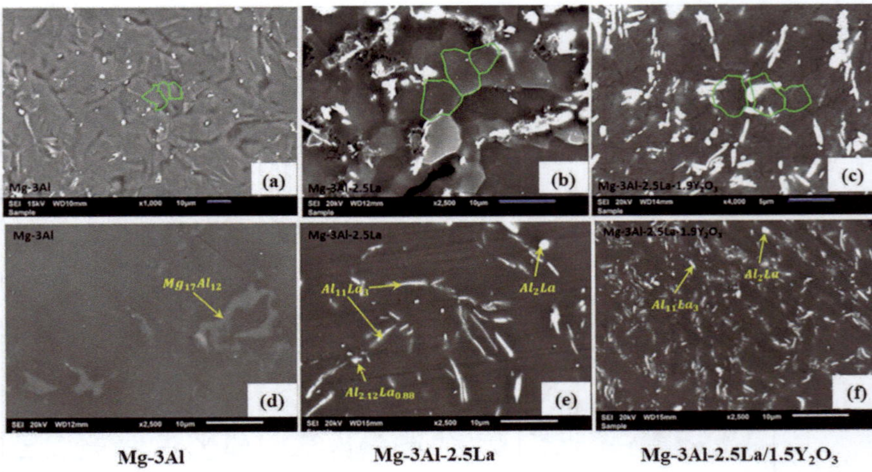

Fig. 5.18 Properties of Mg-3Al-2.5La/1.9Y_2O_3 nanocomposites. (Adapted from [36])

weak basal texture. The composites exhibited minimal tension-compression yield asymmetry and improved ultimate tensile and compressive strengths due to strengthening from fine ZnO particles and grain refinement. Tensile failure strain significantly increased because of non-basal slip activation and microstructure refinement, though compressive failure strain was slightly compromised (Fig. 5.19). XRD analysis revealed weak or random basal texture in both Mg and Mg–ZnO composites, while ZnO particles were distributed at grain boundaries and within grains. These findings demonstrate the potential of Mg–ZnO composites for applications requiring enhanced hardness, strength, and tensile ductility. Mg/ZnO nanocomposites were also developed via DMD method followed by hot extrusion wherein a maximum increment in tensile yield strength by ~55% and compressive yield strength by 90% (with reduced tension–compression asymmetry) was achieved when 0.8 vol.% ZnO nanoparticles were added to Mg (Fig. 5.19). While the fracture strain values under tensile loads were found to increase significantly (by ~95%, in case of Mg–0.48ZnO), it remained largely unaffected under compressive loads [38]. Furthermore, the authors on comparing the experimental properties with that of the theoretical model proposed by Zhang and Chen [39] found that the experimental values were slightly greater than that of the calculated values and attributed it to the presence of fine ZnO clusters in Mg matrix.

In a related study, Tekumalla et al. [40] incorporated ZnO nanoparticles into Mg-3Al-0.4Ce alloy using the DMD technique and evaluated the resulting nanocomposites after hot extrusion for their microstructural and mechanical properties. The addition of ZnO nanoparticles enhanced the thermal expansion coefficient (CTE), microhardness, and tensile strength, with no significant impact on strain to fracture (Fig. 5.20). Among the nanocomposites, Mg-3Al-0.4Ce/2.5ZnO demonstrated the best performance, achieving the highest microhardness (161 Hv), yield strength (180 MPa), and ultimate tensile strength (294 MPa). These improvements

Table 5.3 Properties of Y_2O_3 nanoparticles reinforced Mg composites

References	Matrix	Reinforcement type	Wt.%	Vol.%	Process	Density (g/cc)	Porosity (%)	Average grain size (μm)	Hardness (Hv)	Tensile yield strength (MPa)	Ultimate tensile strength (MPa)	Ductility (%)	Comp yield strength (MPa)	Ultimate comp strength (MPa)	Ductility (%)	Remarks
Hassan and Gupta [1]	Mg	Y_2O_3		0	PM	1.7387	0.08	60 ± 10	37	132 ± 7	193 ± 2	4.2 ± 0.1	–	–	–	–
				0.22		1.7472	0.10	25 ± 3	38	156 ± 1	211 ± 1	15.8 ± 0.7	–	–	–	–
				0.66		1.7616	0.35	13 ± 2	38	151 ± 2	202 ± 2	12.0 ± 1.0	–	–	–	–
				1.11		1.7763	0.49	12 ± 3	51	153 ± 3	195 ± 2	9.1 ± 0.2	–	–	–	–
				0	DMD	1.7397	0.02	49 ± 8	40	97 ± 2	173 ± 1	7.4 ± 0.2	–	–	–	–
				0.22		1.7472	0.00	10 ± 1	51	218 ± 2	277 ± 5	12.7 ± 1.3	–	–	–	–
				0.66		1.7598	0.10	6 ± 1	56	312 ± 4	318 ± 2	6.9 ± 1.6	–	–	–	–
				1.11		1.7730	0.19	6 ± 2	52	–	205 ± 3	1.7 ± 0.5	–	–	–	–
Tun et al. [32–34]	Mg	Y_2O_3		0	PM-MWS	1.74	0.13	20	37	134 ± 7	193 ± 2	7.5 ± 2.5	–	–	–	HR: 49 °C/min; ER: 25:1
				0.2		1.73	0.87	19	38	144 ± 2	214 ± 4	8.0 ± 2.8	–	–	–	
				0.7		1.76	0.35	18	45	157 ± 10	244 ± 1	8.6 ± 1.2	–	–	–	
						1.754	0.49	29	38	111 ± 10	175 ± 8	9.2 ± 0.5	–	–	–	HR: 20 °C/min; ER: 25:1
						1.75	0.8	28	37	105 ± 4	145 ± 4	9.0 ± 0.3	–	–	–	HR: 49 °C/min; ER: 19:1
						1.74	1.1	35	35	90 ± 7	132 ± 2	6.5 ± 0.4	–	–	–	HR: 49 °C/min; ER: 12:1
Kumar et al. [36]	Mg-3Al	–		0	DMD	1.753	0.29	7.7	–	–	–	–	154	481	24	–
	Mg-3Al-2.5La	–		0		1.791	0.17	6.6	–	–	–	–	141	456	18	–
	Mg-3Al-2.5La	Y_2O_3		1.9		1.813	0.16	3.6	–	–	–	–	156	520	18	–

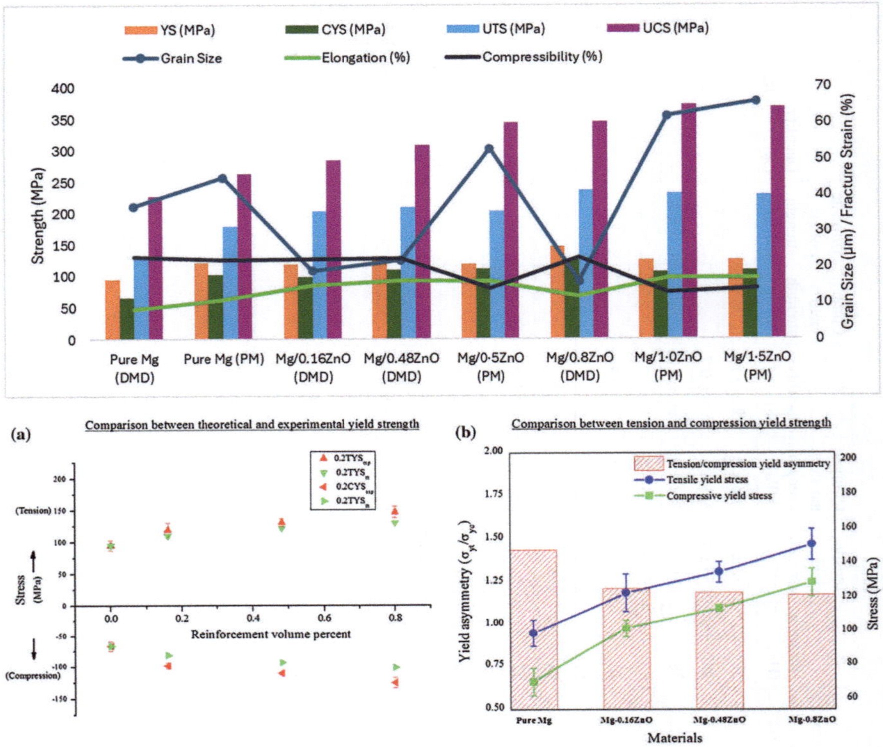

Fig. 5.19 (**i**) Properties of Mg/Zno nanocomposites fabricated. (By Tun et al. [37] and Sankaranarayanan et al. [38]). (**ii**) (**a**) Comparison of experimental vs. theoretical strengths and (**b**) tension-compression asymmetry of Mg/ZnO nanocomposites (from [38])

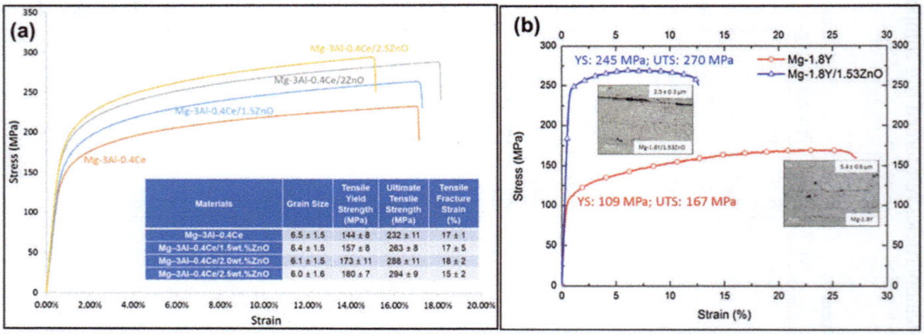

Fig. 5.20 Mechanical properties of (**a**) Mg-3Al-0.4Ce/ZnO and (**b**) Mg-1.8Y/ZnO nanocomposites. (Adapted from [40])

were attributed to the uniform distribution of secondary phases and ZnO reinforcements. Although grain refinement was not observed, ZnO addition contributed to microstructural refinement, dimensional stability, and minimal porosity (~1%), contributing to the improvement in alloy performance.

In a similar study, Chen et al. [41] developed a Mg-4Zn-3Gd-1Ca/ZnO nanocomposite using the DMD method followed by hot extrusion. The incorporation of 2 wt.% ZnO nanoparticles into the magnesium alloy enhanced the compressive yield stress to 355 MPa, ultimate compressive stress to 703 MPa, and failure strain to 10.6% (Fig. 5.21). These properties significantly surpassed those of conventional magnesium alloys, such as WE43 and ZK60, and matched or exceeded mild steel grades S275 and S355. The ~100 MPa strength improvement over the base alloy was attributed to substantial grain refinement, reducing the grain size from 24.6 μm in pure magnesium to approximately 0.9 μm in the nanocomposite. X-ray diffraction analysis identified tension twinning as the dominant deformation mechanism, highlighting the effectiveness of ZnO reinforcement in enhancing both strength and deformation behavior.

Mg-based nanocomposites reinforced with nano-sized zirconia (ZrO_2) particles were developed using ingot metallurgy (DMD) and powder metallurgy (PM) techniques [1]. The addition of 0.2 and 0.7 vol.% ZrO_2 significantly refined the grain structure, improved hardness, and enhanced mechanical properties. DMD-processed composites exhibited superior grain refinement, lower porosity, and higher strength, achieving a 128% increase in yield strength and a 57% rise in ultimate tensile strength with 0.7 vol.% ZrO_2. In contrast, PM-processed composites demonstrated better ductility and fracture resistance, with an 87% increase in elongation-to-fracture at the same reinforcement level. Fracture analysis revealed that DMD-processed composites primarily failed through intergranular cracking, while PM-processed composites exhibited ductile failure mechanisms, explaining the observed differences in mechanical performance.

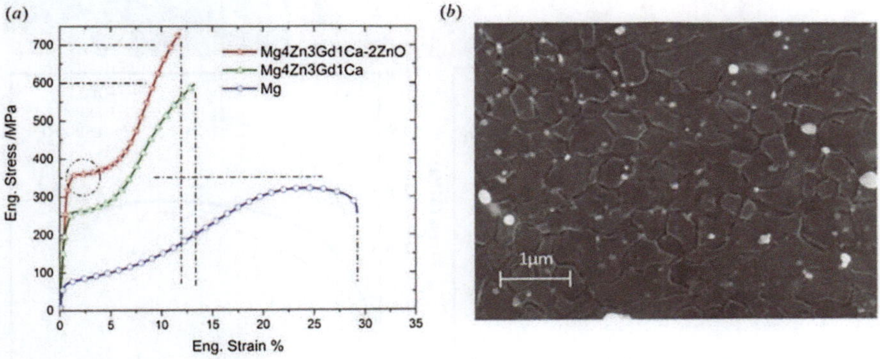

Fig. 5.21 Compressive properties and microstructure of Mg-4Zn-3Gd-1Ca/2ZnO nanocomposite. (Adapted from [41])

In another study, Qiao et al. [42] investigated the fabrication of Mg/ZrO_2 nanocomposites using friction stir processing (FSP) to enhance their microstructure, mechanical properties, and corrosion resistance for biomedical applications. The addition of ZrO_2 nanoparticles refined the grain size from 10 μm in the base material to 3.2 μm in the composites, resulting in improved mechanical performance, including an increase in ultimate tensile strength from 283 to 427 MPa and yield strength from 137 to 217 MPa (Fig. 5.22). These improvements were attributed to grain refinement and Orowan strengthening, although the elongation decreased due to increased brittleness. Microstructural analysis revealed a uniform distribution of ZrO_2 particles, which contributed to strengthening and modifying grain orientation.

The summary of processing type, grain size, and tensile response of Mg matrices reinforced with ZnO and ZrO_2 nanoparticles are summarized in Table 5.4.

Ferkel and Mordike [43] synthesized sub-microcrystalline Mg composites containing nano-SiC particles using powder metallurgy and mechanical alloying. The tensile tests conducted on extruded rods from ball-milled composites revealed significant strength gains, attributed to grain refinement (Hall-Petch strengthening), although this came at the expense of ductility. Composites processed via blending and mixing displayed a minor decrease in strength and adverse effects on ductility (Fig. 5.23).

Wong and Gupta [44] employed the microwave sintering method to fabricate Mg/SiC nanocomposites, incorporating different volume fractions of nano-SiC. The properties of extruded Mg/SiC composites were evaluated to assess the efficiency of

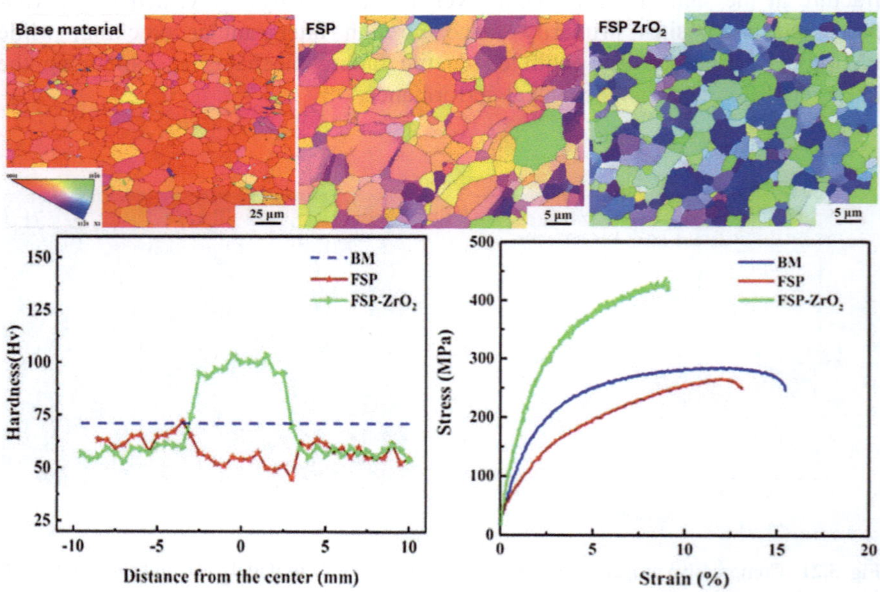

Fig. 5.22 Properties of FSP processed Mg/ZrO_2 nanocomposites (from [42])

Table 5.4 Properties of Mg nanocomposites reinforced with other oxide nanoparticles

References	Matrix	Reinforcement type	Wt.%	Vol.%	Process	Density (g/cc)	Porosity (%)	Average grain size (μm)	Hardness (Hv)	Tensile yield strength (MPa)	Ultimate tensile strength (MPa)	Ductility (%)	Comp yield strength (MPa)	Ultimate comp strength (MPa)	Ductility (%)
Hassan and Gupta [1]	Mg	ZrO_2		0	PM	1.7387	0.08	60 ± 10	37	132 ± 7	193 ± 2	4.2 ± 0.1	–	–	–
				0.22		1.7436	0.32	13 ± 3	38	140 ± 3	160 ± 8	6.4 ± 1.5	–	–	–
				0.66		1.7599	0.42	13 ± 2	44	163 ± 3	202 ± 6	11.4 ± 0.9	–	–	–
				1.11		1.7829	0.18	11 ± 3	46	146 ± 1	199 ± 5	10.8 ± 1.3	–	–	–
				0	DMD	1.7397	0.02	49 ± 8	40	97 ± 2	173 ± 1	7.4 ± 0.2	–	–	–
				0.22		1.7466	0.15	8 ± 2	47	186 ± 2	248 ± 4	4.7 ± 0.2	–	–	–
				0.66		1.7674	0.17	5 ± 2	51	221 ± 5	271 ± 6	4.8 ± 0.7	–	–	–
				1.11		1.7861	0.03	2 ± 1	55	216 ± 4	250 ± 6	3.0 ± 0.2	–	–	–
Tun et al. [37]	Mg	ZnO	0		PM	–	–	20.4 ± 3.4	45	121 ± 5	179 ± 6	11 ± 1	103 ± 4	263 ± 5	22 ± 2
			0.5			–	–	8.9 ± 4.9	53	119 ± 9	203 ± 17	16 ± 2	111 ± 8	344 ± 15	14 ± 2
			1.0			–	–	8.8 ± 5.4	62	125 ± 4	231 ± 13	17 ± 2	106 ± 11	372 ± 15	13 ± 1
			1.5			–	–	8.8 ± 5.1	66	125 ± 4	229 ± 4	17 ± 2	109 ± 3	368 ± 21	14 ± 1
Sankaran-arayanan et al. [38]	Mg	ZnO		0	DMD	–	–	37 ± 11	–	95 ± 8	133 ± 7	8.3 ± 2.9	66 ± 8	228 ± 7	22.2 ± 0.9
				0.16		–	–	19 ± 9	–	119 ± 11	204 ± 9	15 ± 1.4	98 ± 5	284 ± 6	22.3 ± 0.5
				0.48		–	–	22 ± 10	–	131 ± 6	210 ± 8	16.3 ± 1.4	110 ± 2	308 ± 9	22.5 ± 1.2
				0.8		–	–	16 ± 7	–	147 ± 9	237 ± 8	11.88 ± 1.7	125 ± 8	345 ± 4	21.7 ± 2.1
Tekumalla et al. [40]	Mg-3Al-0.4Ce	ZnO	0		DMD	1.759	0.20	6.5	105 ± 4	144 ± 8	232 ± 11	17 ± 1	–	–	–
			1.5			1.779	0.12	6.4	113 ± 3	157 ± 8	263 ± 8	17 ± 5	–	–	–
			2			1.783	0.24	6.1	144 ± 4	173 ± 11	288 ± 11	18 ± 2	–	–	–
			2.5			1.774	1.08	6.0	161 ± 5	180 ± 7	294 ± 9	15 ± 2	–	–	–

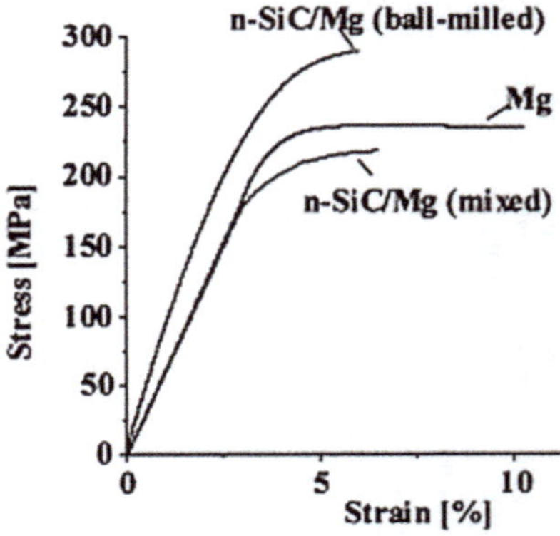

Fig. 5.23 Properties of Mg-SiC nanocomposites developed by Ferkel and Mordike. (Adapted from [43])

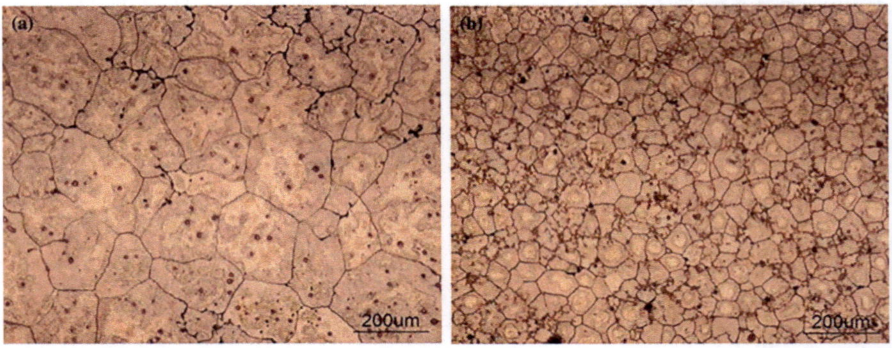

Fig. 5.24 Microstructure of (**a**) Mg-6Zn and (**b**) Mg-6Zn/1.5%SiC nanocomposites fabricated by ultrasonic cavitation-based solidification processing. (Adapted from Cao et al. [46])

microwave sintering. Results showed a simultaneous increase in microhardness, 0.2% yield strength (YS), ultimate tensile strength (UTS), and ductility due to nano-SiC addition. Among the microwave-sintered composites, Mg/1SiC exhibited the best improvements: microhardness increased by ~10%, 0.2% YS by ~26%, UTS by ~18%, and ductility by ~31%. In a similar study, Cao et al. [45, 46] utilized ultrasonic cavitation to disperse nano-SiC particles in pure Mg and Mg-6Zn alloy (Fig. 5.24). The addition of nano-SiC reduced grain size and improved strength (201 MPa, improved by ~40%) without significantly affecting ductility. Cicco et al. [47] also observed enhanced strength and ductility in Mg-4Zn (199 MPa and 10% compared to 112 MPa and 5%, respectively), Mg-6Zn (194 MPa and 7% compared to 136 MPa and 5%, respectively), and Mg-8Zn (222 MPa and 7% compared to 152 MPa and 3%, respectively) alloys with ultrasonically dispersed nano-SiC particles. Similar findings were reported by Wang et al. [48] for AZ91 alloys. Shiying et al. [49], using up to 2.0 wt.% SiC nanoparticles, reported improved strength of AZ91D alloy (105 MPa vs. 65 MPa for the unreinforced alloy), although ductility decreased when more than 1 wt.% of nano-SiC was added.

Nie et al. [50] investigated the influence of extrusion temperature and processing parameters on the microstructure and tensile properties of Mg/SiC nanocomposites produced using ultrasonic vibration-assisted semi-solid stirring. SiCp/AZ91 nanocomposites were homogenized at 415 °C for 24 h (T4 treatment) and extruded at temperatures ranging from 250 to 350 °C with an extrusion ratio of 12:1 and speed of 15 mm/s. The results indicated that increasing the SiC nanoparticle content refined the grain structure, as did higher extrusion temperatures and shorter stirring times. Mechanical property testing further confirmed improvements in yield and ultimate tensile strength with these processing adjustments (Fig. 5.25).

The summary of mechanical response of SiC reinforced magnesium-based matrices as a function of processing type and the amount of SiC particulates is presented in Table 5.5.

Magnesium composites containing B_4C nanoparticles were developed using DMD and PM methods, followed by hot extrusion [51, 52]. Incorporating B_4C nanoparticles enhanced the strength properties of Mg composites; however, ductility improvements were limited to a critical volume fraction of B_4C nanoparticles (0.66 vol.% for PM and 1.04 vol.% for DMD) (Fig. 5.26). The PM-processed Mg/B_4C nanocomposites demonstrated increased strength under compressive loads, though there was a reduction in compressive failure strain. The improved strength was attributed to grain refinement and dispersion strengthening due to the nano-B_4C addition, while the ductility enhancement was linked to grain refinement and texture weakening, promoting the activation of multiple basal and non-basal slip systems. Similarly, when 1.5% BC was added to a hybrid AZ91-ZK60 alloy using the DMD method, a balance between improved tensile ductility and reduced compressive failure strain was observed [53].

In another study, Tekumalla et al. [54] explored the impact of nano-B_4C addition on the properties of Elektron 21 (E21) magnesium alloys to expand their applications in the aerospace sector. E21 alloys reinforced with varying amounts of nano-sized B_4C particulates were synthesized using the disintegrated melt deposition technique followed by hot extrusion. The addition of B_4C nanoparticles refined grain sizes by 29% and resulted in uniform particle distribution. Mechanical testing showed significant enhancements in compressive yield strength and ultimate

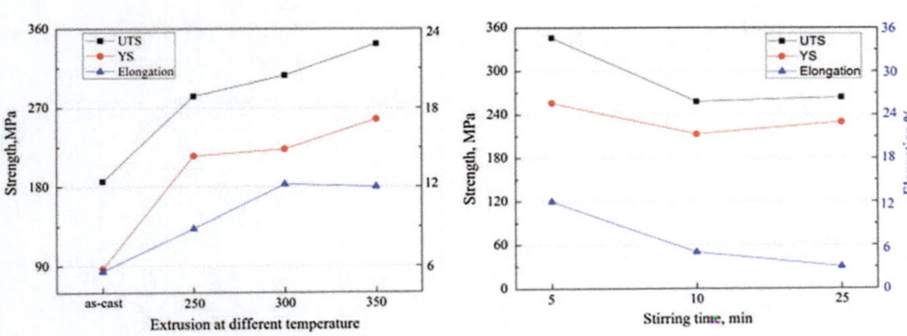

Fig. 5.25 Effect of extrusion temperature and stirring time on the mechanical properties of AZ91/SiC nanocomposites as reported by Nie et al. (Adapted from [50])

Table 5.5 Tensile properties of Mg/SiC nanocomposites

References	Matrix	Reinforcement type	Wt.%	Vol.%	Process	Density (g/cc)	Porosity (%)	Average grain size (μm)	Hardness (Hv)	Tensile yield strength (MPa)	Ultimate tensile strength (MPa)	Ductility (%)	Comp yield strength (MPa)	Ultimate comp strength (MPa)	Ductility (%)	Remarks
Wong & Gupta [44]	Mg	SiC		0	PM-HE	1.74	0.86	–	35	106 ± 7	160 ± 8	5.8 ± 1.8	–	–	–	–
				0.35		1.745	1.39	–	38	116 ± 12	169 ± 17	5.2 ± 1.4	–	–	–	–
				0.5		1.747	1.41	–	40	107 ± 10	161 ± 11	6.5 ± 0.2	–	–	–	–
				1.0		1.755	2.02	–	41	125 ± 2	181 ± 4	6.1 ± 0.9	–	–	–	–
				0	PM-MWS-HE	1.740	0.52	–	39	125 ± 15	172 ± 12	5.8 ± 0.9	–	–	–	–
				0.35		1.745	0.58	–	40	132 ± 14	194 ± 11	6.3 ± 1.8	–	–	–	–
				0.5		1.747	0.48	–	42	144 ± 12	194 ± 10	7.0 ± 2.0	–	–	–	–
				1.0		1.755	0.11	–	43	157 ± 22	203 ± 22	7.6 ± 1.5	–	–	–	–
Cao et al. [45]	Mg	SiC	0		Ultrasonic casting	–	–	–	–	20	89.6	14.0	–	–	–	–
			0.5			–	–	–	–	28.3	120.7	15.5	–	–	–	–
			1.0			–	–	–	–	30.3	124.1	14.2	–	–	–	–
			2.0			–	–	–	–	35.9	131.0	12.6	–	–	–	–
			4.0			–	–	–	–	47.6	106.9	5.5	–	–	–	–
De Cicco et al. [47]	Mg-4Zn	SiC	0		Ultrasonic casting	–	–	–	–	44 ± 2	112 ± 14	5 ± 1	–	–	–	–
			1.5			–	–	–	–	67 ± 4	199 ± 6	10 ± 1	–	–	–	–
	Mg-6Zn		0			–	–	–	–	51 ± 4	136 ± 19	5 ± 1	–	–	–	–
			1.5			–	–	–	–	79 ± 5	194 ± 15	7 ± 1	–	–	–	–
	Mg-8Zn		0			–	–	–	–	81	152	3	–	–	–	–
			3.0			–	–	–	–	111	222	7	–	–	–	–
Cao et al. [46]	Mg-6Zn	SiC			Ultrasonic casting	–	–	–	–	125	200	3.5	–	–	–	T5
			1.5			–	–	–	–	165	230	3.0	–	–	–	–

Wang et al. [48]	AZ91	SiC	0		Ultrasonic casting	–	–	–	–	100	175	3	–	–	–	–
			0.1			–	–	–	–	120	190	5	–	–	–	–
			0.3			–	–	–	–	120	200	5.8	–	–	–	–
			0.5			–	–	–	–	120	210	6.2	–	–	–	–
Shiying et al. [49]	AZ91	SiC	0		Ultrasonic casting	–	–	–	–	65	133	2.2	–	–	–	–
			0.5			–	–	–	–	140	180	2.1	–	–	–	–
			1.0			–	–	–	–	141	191	2.4	–	–	–	–
			2.0			–	–	–	–	117	137	1.3	–	–	–	–

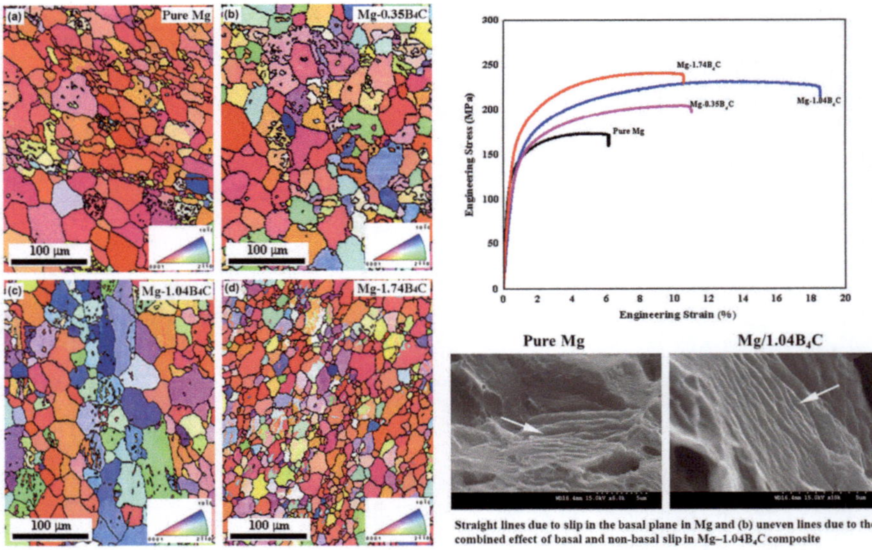

Fig. 5.26 Microstructure and tensile properties of DMD-processed Mg/B_4C nanocomposites. (Adapted from [51])

strength, particularly in the E21–2.5B_4C nanocomposite, without compromising ductility (Fig. 5.27). This composite also exhibited the highest damping capacity, fastest vibration suppression, and superior ignition resistance, with an ignition temperature of 798 °C, 57 °C higher than the base alloy. These improvements are attributed to refined grain structure, thermal insulation, and the uniform dispersion of B_4C nanoparticles.

ZK60A nanocomposite reinforced with 1.5 vol.% TiC nanoparticles was fabricated using solidification processing followed by hot extrusion [55]. The nanocomposite exhibited a similar grain size to monolithic ZK60A but showed a reduced presence of intermetallic phases, a reasonable TiC nanoparticle distribution, and a nondominant (0 0 0 2) texture in the longitudinal direction. Despite a 16% lower hardness than monolithic ZK60A, the nanocomposite demonstrated superior mechanical performance in tension, with increases in 0.2% yield strength (13%), ultimate tensile strength (15%), and failure strain (76%). Under compression, the nanocomposite exhibited a 17% reduction in 0.2% compressive yield strength (CYS) but improved ultimate compressive strength (UCS) (11%), and failure strain (29%). These enhancements were attributed to the uniform distribution of TiC nanoparticles, refined intermetallic phases, and mechanisms like compressive shear buckling of Mg-Zn-based nanoscale intermetallics (Fig. 5.28). While the tensile response improved due to reinforcement effects, the reduction in compressive yield strength was linked to smaller intermetallic precipitates and buckling of nanoscale intermetallics. Overall, the ZK60A/1.5 vol.% TiC nanocomposite demonstrated a favorable balance of tensile and compressive properties, with significantly enhanced ductility and energy absorption capacity compared to monolithic ZK60A.

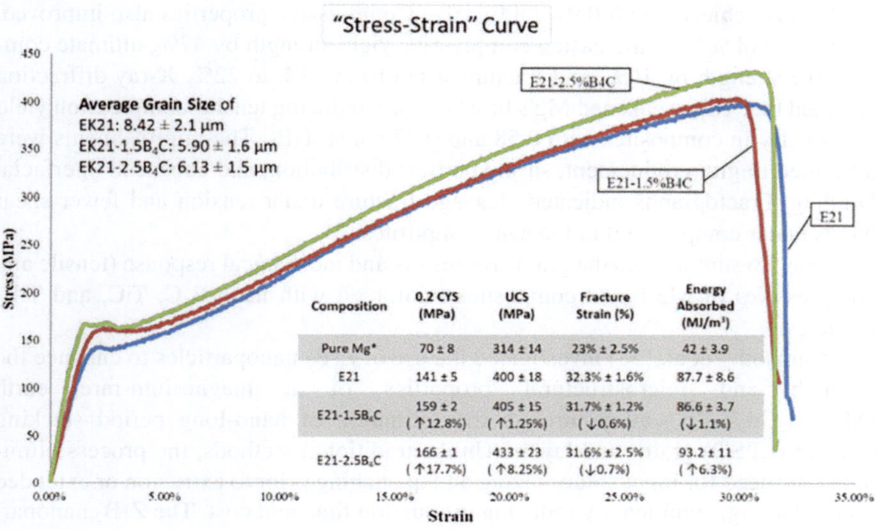

Fig. 5.27 Properties of Elektron 21/B4C nanocomposites. (Adapted from [54])

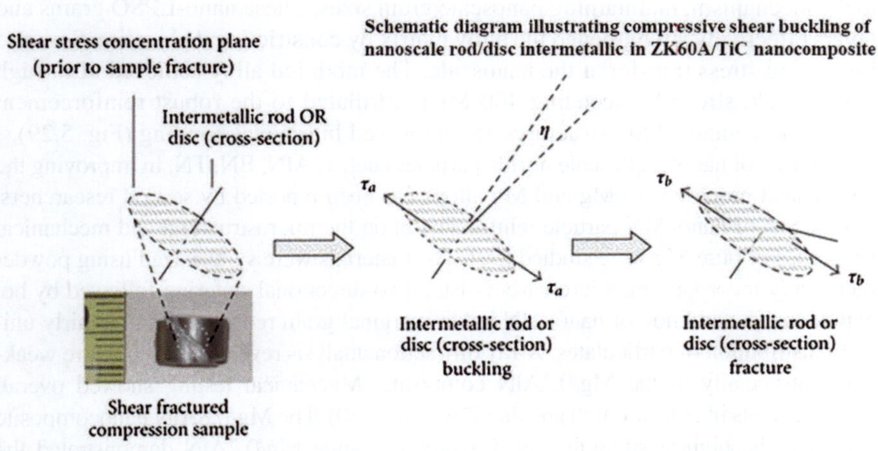

Fig. 5.28 Compressive properties and the schematic of compressive shear buckling of nano-rod/disc intermetallics in ZK60A/TiC nanocomposite [55]

TiB_2 nanoparticles reinforced Mg composites were successfully synthesized using the DMD method followed by hot extrusion [56]. The addition of up to 2 vol.% nano-TiB_2 significantly improved dimensional stability by reducing the coefficient of thermal expansion and refining grain size. Mechanical testing showed substantial enhancements, with Mg-1.98 vol.% TiB_2 exhibiting a 54% increase in tensile yield strength, a 15% rise in ultimate tensile strength, and an 83% improvement in fracture strain compared to pure Mg. A maximum tensile fracture strain of

~16% was achieved with 0.97 vol.% TiB_2. Compressive properties also improved, with 1.98 vol.% TiB_2 increasing compressive yield strength by 47%, ultimate compressive strength by 10%, and fracture strain from 18% to 22%. X-ray diffraction revealed that TiB_2 weakened Mg's basal texture, reducing tensile-compression yield asymmetry in composites with 0.58 and 0.97 vol.% TiB_2. The improvements were attributed to grain refinement, strong particle distribution, and enhanced interfacial bonding. Fractographs indicated cleavage fracture under tension and fewer shear bands under compression in the nanocomposites.

Table 5.6 summarizes the grain size results and mechanical response (tensile and compressive) of Mg-based composites reinforced with nano-B_4C, TiC, and TiB_2 particles.

Paramsothy et al. [58] investigated the use of ZrB_2 nanoparticles to enhance the strength and microstructural properties of a magnesium-rare earth ($Mg_{94}Al_2Gd_2Y_1Zn_1$) alloy through the formation of nano-long period stacking ordered (LPSO) grains and layers. Unlike traditional methods, the process eliminated the need for ingot solutionizing and quenching prior to extrusion or extended thermal aging, significantly reducing production time and cost. The ZrB_2 nanoparticles induced fine LPSO phase formation, leading to nanoscale grain refinement and increased thermal stability due to the high melting point of rare earth components. During thermal aging at 200 °C for 24 h, the LPSO phase acted as an "auto-lock" mechanism, maintaining nanoscale grain sizes. These nano-LPSO grains and layers effectively strengthened the alloy matrix by constricting dislocation flow and improving stress transfer at the nanoscale. The modified alloy achieved ultra-high tensile yield strength exceeding 400 MPa, attributed to the robust reinforcement provided by nano-LPSO structures and improved interfacial bonding (Fig. 5.29).

The role of nanolength scale nitride particles such as AlN, BN, TN, in improving the mechanical properties of Mg and Mg alloys has been reported by several researchers. The effects of nano-AlN particle reinforcement on the microstructural and mechanical properties of pure Mg were studied [59]. The materials were synthesized using powder metallurgy incorporating microwave-assisted two-directional sintering followed by hot extrusion. The addition of nano-AlN led to marginal grain refinement and a fairly uniform distribution of particulates. X-ray diffraction analysis revealed basal texture weakening, especially in the Mg/0.2AlN composite. Mechanical testing showed overall improvements in both strength and ductility (Fig. 5.30). The Mg/0.8AlN nanocomposite exhibited the highest strength (~30% increase), while Mg/0.2AlN demonstrated the greatest ductility enhancement (~85%). Under compressive loading, the addition of nano-AlN particles improved strength by ~30% without significantly affecting ductility, which was attributed to reduced twinning in the material.

Maung et al. [60] synthesized nano-AlN reinforced Mg nanocomposites by ball milling the Mg chips together with 1%, 2%, and 5% nano-sized AlN particulates. The composite powder mixture was milled for varying durations to study the effects of grain refinement on mechanical properties, including tensile strength, ductility, and hardness. Results showed that refined grain size significantly influenced ductility, with 40-h milled specimens achieving the highest elongation (37%) in the Mg–5 wt% Al–1% AlN system. Strength was maximized in 20-h milled specimens, which exhibited strain softening behavior, while shorter milling durations (0 and 10 h) showed strain hardening (Fig. 5.31). The transition from strain hardening to

Table 5.6 Properties of nano-B_4C, TiC, and TiB_2 reinforced Mg composites

References	Matrix	Reinforcement type	Wt.%	Vol.%	Process	Density (g/cc)	Porosity (%)	Average grain size (μm)	Hardness (Hv)	Tensile yield strength (MPa)	Ultimate tensile strength (MPa)	Ductility (%)	Comp yield strength (MPa)	Ultimate comp strength (MPa)	Ductility (%)
Habibi et al. [52]	Mg	B_4C		0	PM-MWS-HE	–	–	20 ± 3	–	109 ± 09	161 ± 16	8.9 ± 1.1	93 ± 02	246 ± 23	19.4 ± 0.8
				0.22		–	–	11 ± 2	–	110 ± 15	159 ± 14	9.9 ± 0.6	97 ± 04	337 ± 14	10.0 ± 1.6
				0.66		–	–	10 ± 3	–	120 ± 05	164 ± 06	10.0 ± 0.3	100 ± 05	335 ± 11	11.8 ± 1.8
				1.11		–	–	10 ± 2	–	82 ± 11	119 ± 17	5.5 ± 1.2	105 ± 03	331 ± 10	13.3 ± 1.4
Sankaranarayanan et al. [51]	Mg	B_4C		0	DMD-HE	1.7397	0.02	28 ± 9	48	120 ± 9	169 ± 11	06.4 ± 0.7	–	–	–
				0.35		1.7405	0.04	21 ± 5	51	127 ± 6	202 ± 6	11.8 ± 1.6	–	–	–
				1.04		1.7443	0.08	18 ± 9	55	137 ± 5	215 ± 8	17.4 ± 2.0	–	–	–
				1.74		1.7507	0.07	11 ± 6	57	160 ± 2	240 ± 5	12.4 ± 1.7	–	–	–
Paramsothy et al. [53]	AZ91-ZK60A	B_4C		0	DMD-HE	–	–	5 ± 1	–	225 ± 4	321 ± 4	16.1 ± 0.3	106 ± 5	508 ± 17	19.5 ± 1.7
				1.5		–	–	4.2	–	232 ± 2	345 ± 2	25.1 ± 1.2	140 ± 20	498 ± 2	12.1 ± 1.1
Tekumalla et al. [54]	E21	B_4C	0	0	DMD-HE	1.798	0.18	8 ± 2					141 ± 5	400 ± 18	87.7 ± 8.6
				1.5		1.808	0.53	6 ± 2					159 ± 2	405 ± 15	86.6 ± 3.7
				2.5		1.813	0.23	6 ± 2					166 ± 1	433 ± 23	93.2 ± 11
Paramsothy et al. [55]	ZK60A	TiC	0		DMD-HE	–	–	8.9	138	163	268	6.6	128	522	19.6
			1.5			–	–	7.0	116	184	309	11.6	106	577	25.3
Meenashisundaram et al. [56]	Mg	TiB_2		0	DMD-HE	1.7362	0.20	45 ± 2.4	52	92 ± 5	157 ± 5	8.2 ± 0.2	57 ± 3	332 ± 10	18
				0.58		1.7530	0.16	37 ± 2.1	65	93 ± 7	149 ± 9	13 ± 0.5	78 ± 3	352 ± 4	17 ± 1
				0.97		1.7630	0.20	28 ± 3.6	69	110 ± 3	173 ± 8	16 ± 0.5	75 ± 7	333 ± 9	21 ± 1
				1.98		1.7910	0.19	16.5 ± 2	76	140 ± 9	186 ± 1	14 ± 1	81 ± 7	365 ± 10	19 ± 1
Choi et al. [57]	AZ91D	TiB_2		0	Ultrasonic casting	–	–	100	–	88 ± 3	162 ± 5	2.88 ± 0.12	–	–	–
				1		–	–	50	–	104 ± 1	180 ± 0	3.33 ± 0.05	–	–	–
				2		–	–	50	–	107 ± 1	188 ± 7	4.27 ± 0.61	–	–	–

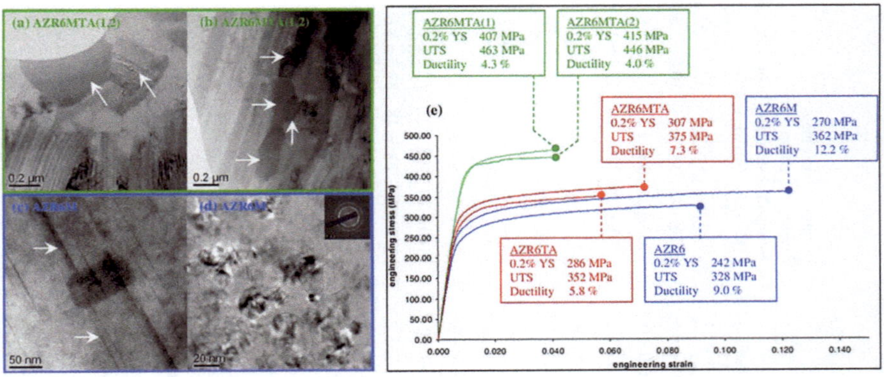

Fig. 5.29 Microstructure and properties of $Mg_{94}Al_2Gd_2Y_1Zn_1/ZrB_2$ (AZR6M) nanocomposites as reported by Paramsothy and Gupta. (Adapted from [58])

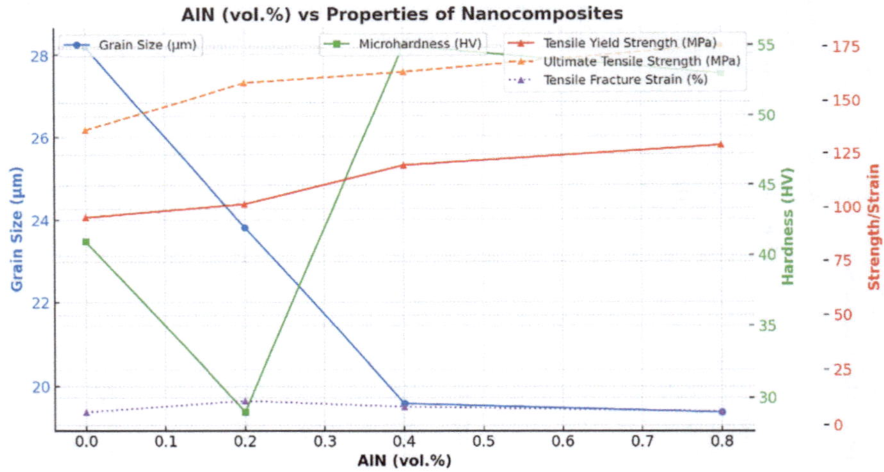

Fig. 5.30 Effects of AlN content on the properties of Mg/AlN nanocomposites (from [59])

strain softening after 20 h of milling marked a shift in the deformation mechanism from strengthening by grain refinement and particle dispersion to grain boundary sliding and diffusion. The variation in AlN reinforcement percentage (1%, 2%, or 5%) had minimal impact on composite strength.

In another study, AlN nanoparticles were introduced into AZ91D and AM60 alloys using ultrasonic stir casting and PM methods [61, 62] (Figs. 5.32 and 5.33). The AZ91D/AlN nanocomposites displayed significant improvements in tensile strength without compromising ductility. Meanwhile, AM60/AlN composites exhibited both strength and ductility enhancements, with grain size reduction from 1227 ± 301 to 85 ± 6 μm. However, the addition of AlN particles to Elektron 21 did not significantly change its mechanical properties, though marginal grain refinement was noted [63].

The effects of nano-BN reinforcements on pure Mg were investigated utilizing the melt casting [64] and powder metallurgy methods [65]. Microstructural examination of the DMD-processed composites followed by hot extrusion revealed significant grain refinement from 28 μm in pure Mg to 9 μm in Mg-1.2BN due to grain boundary pinning and dynamic recrystallization. This resulted in notable improvements in mechanical properties, with microhardness increasing from 48 to 67 HV, ultimate tensile strength rising from 169 to 255 MPa, and compressive yield strength improving from 70 to 109 MPa. The tensile fracture strain showed a 140% increase at 0.6 vol% nano-BN before declining at higher contents, attributed to changes in crystallographic texture, including increased fiber texture intensity. In PM-processed

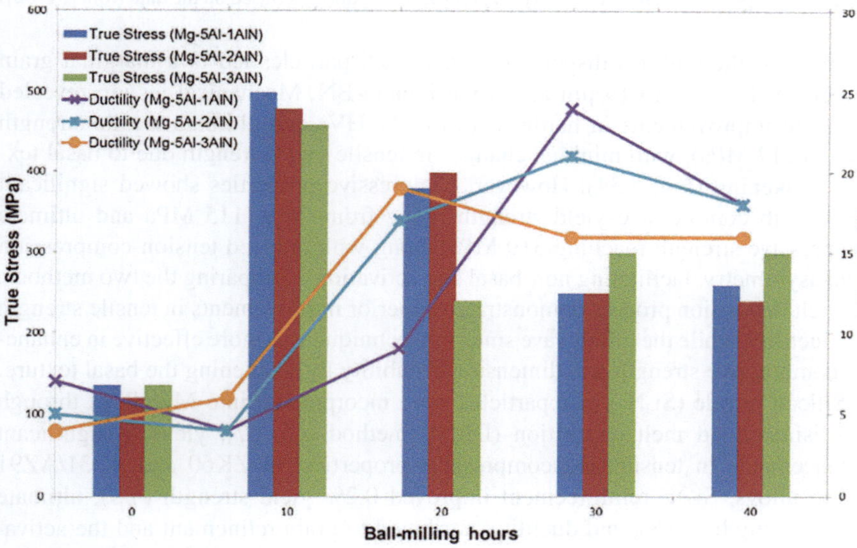

Fig. 5.31 Effect of ball milling on the properties of Mg-5Al/AlN nanocomposites. (Based on the data from [60])

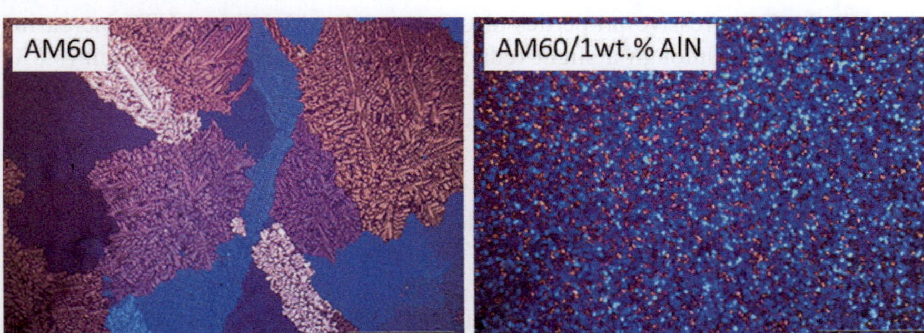

Fig. 5.32 Microstructure of AM60 and AM60/1 wt.% AlN nanocomposite. (Adapted from Dieringa et al. [62])

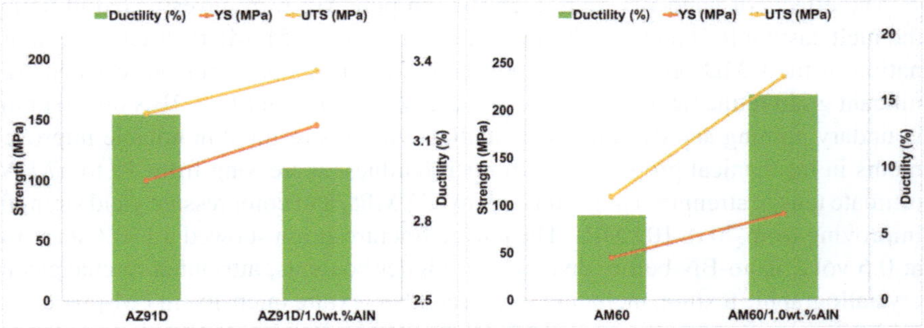

Fig. 5.33 Properties of nano-AlN added AZ91D & AM60 alloys. (Based on the data from [61, 62])

composites, the uniform dispersion of nano-BN particles led to a marginal grain refinement from 28 to 19 μm at 2.5 vol% nano-BN. Mechanical testing revealed moderate improvements in hardness (48 to 57 HV) and ultimate tensile strength (170 to 217 MPa), with minimal changes in tensile yield strength due to basal texture weakening (Fig. 5.34). However, compressive properties showed significant gains, with compressive yield strength rising from 70 to 115 MPa and ultimate compressive strength reaching 319 MPa, along with reduced tension-compression yield asymmetry, facilitating non-basal slip activation. Comparing the two methods, the melt deposition process demonstrated superior improvements in tensile strength and ductility, while the microwave sintering technique was more effective in enhancing compressive strength and dimensional stability by weakening the basal texture.

Silicon nitride (Si_3N_4) nanoparticles were incorporated into Mg alloys through the disintegrated melt deposition (DMD) method [66, 67], yielding significant enhancements in tensile and compressive properties. In ZK60 and AZ31/AZ91 hybrid alloys, Si_3N_4 reinforcement improved 0.2% yield strength (YS), ultimate tensile strength (UTS), and ductility, attributed to grain refinement and the activation of non-basal slip systems [66] (Fig. 5.35). The AZ31/AZ91 hybrid nanocomposite achieved a slight hardness improvement (~13%), with tensile YS and UTS increasing by 12% and 5%, respectively [67]. Substantial gains were observed in failure strain and work of fracture (WOF), which increased by 64% and 71%, respectively. Compressive properties also improved, with YS and UTS increasing by 35% and 4%, although a slight reduction in compressive failure strain (−6%) was reported. Similarly, Mg-TiN composites were extensively studied [68, 69], demonstrating notable mechanical and microstructural improvements (Fig. 5.35). Composites synthesized using DMD and powder metallurgy exhibited grain refinement, with grain size reduced by 57% to ~11 μm for 2.5 vol.% TiN. Microhardness increased by 26% to 67 HV, while tensile properties were enhanced, with YS and UTS reaching 135 MPa and 196 MPa, reflecting 26% and 17% improvements, respectively [68]. Compression strength peaked at 1.98 vol.% TiN, with ultimate compressive strength increasing by 11% to 385 MPa. The processing method influenced reinforcement dispersion, with powder metallurgy providing better uniformity and reduced agglomeration, leading to improved load transfer efficiency. Strengthening mechanisms such as Hall-Petch and Orowan effects, along with

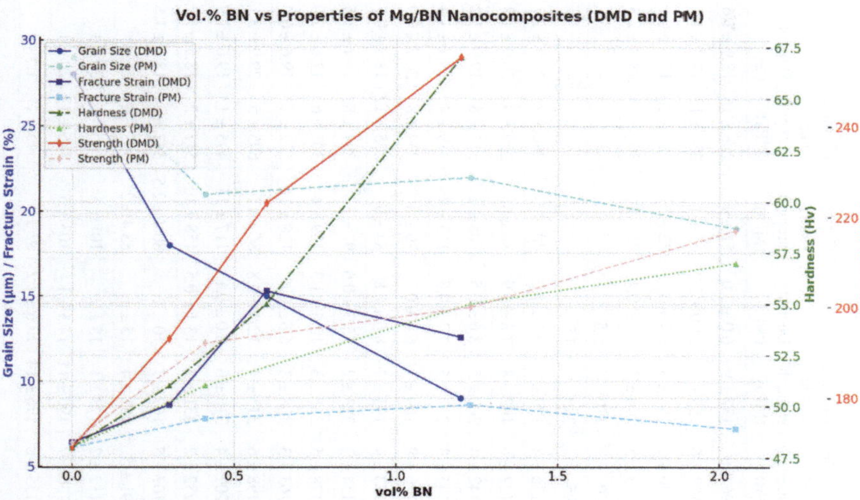

Fig. 5.34 Properties of Mg/BN nanocomposites [64, 65]

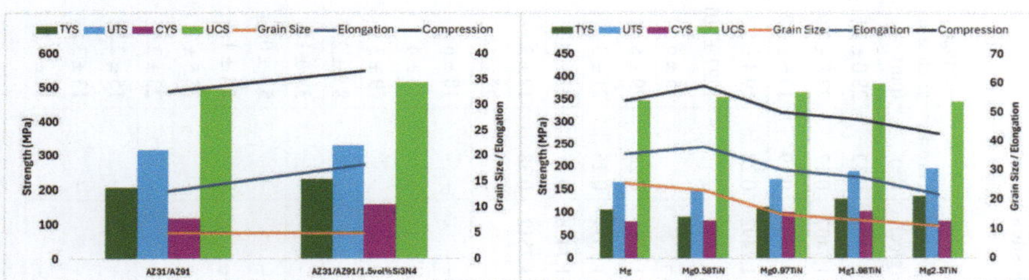

Fig. 5.35 Properties of Si_3N_4 and TiN nanoparticles reinforced Mg nanocomposites [66, 68]

thermal and modulus mismatch, were identified as contributors to the observed enhancements. However, reinforcement levels exceeding 2.5 vol.% TiN caused reduced ductility due to particle clustering. In machinability studies, TiN-reinforced Mg composites displayed enhanced cutting resistance, surface quality, and tool wear resistance, with 2.5 vol.% TiN providing an optimal balance between machinability and mechanical performance. Building on previous findings, the incorporation of upto 12 wt.% TiN into ZK60A Mg alloys was carried out using powder metallurgy method [70]. In this study. The microhardness increased by 62% to 80 HV, while compressive strength rose by 33% to 160 MPa when 12 wt.% was added. Corrosion resistance improved significantly, with the corrosion rate reduced to 0.00032 mm/year due to the protective barrier effect of TiN particles. Biocompatibility studies validated the biomedical potential of these composites, with high cell viability, osteoblast proliferation, and minimal cytotoxicity observed.

The grain size, processing type, and mechanical response of magnesium and its matrices as a result of the presence of AlN, Si_3N_4, BN, and TiN reinforcements are summarized in Table 5.7.

Table 5.7 Properties of nitride nanoparticles reinforced Mg composites

References	Matrix	Reinforcement type	Wt.%	Vol.%	Process	Density (g/cc)	Porosity (%)	Average grain size (μm)	Hardness (Hv)	Tensile yield strength (MPa)	Ultimate tensile strength (MPa)	Ductility (%)	Comp yield strength (MPa)	Ultimate comp strength (MPa)	Ductility (%)
Sankaranarayanan et al. [59]	Mg	AlN		0	PM-MWS-HE	1.7379	0.12	28.0 ± 7.7	41	96 ± 6	137 ± 9	6.0 ± 3.0	51 ± 9	268 ± 16	18.9 ± 1.6
				0.1		1.7407	0.17	23.8 ± 7.9	49	102 ± 6	159 ± 8	11.0 ± 2.2	65 ± 1	284 ± 12	19.3 ± 1.5
				0.4		1.7421	0.19	19.6 ± 6.2	55	120 ± 1	164 ± 3	8.4 ± 0.9	72 ± 5	314 ± 20	17.5 ± 0.6
				0.8		1.7492	0.28	19.4 ± 5.3	53	129 ± 5	176 ± 3	6.3 ± 0.4	71 ± 3	307 ± 17	18.3 ± 2.3
Dieringa et al. [62]	AM60	AlN		0	Ultrasonic casting	–	–	1227 ± 301	–	50	110	6.5	–	–	–
				1		–	–	85 ± 6	–	95	240	15	–	–	–
Seetharaman et al. [65]	Mg	BN	0		PM-MWS-HE	1.7365	0.20	29 ± 6	48	136 ± 8	170 ± 7	6.1 ± 1.2	70 ± 2	250 ± 7	24.5 ± 2.7
			0.29			1.7391	0.26	21 ± 3	51	127 ± 6	192 ± 8	7.8 ± 0.9	88 ± 6	290 ± 9	20.9 ± 1.8
			0.86			1.7458	0.21	22 ± 2	55	142 ± 4	200 ± 5	8.6 ± 0.5	108 ± 2	312 ± 8	19.9 ± 1.2
			1.44			1.7512	0.38	19 ± 3	57	145 ± 3	217 ± 5	7.2 ± 0.8	115 ± 4	319 ± 4	19.7 ± 1.4
Sankaranarayanan et al. [64]	Mg	BN	0		DMD-HE	–	–	28 ± 11	48	120 ± 9	169 ± 11	6.4 ± 0.7	70 ± 8	234 ± 8	20.7 ± 0.9
			0.3			–	–	18 ± 6	51	133 ± 4	193 ± 7	8.6 ± 0.8	84 ± 9	275 ± 12	18.9 ± 0.7
			0.6			–	–	15 ± 4	55	154 ± 2	223 ± 2	15.3 ± 0.6	97 ± 3	297 ± 8	17.1 ± 1.5
			1.2			–	–	9 ± 3	67	178 ± 5	255 ± 3	12.6 ± 1.3	109 ± 4	307 ± 6	17.6 ± 2.0
Paramsothy et al. [67]	ZK60A	Si_3N_4	0		DMD-HE	–	–	8.9 ± 2.0	138	163 ± 3	268 ± 3	6.6 ± 0.6	128 ± 11	522 ± 11	19.6 ± 0.9
			1.5			–	–	3.4 ± 1.7	121	198 ± 6	313 ± 4	12.2 ± 0.8	86 ± 11	605 ± 5	29.9 ± 3.1
Paramsothy et al. [66]	AZ31-AZ91	Si_3N_4	0		DMD-HE	–	–	5.1 ± 0.7	111	207 ± 4	316 ± 6	8.0 ± 0.1	117 ± 15	495 ± 13	19.6 ± 1.9
			1.5			–	–	5.1 ± 1.0	125	232 ± 2	331 ± 2	13.1 ± 0.5	158 ± 10	517 ± 2	18.5 ± 1.7
Meenashis-undaram et al. [68]	Mg	TiN		0	DMD-HE	–	–	25.5 ± 2	53	107 ± 5	167 ± 7	10 ± 1	80.4 ± 2.5	347 ± 4	18.5 ± 1.5
				0.58		–	–	23 ± 2.5	56	91 ± 5	151 ± 4	15 ± 1	83.4 ± 2	355 ± 8	21
				0.97		–	–	15 ± 2.5	60	112 ± 2	173 ± 1	15 ± 2	101 ± 3	365.5	20 ± 1
				1.98		–	–	13 ± 3.5	65	130 ± 7	190 ± 11	14.5 ± 1	103 ± 5	385 ± 13	20 ± 1
				2.5		–	–	11 ± 3.5	67	135 ± 8	196 ± 14	10.6 ± 1.2	82 ± 3	345 ± 1	1.5

5.3 Carbon Nanotubes and Graphene Nanoparticles

The incorporation of carbonaceous nanomaterials, such as graphene and carbon nanotubes (CNTs), into magnesium matrix composites (MMCs) has gained substantial interest due to their outstanding mechanical properties. CNTs are broadly classified into single-walled nanotubes (SWNTs) and multi-walled nanotubes (MWNTs). SWNTs consist of a single sheet of carbon atoms wrapped into a cylinder with a diameter of 1–2 nm, whereas MWNTs are composed of multiple concentric single walls with diameters ranging from 4 to 20 nm. A schematic of these structures is provided in Fig. 5.36.

Despite the remarkable properties of graphene and CNTs, challenges such as agglomeration and poor wetting limit their use in MMCs. To address these, various techniques including powder metallurgy, mechanical alloying, thermomechanical processing, metal casting, and additive manufacturing have been developed to produce CNT- or graphene-reinforced magnesium matrix nanocomposites (MMNCs).

5.3.1 Carbon Nanotubes (CNTs)

Goh et al. [72] examined the incorporation of multi-walled carbon nanotubes (MWCNTs) into magnesium (Mg) to fabricate nanocomposites using the powder metallurgy technique, followed by hot extrusion. The research demonstrated significant improvements in mechanical and thermal properties with the addition of CNTs up to 0.3 wt%. The coefficient of thermal expansion was reduced by approximately

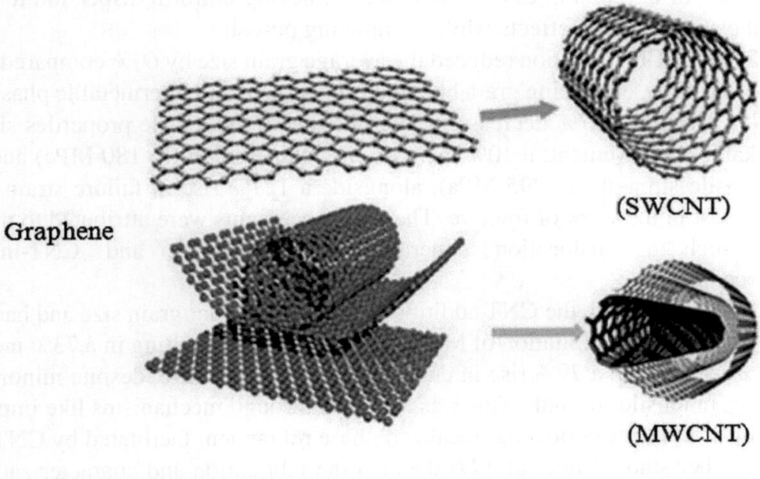

Fig. 5.36 Graphene, single-walled carbon nanotubes (SWCNT) and multiwalled carbon nanotubes (MWCNT) structures. (Adapted from [71])

9%, and the nanocomposites exhibited enhanced yield strength, ductility, and work of fracture due to effective load transfer and strong interfacial bonding between the CNTs and Mg matrix. The authors observed an increase in porosity with higher CNT content, reaching 1.15 vol% at 0.3 wt%, which slightly reduced the density but remained within acceptable limits.

In another study, Hou et al. [73] investigated the effects of adding multi-walled carbon nanotubes (MWCNTs) to Mg-9Al alloy, focusing on the resultant changes in microstructure and mechanical properties. The composites were fabricated using powder metallurgy and hot extrusion, with the addition of 0.4 wt% MWCNTs resulting in significant enhancements. The ultimate tensile strength and elongation increased by 18% and 150%, respectively, compared to the Mg-9Al alloy. These improvements were attributed to the uniform dispersion of MWCNTs and their role in refining the $Mg_{17}Al_{12}$ phase from the micron to the nanoscale, as well as the excellent interfacial bonding between the CNTs and matrix. The authors also highlighted that the ionic gemini dispersant used for dispersing the CNTs ensured structural integrity and avoided the clustering of CNTs, which was critical for achieving the improved properties.

In a similar study, Zhou et al. [74] explored the effects of CNT reinforcement on the microstructure and mechanical properties of AZ31 magnesium alloy using powder metallurgy and hot extrusion. The addition of CNTs, up to 2.0 wt%, led to a 28.2% improvement in yield strength (to 241 MPa) and a 6.1% increase in ultimate tensile strength (to 297 MPa), primarily due to mechanisms such as load transfer, Orowan strengthening, and thermal mismatch-induced dislocations. The CNTs also contributed to grain refinement and weakened the basal plane texture, which facilitated easier slip during deformation. However, increasing CNT content beyond 2.0 wt% resulted in higher porosity levels, reaching 5.3% at 4.0 wt%, which adversely affected the composite's properties (Fig. 5.37). The study highlighted the importance of optimizing CNT content and achieving uniform dispersion to maximize the strengthening effects while minimizing porosity.

In ZK60A, CNT addition reduced the average grain size by 60% compared to the monolithic alloy, enhancing grain boundary refinement and intermetallic phase control [75]. Despite a 17% decrease in microhardness, the tensile properties showed remarkable improvement: a 10% increase in yield strength (to 180 MPa) and ultimate tensile strength (to 295 MPa), alongside a 127% rise in failure strain and a 156% boost in the work of fracture. These enhancements were attributed to mechanisms such as dislocation generation, load transfer, and CNT-induced reinforcement.

In AZ81 alloy [76], the CNT addition maintained similar grain size and hardness but regulated the precipitation of Mg-Al second phases, resulting in a 73% increase in failure strain and a 79% rise in energy absorbed to fracture, despite minor compromises in tensile strength. This was achieved through mechanisms like improved crack tip stress distribution and localized phase refinement facilitated by CNTs.

In another study, Liu et al. [77] detailed the fabrication and characterization of AZ91D magnesium matrix composites reinforced with 1.5% multi-walled CNTs using mechanical stirring combined with high-intensity ultrasonic dispersion. The

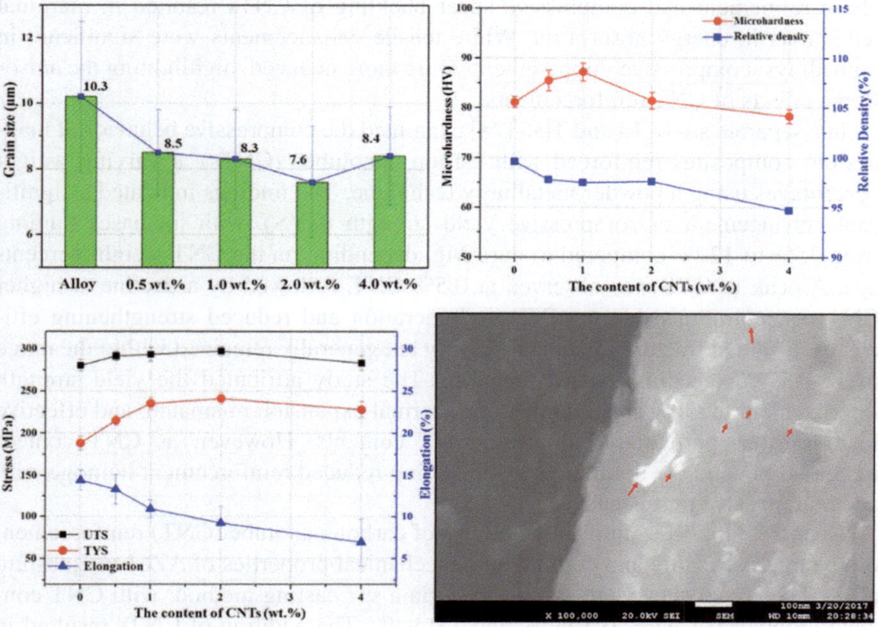

Fig. 5.37 Microstructural and mechanical properties of CNT reinforced Mg nanocomposites [74]

introduction of CNTs resulted in a 72% reduction in grain size compared to unreinforced AZ91D alloy, with the primary α-Mg phase and eutectic β-$Mg_{17}Al_{12}$ phases being significantly refined. The dispersion of CNTs was homogeneous and free from clustering due to the cooperative effects of acoustic cavitation and acoustic streaming during ultrasonic processing. Mechanical property enhancements were notable, with UTS, YS, and elongation increasing by 22%, 21%, and 42%, respectively. Specifically, UTS increased from 128.3 MPa in the unreinforced AZ91D to 157.0 MPa in the composite, while YS rose from 86.0 to 104.0 MPa, and elongation improved from 0.9% to 1.28%. The improved properties were attributed to strong CNT-matrix bonding, effective load transfer, and the prevention of crack propagation by CNTs. Despite the enhancements, tensile fracture analysis revealed brittle failure mechanisms, with CNTs playing a role in bridging cracks and mitigating local crack propagation.

Under compressive loading, ZK60A with CNT reinforcement exhibited a 14% reduction in compressive yield strength (to 110 MPa) but improvements in ultimate compressive strength (+5%) and failure strain (+69%), with a significant 58% increase in the work of fracture [75]. These improvements were linked to CNT-induced strain energy dissipation and dislocation dynamics.

For AZ81, CNT addition led to an 18% reduction in compressive yield strength (to 129 MPa), no significant change in ultimate compressive strength, and a slight 6% decrease in failure strain [76]. The interplay between CNT-induced second phase refinement and compressive shear buckling of CNTs resulted in marginal reductions in energy absorption. While tensile enhancements were significant in both alloys, compressive improvements were more nuanced, highlighting the anisotropic effects of CNT reinforcements.

In a separate study, Li and Tian [78] examined the compressive behavior of magnesium composites reinforced with carbon nanotubes (CNTs) at varying weight percentages using a powder metallurgy technique. The findings indicated a significant enhancement in compressive yield strength (CYS), with increases ranging from 35% to 129% compared to pure Mg, depending on the CNT weight percentage. A peak in CYS was observed at 0.5% CNT, followed by a decline at higher CNT concentrations due to CNT agglomeration and reduced strengthening efficiency. Strain at fracture fluctuated slightly but generally remained within the range of 12–18%, exceeding that of pure Mg. The study attributed the yield strength improvement to Orowan strengthening, thermal expansion mismatch, and effective load transfer, particularly at lower CNT contents. However, as CNT content increased beyond the optimal level, clustering reduced reinforcement homogeneity and compressive performance.

Huang et al. [79] examined the effects of carbon nanotube (CNT) reinforcement on the microstructure and compressive mechanical properties of AZ61 magnesium alloy. The composites were synthesized via a stir casting method, with CNT concentrations of 0.1 wt%, 0.5 wt%, and 1.0 wt%. The addition of CNTs resulted in significant grain refinement and modification of the microstructure, particularly the redistribution of secondary phases such as $Mg_{17}Al_{12}$ along grain boundaries. The grain size decreased with increased CNT content due to grain boundary pinning and heterogeneous nucleation facilitated by the CNTs. Under compressive loading, the ultimate compressive strength (UCS) improved significantly with CNT addition, peaking at 488.77 MPa for the aged 1.0 wt% CNT composite—a 35.5% increase over the unreinforced aged AZ61 alloy. The energy absorbed until fracture also increased substantially with CNT reinforcement, reflecting enhanced toughness. These improvements were attributed to mechanisms such as load transfer, dislocation generation from thermal expansion mismatch, and Orowan strengthening. However, a marginal reduction in ductility at higher CNT concentrations was noted, likely due to increased porosity and CNT agglomeration. The study demonstrated that the aging process further enhanced compressive properties by dissolving low-strength $Mg_{17}Al_{12}$ phases and promoting finer grain structures.

Nai et al. [80] focused on enhancing the mechanical properties of magnesium (Mg) composites through the incorporation of carbon nanotubes (CNTs) and nickel-coated carbon nanotubes (Ni-CNTs) via powder metallurgy and microwave-assisted sintering. The addition of 0.3 wt.% Ni-CNTs resulted in significant microstructural refinement, with a grain size reduction of over 62% compared to pure Mg, attributed to the dual effects of grain boundary pinning and recrystallization nucleation. The Ni coating facilitated interfacial bonding through the formation of Mg_2Ni intermetallics, enhancing load transfer and preventing crack initiation (Fig. 5.38). Mechanically, the Mg/Ni-CNT composite exhibited superior properties, with a 41% increase in microhardness, a 64% improvement in 0.2% yield strength (to 206 MPa), and a 39% rise in ultimate tensile strength (to 237 MPa), compared to pure Mg. This

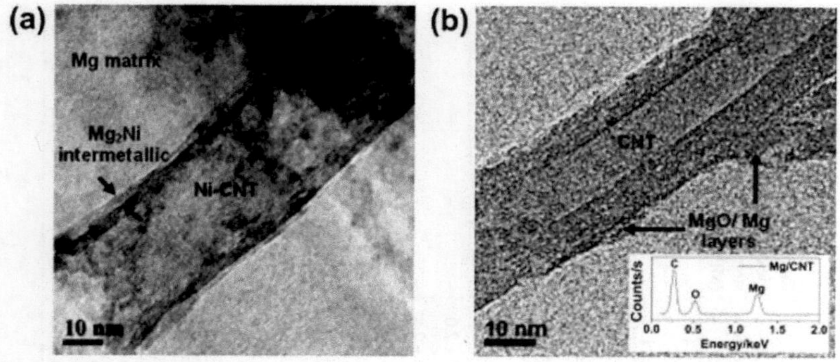

Fig. 5.38 TEM micrographs of Mg/Ni-CNT composites showing the presence of (**a**) intermetallics and (**b**) oxide layer (EDX inset shows the elemental distribution on the surface of the CNT) at the interface of a Ni-CNT embedded in the Mg matrix. (Adapted from [80])

contrasted sharply with the Mg/CNT composite, where inadequate bonding and CNT clustering led to reduced mechanical performance. Furthermore, porosity was minimized in the Ni-CNT composite, achieving 0.46 vol%, compared to 0.75 vol% in the CNT composite.

Summary of findings illustrating the effect of CNTs on grain size and mechanical properties of Mg, Mg-Al, AZ81 and ZK 60A alloy matrices is presented in Table 5.8.

5.3.2 Graphene Nanoplatelets (GNPs)

The introduction of graphene nanoplatelets (GNPs) into Mg alloys via PM has been shown to improve mechanical properties. Chen et al. [81] developed a novel processing method combining liquid state ultrasonic processing and solid state stirring to effectively incorporate and uniformly disperse graphene nanoplatelets into a magnesium metal matrix. The resulting Mg-graphene nanoplatelet nanocomposites exhibited significantly enhanced properties (~78% increased hardness compared to pure Mg) due to the uniform distribution of the graphene reinforcement. Using DMD, Xiang et al. [82] also produced Mg composites reinforced with lower volume fractions of graphene nanoplatelets. They examined two compositions with GNP contents 0.10 wt.% and 0.25 wt.%. The 0.25 wt.% GNP composite exhibited the most significant improvements, with a 27% increase in yield strength (to 122 MPa), a 24% increase in ultimate tensile strength (to 202 MPa), and a nearly doubled fracture elongation (to 14.5%) compared to pure Mg (Fig. 5.39). These enhancements were attributed to the uniform dispersion of GNPs, grain refinement via twinning, and a weakening of the basal texture, which facilitated basal slip. Mechanisms contributing to the enhanced properties included the load transfer effect of the wrinkled GNP surfaces, Orowan strengthening, and dislocation generation due to mismatched thermal expansion coefficients between the GNPs and Mg matrix. The introduction

Table 5.8 Effect of CNTs on grain size and mechanical properties of Mg, Mg-Al, AZ81, and ZK 60A alloys

References	Matrix	Reinforcement type	Wt.%	Vol.%	Process	Density (g/cc)	Porosity (%)	Average grain size (μm)	Hardness (Hv)	Tensile yield strength (MPa)	Ultimate tensile strength (MPa)	Ductility (%)	Comp yield strength (MPa)	Ultimate comp strength (MPa)	Ductility (%)
Goh et al. [72]	Mg	CNT	0		PM-HE	1.738	0.12	–	45	127 ± 5	205 ± 4	9 ± 2	–	–	–
			0.06			1.738	0.25	–	44	133 ± 2	203 ± 1	12 ± 1	–	–	–
			0.18			1.737	0.83	–	44	138 ± 4	206 ± 7	11 ± 1	–	–	–
			0.30			1.736	1.15	–	44	146 ± 5	210 ± 6	8 ± 1	–	–	–
Zhou et al. [74]	AZ31	CNT	0		DMD	1.7637	0.4	10.3	82	188 ± 5.2	280 ± 4.5	14.4 ± 1.2	–	–	–
			0.5			1.7327	2.2	8.5	85	215 ± 3.5	291 ± 3.8	13.3 ± 1.8	–	–	–
			1.0			1.7291	2.5	8.3	87	235 ± 4.8	293 ± 5.5	10.9 ± 1.1	–	–	–
			2.0			1.7298	2.6	7.6	82	241 ± 4.3	297 ± 3.6	9.2 ± 2.0	–	–	–
			4.0			1.7173	5.3	8.4	79	228 ± 6.5	283 ± 5.8	6.9 ± 2.8	–	–	–
Paramsothy et al. [76]	AZ81	CNT		0	DMD	–	–	6 ± 1	119	225 ± 14	336 ± 10	7.9 ± 0.5	157 ± 17	487 ± 14	17.0 ± 0.1
				1.5		–	–	6 ± 1	114	209 ± 9	328 ± 4	13.7 ± 2.2	129 ± 19	488 ± 13	16.0 ± 1.8
Paramsothy et al. [75]	ZK60A	CNT		0	DMD	–	–	9 ± 2	138	163 ± 3	268 ± 3	6.6 ± 0.6	128 ± 11	522 ± 11	19.6 ± 0.9
				1.0		–	–	3 ± 1	114	180 ± 6	295 ± 8	15.0 ± 0.7	110 ± 7	547 ± 3	33.2 ± 6.2
Hou et al. [74]	Mg-9Al	CNT	0		PM	–	–	12.9 ± 2.1	80.6	235 ± 3	301 ± 5	6 ± 2	–	–	–
			0.2			–	–	10.9 ± 1.5	91.8	242 ± 3	346 ± 4	14 ± 1	–	–	–
			0.4			–	–	11.1 ± 1.7	94.2	248 ± 5	355 ± 7	15 ± 3	–	–	–
			0.6			–	–	11.8 ± 1.2	89.6	230 ± 2	329 ± 1	13 ± 1	–	–	–
Nai et al. [80]	Mg	CNT	0		PM-MWS-HE	1.734	0.1	17.9 ± 8.2	39	126 ± 1	171 ± 2	7.9 ± 0.3	–	–	–
			0.3			1.737	0.75	15.6 ± 6.4	36	119 ± 4	163 ± 7	5.7 ± 0.2	–	–	–
		CNT-Ni	0.3			1.743	0.46	6.1 ± 1.4	55	206 ± 2	237 ± 1	6.4 ± 0.3	–	–	–

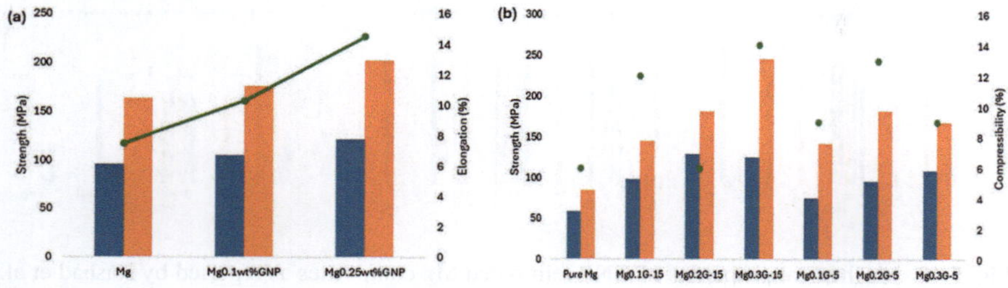

Fig. 5.39 Mechanical properties of GNP reinforced Mg composites reported by Xiang et al. [82] and Munir et al. [83]

of GNPs also promoted extension twinning, which refined grains and aligned them in orientations favorable for slip, further enhancing ductility.

Munir et al. [83] also investigated the properties of Mg composites reinforced with low volume fraction GNPs with varying thicknesses (5 and 9 nm) and particle sizes (5 and 15 μm). Here, the composites with different amounts (0.1–0.3 wt.%) of GNPs were fabricated via powder metallurgy method. The 0.2 wt.% GNP composite with finer GNPs (5 μm, 9 nm) achieved the best mechanical performance, with a 120% increase in compressive yield strength (to 130 MPa) and a 114% improvement in ultimate compressive strength (to 246 MPa), while strain to failure reached 14%. These improvements were attributed to effective load transfer, grain refinement, and thermal mismatch strengthening, with GNPs uniformly distributed along grain boundaries (Fig. 5.39).

Rashad et al. [84–87] conducted extensive studies on magnesium-based composites processed via a semi-powder metallurgy method, demonstrating significant improvements in mechanical properties through the incorporation of graphene nanoplatelets (GNPs). In one study [84], the addition of 0.30 wt.% GNPs to an Mg-1Al alloy resulted in a 131% increase in Young's modulus, a 49.5% rise in yield strength (208 MPa), and a 74.2% improvement in fracture strain (Fig. 5.40). These enhancements were attributed to the uniform dispersion and high surface area of GNPs, which facilitated effective load transfer and mitigated crack propagation. Microstructural analysis confirmed grain refinement and improved crystallographic texture due to GNP incorporation. In another study [85], the researchers varied the aluminum content (0.5–1.5 wt.%) while maintaining a fixed GNP content of 0.18 wt.%. The Mg–1.0Al–0.18GNP composite exhibited optimal mechanical performance, achieving a 17% increase in 0.2% yield strength (190 MPa), a 30% improvement in ultimate tensile strength (254 MPa), and a failure strain of 15.5%, nearly four times that of pure Mg. A Vickers hardness of 60 HV was recorded with 1.5 wt.% Al. These improvements were driven by mechanisms such as load transfer, Orowan strengthening, and dislocation generation due to thermal expansion and elastic modulus mismatches. However, excessive aluminum content led to reinforcement clustering, reducing ductility. In a related study [86], an Mg–1%Al–1%Sn alloy reinforced with 0.18 wt.% GNPs achieved a 29% increase in yield strength (208 MPa) and a 14% rise in ultimate tensile strength (269 MPa). However, failure strain decreased from 16.7% to 10.9% due to GNP clustering. Uniform GNP dispersion was critical in enhancing strength through Orowan looping and interfacial load transfer while constraining deformation.

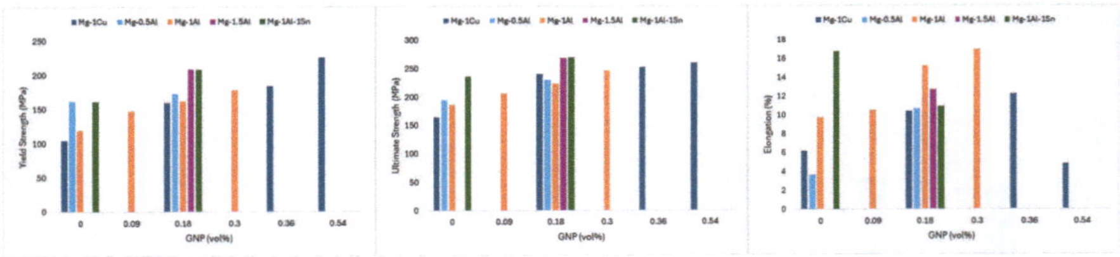

Fig. 5.40 Mechanical properties of GNP reinforced Mg composites as reported by Rashad et al. [84, 86, 87]

The superior performance of GNPs over carbon nanotubes (CNTs) in magnesium composites was underscored by their two-dimensional structure, enabling simultaneous improvements in strength and ductility. In a separate study [87], the introduction of Cu-GNP hybrids into Mg composites led to a 77% improvement in ultimate tensile strength (252 MPa) and a 96% increase in failure strain. Compressive properties also improved, with a 34% rise in compressive yield strength and a 59% improvement in ultimate compressive strength. The smallest grain size (11 μm) was observed in the Mg–1Cu–0.54GNP composite. These enhancements were attributed to Hall-Petch effects, thermal expansion mismatch, and efficient load transfer, although excessive GNP content caused aggregation and reduced ductility. Collectively, these studies highlight the exceptional potential of GNPs and Cu-GNP hybrids in enhancing the mechanical performance of magnesium composites.

Meng et al. [88] developed GNP reinforced Mg composites using a laminated structure approach to enhance mechanical properties. The addition of 0.25 vol% and 0.75 vol% GNPs resulted in ultimate tensile strength increases from 136 MPa to 160 MPa and 179 MPa, respectively, while maintaining reasonable elongation to failure (2.7% for 0.75 vol% GNPs compared to 5.5% for pure Mg). The laminated structure altered crack propagation pathways, reduced strain localization, and dissipated energy during deformation, enhancing toughness while maintaining ductility. Uniform dispersion of GNPs and strong interfacial bonding, aided by the formation of MgO and GNP exfoliation during rolling, contributed to strengthening through load transfer and inhibition of dislocation motion. This study demonstrated the potential of laminated GNP/Mg composites for lightweight structural applications requiring high strength and toughness.

Du et al. [89] fabricated ZK60 alloy composites reinforced with low-content graphene nanoplatelets (GNPs) using a combination of melt stirring and hot extrusion, ensuring excellent dispersion and interfacial bonding through pre-dispersion with Mg powders. The incorporation of 0.05 and 0.10 wt% GNPs led to remarkable mechanical improvements. The 0.10 wt% GNP composite exhibited a 79% increase in tensile yield strength (TYS) to 283 MPa, a 21% rise in ultimate tensile strength (UTS) to 343 MPa, and a 55% improvement in elongation (Fig. 5.41). Additionally, compressive yield strength (CYS) increased by 121% to 279 MPa, and failure strain improved by 33%. Microstructural analysis showed a significant reduction in grain size from 92 μm in pure ZK60 to 41 μm in the composite, along with uniform GNP distribution and robust interfacial integrity. The strengthening mechanisms included load transfer (contributing 72%), thermal mismatch-induced dislocation generation, and grain refinement. The study achieved an ultra-high strengthening efficiency of up to 1550, highlighting the potential of low GNP contents to significantly enhance the mechanical properties of magnesium alloys for lightweight structural applications.

Chen et al. [90] developed AZ91D magnesium alloy composites reinforced with graphene nanoplatelets (GNPs) using a novel thixomolding process, which demonstrated significant mechanical and microstructural improvements. The optimal composite, containing 0.6 wt.% GNPs at a barrel temperature of 603 °C, achieved a 38.4% increase in ultimate tensile strength (UTS) to 245 MPa, an 85.7% rise in elongation to 3.9%, a 29.9% improvement in Vickers hardness (HV) to 92.1 HV, and a 33.3% reduction in porosity to 5.2%, compared to unreinforced AZ91D. The enhancements were attributed to grain refinement, effective load transfer, and strong interfacial bonding between the GNPs and the magnesium matrix. GNPs were well-dispersed in the matrix without clustering, and their structural integrity was preserved during processing. Microstructural analysis showed that GNPs acted as nucleation sites, facilitating grain refinement and increasing nucleation density.

Ramezanzade et al. [91] investigated the microstructural and mechanical improvements in Mg-0.3Sr-0.3Ca composites reinforced with graphene nanoplatelets (GNPs), fabricated using stir casting followed by homogenization and hot extrusion. Composites with 0.1, 0.2, and 0.4 wt.% GNPs were analyzed, with the 0.2 wt.% GNP composite demonstrating optimal performance. This composite exhibited a 22.8% increase in tensile yield strength (TYS) to 210 MPa, a 66.7% rise in compressive yield strength (CYS) to 93 MPa, and significant improvements in tensile and compressive fracture strains by 43.1% and 28%, respectively (Fig. 5.42). Grain refinement was achieved, with the grain size reducing to 14 μm, and anisotropy decreased by 14%. These enhancements were attributed to mechanisms such as effective load transfer, dynamic recrystallization, and the activation of non-basal slip systems induced by the uniformly dispersed GNPs. Microstructural analysis revealed uniform GNP dispersion up to 0.2 wt.%, while higher contents caused

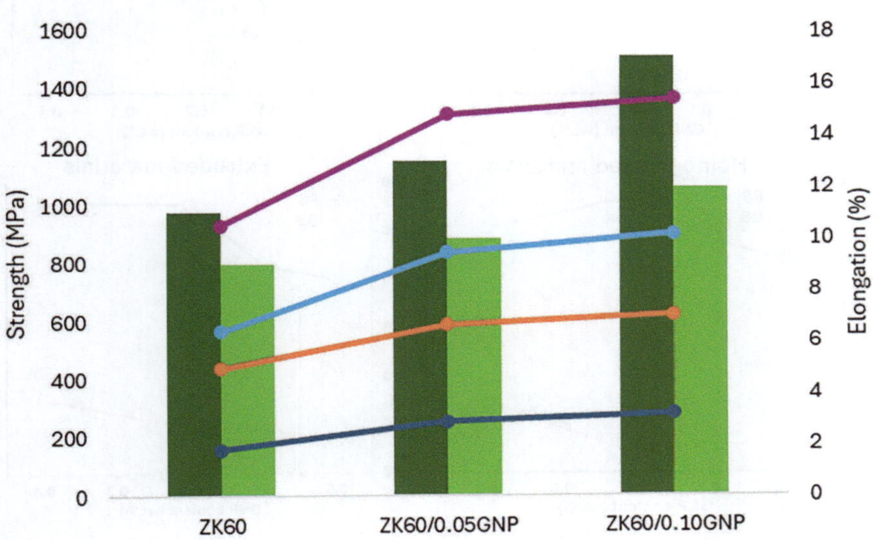

Fig. 5.41 Mechanical properties of GNP reinforced Mg composites as reported by Du et al. [89]

agglomeration and reduced performance. Additionally, GNPs facilitated dynamic precipitation and limited twin formation, further contributing to mechanical stability. Biocorrosion resistance was also improved, with the 0.2 wt.% GNP composite showing an 11% reduction in the corrosion rate compared to the base alloy.

Table 5.9 summarizes the effects of GNP on the grain size and mechanical response of Mg, Mg-Al, ZK80, and Mg-Sr-Ca matrices.

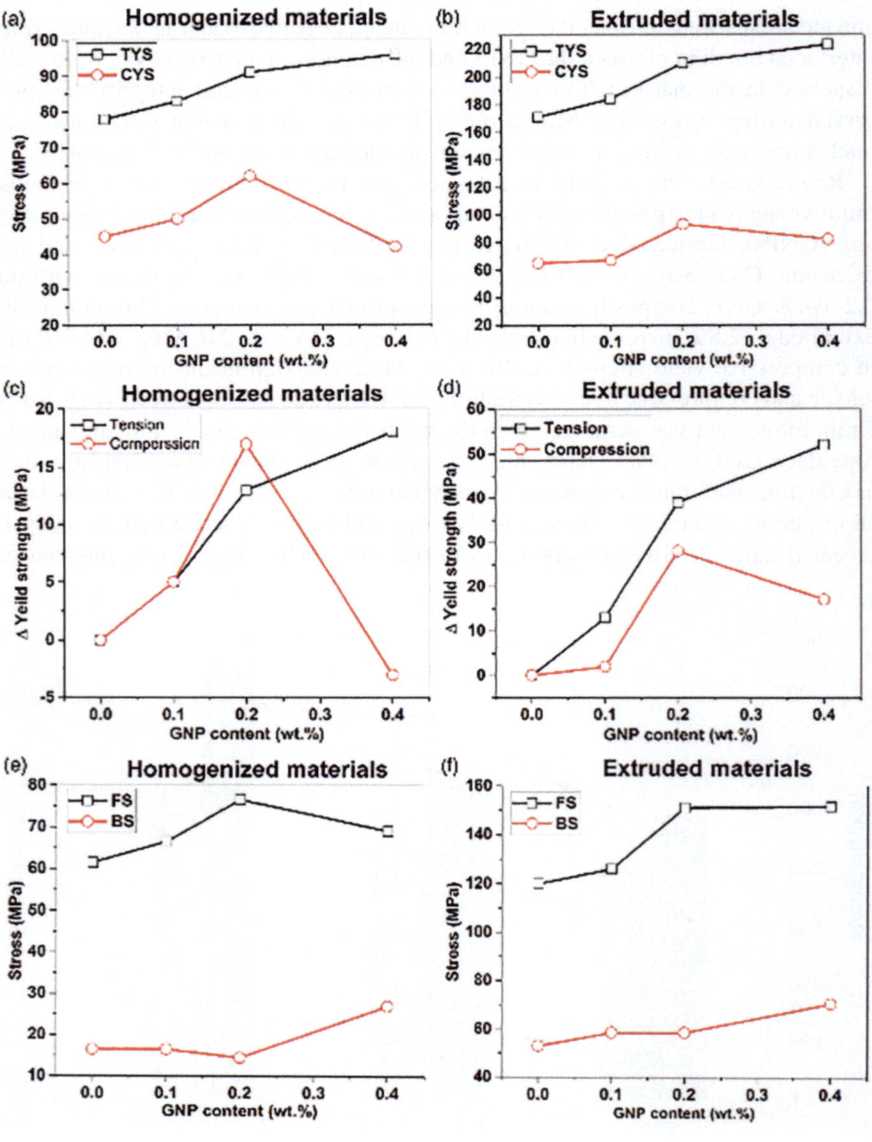

Fig. 5.42 Mechanical properties of Mg-0.3Sr-0.3Ca composites reinforced with graphene nanoplatelets [91]

Table 5.9 Effects of GNP on the grain size and mechanical response of Mg, Mg-Al, ZK80, and Mg-Sr-Ca alloys

References	Matrix	Reinfor-cement type	Wt.%	Vol.%	Process	Density (g/cc)	Porosity (%)	Average grain size (μm)	Hardness (Hv)	Tensile yield strength (MPa)	Ultimate tensile strength (MPa)	Ductility (%)	Comp yield strength (MPa)	Ultimate comp strength (MPa)	Ductility (%)	Remarks
Rashad et al. [84–86]	Mg	GNP	0		PM	–	–	–	41 ± 3.5	119 ± 5	186 ± 6	9.7 ± 3	–	–	–	–
	Mg-1Al		0.09			–	–	–	48 ± 2.9	148 ± 3	206 ± 4	10.5 ± 3.4	–	–	–	–
			0.18			–	–	–	51 ± 3	162 ± 4.1	223 ± 5	15.2 ± 2	–	–	–	–
			0.3			–	–	–	55 ± 4	178 ± 2.9	246 ± 3.5	16.9 ± 3	–	–	–	–
	Mg		0			1.7385	–	–	41 ± 4	162 ± 5	195 ± 4	3.7 ± 2.5	–	–	–	–
	Mg-0.5Al		0.18			1.7335	–	–	55 ± 2	173 ± 4	230 ± 5.1	10.7 ± 3	–	–	–	–
	Mg-1Al		0.18			1.7360	–	–	58 ± 3.5	190 ± 5.3	254 ± 3	15.5 ± 3.4	–	–	–	–
	Mg-1.5Al		0.18			1.7382	–	–	60 ± 3	209 ± 3.9	268 ± 4.5	12.7 ± 2	–	–	–	–
	Mg-1Al-1Sn		0			–	–	–	–	161 ± 04	236 ± 5.1	16.7 ± 03	–	–	–	–
			0.18			–	–	–	–	208 ± 5.3	269 ± 03	10.9 ± 3.4	–	–	–	–
Xiang et al. [82]	Mg	GNP	0		DMD	–	–	7.2	–	96 ± 4	163 ± 3	7.5 ± 1.5	–	–	–	–
			0.1			–	–	5.8	–	105 ± 1	176 ± 2	10.3 ± 0.6	–	–	–	–
			0.25			–	–	4.8	–	122 ± 2	202 ± 3	14.5 ± 1.2	–	–	–	–
Du et al. [89]	ZK60	GNP	0		Melt stirring & HE	–	–	–	68	282 ± 3.0	158 ± 2.0	11 ± 0.8	364 ± 2.8	126 ± 3.0	9 ± 0.3	–
			0.05			–	–	–	78	336 ± 4.0	256 ± 4.0	13 ± 1.2	473 ± 6.2	249 ± 4.0	10 ± 1.0	–
			0.1			–	–	–	75	343 ± 3.8	283 ± 3.5	17 ± 2.0	463 ± 5.0	279 ± 3.4	12 ± 1.1	–
Ramezanzade et al. [91]	Mg-0.3Sr-0.3Ca	GNP	0		Stir casting	–	–	–	–	78 ± 4	102 ± 5	5.4 ± 0.27	45 ± 2	208 ± 10	31 ± 1.6	Homogenized
						–	–	–	–	171 ± 9	228 ± 11	6 ± 0.3	65 ± 3	339 ± 17	14.3 ± 0.7	Extruded
			0.1			–	–	–	–	83 ± 4	108 ± 5	7.2 ± 0.4	50 ± 3	213 ± 11	29.4 ± 1.5	Homogenized
						–	–	–	–	184 ± 9	232 ± 11	8.1 ± 0.4	67 ± 3	335 ± 16	15.5 ± 0.8	Extruded
			0.2			–	–	–	–	91 ± 5	117 ± 6	8 ± 0.4	62 ± 3	218 ± 10	30.2 ± 1.5	Homogenized
						–	–	–	–	210 ± 11	231 ± 12	10 ± 1	93 ± 5	339 ± 18	18.3 ± 1.0	Extruded
			0.4			–	–	–	–	96 ± 7	129 ± 9	7.8 ± 0.6	42 ± 3	227 ± 11	29.1 ± 2.3	Homogenized
						–	–	–	–	223 ± 13.3	245 ± 15	8.8 ± 0.6	82 ± 5	309 ± 17	14.3 ± 0.9	Extruded

5.4 Biodegradable Particle Reinforcements

Calcium phosphates (CaP) are the major constituents of bone and tooth enamel. Their high chemical stability and low solubility make them ideal for applications in regenerative medicine and orthopedics [76, 92]. Specifically, apatites, which are CaP compounds with a Ca/P atomic ratio between 1.5 and 1.67, are widely used as bone substitutes, such as in maxillofacial surgery for restoring cavities, as components in toothpaste, and as implant coatings for dental applications [76, 92]. Due to these properties, calcium phosphate compounds are increasingly being explored as reinforcements in biodegradable magnesium composites [77, 93].

Khalil et al. [94] demonstrated the potential of Mg/nHA composites fabricated using high-frequency induction heat sintering (HFIHS). Composites with 1–3 wt.% nHA achieved optimal properties, including a 99.7% relative density, a hardness of 70 HV, and a 16% improvement in compressive strength due to grain refinement and uniform dispersion of nHA. However, higher nHA content caused agglomeration, adversely affecting mechanical properties.

In a similar study, Khodaei et al. [95] further advanced the understanding of Mg/nHA composites by comparing conventional sintering (CS) and spark plasma sintering (SPS). The SPS method proved superior, producing composites with a 27% higher ultimate compressive strength (221 MPa) than CS (136 MPa). SPS composites exhibited reduced porosity and minimized chemical reactions between Mg and nHA, avoiding the formation of detrimental phases. Additionally, SPS-fabricated composites demonstrated significantly lower corrosion rates due to enhanced interfacial bonding and the protective effects of nHA, making SPS an optimal method for processing Mg/nHA composites.

In another study, Cao et al. [96] explored the use of friction stir processing (FSP) to fabricate WE43 Mg alloy composites reinforced with nHA. The addition of nHA reduced the average grain size in the stir zone from 5.7 to 3.3 μm, significantly enhancing corrosion resistance in simulated body fluid. However, slight reductions in tensile and yield strength were observed due to localized agglomeration and poor interfacial bonding. Despite this, FSP enabled uniform nHA dispersion and grain refinement, underscoring its potential for improving the corrosion resistance and microstructural properties of Mg-based biomaterials (Fig. 5.43).

Razavi et al. [97–99] conducted a series of studies to evaluate the effects of fluorapatite (FA) nanoparticles as reinforcements in AZ91 magnesium alloy composites, focusing on mechanical properties, corrosion resistance, and bioactivity for biomedical applications. In these studies, AZ91 composites with 10, 20, and 30 wt.% FA were fabricated using a powder metallurgy process. The AZ91-20FA composite demonstrated the best mechanical performance, achieving a compressive yield strength of 123 MPa (+16.9%) and a hardness of 93 HV (+16.2%) compared

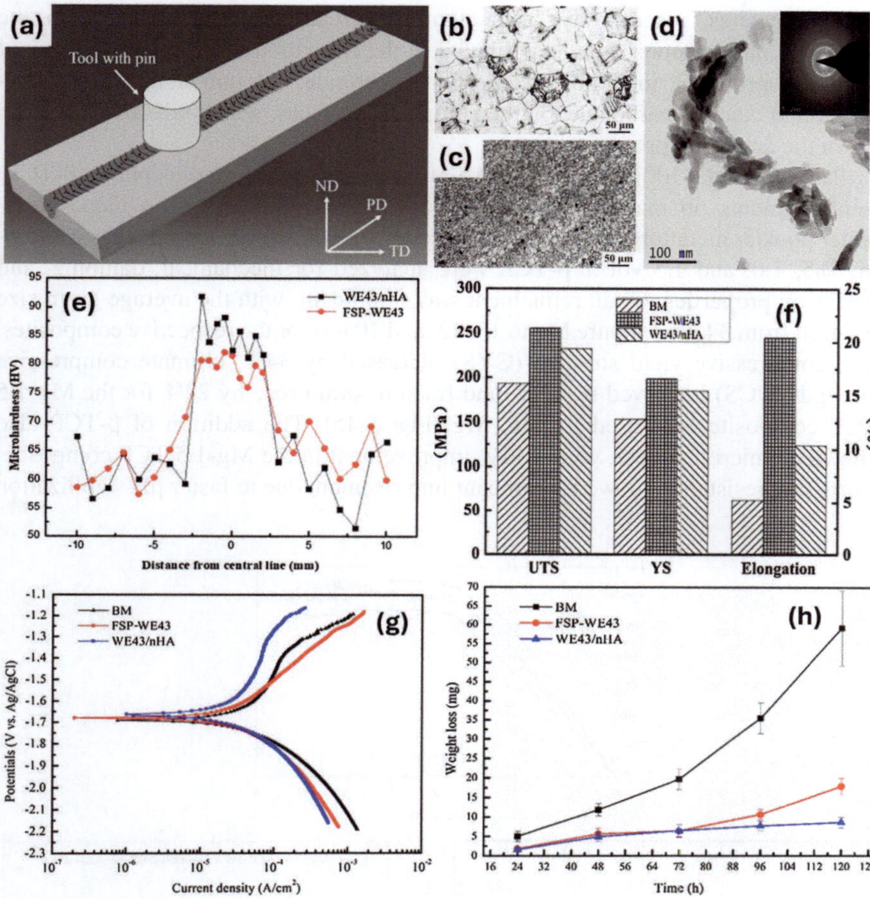

Fig. 5.43 Microstructural, mechanical, and corrosion response of WE43-nHA composites [96]

to pure AZ91. This improvement was attributed to the uniform dispersion of FA nanoparticles, which refined the grains and strengthened the matrix through dislocation pinning. Corrosion tests conducted in Ringer's solution revealed a significant reduction in corrosion rates, with the corrosion current density decreasing from 8.5×10^{-4} A/cm^2 for AZ91 to 2.3×10^{-6} A/cm^2 for AZ91-20FA and 3.5×10^{-7} A/cm^2 for AZ91-30FA (Fig. 5.44). Additionally, the total corrosion rate for AZ91 after 5 h of immersion was approximately 1.3 mg/cm^2/h, compared to 1.11 mg/cm^2/h, 0.8 mg/cm^2/h, and 0.44 mg/cm^2/h for AZ91-10FA, AZ91-20FA, and AZ91-30FA nanocomposites, respectively. This significant retardation of corrosion in the AZ91-FA nanocomposites was attributed to the formation of a protective layer of corrosion products, including magnesium hydroxide, calcium magnesium phosphate, and calcium carbonate, on the composite surfaces, which effectively slowed down the corrosion process.

Parande et al. [100] investigated the effects of β-tricalcium phosphate (β-TCP) reinforcements on magnesium (Mg) composites synthesized via a blend-press-sinter powder metallurgy technique followed by hot extrusion. Composites containing 0.5, 1.0, and 1.5 vol.% β-TCP were analyzed for mechanical, damping, and corrosion properties. Grain refinement was significant, with the average grain size reduced from 34 μm for pure Mg to 18, 13, and 10 μm for the respective composites. The compressive yield strength (CYS) increased by 34%, ultimate compressive strength (UCS) improved by 53%, and fracture strain rose by 22% for the Mg-1.5 TCP composite compared to pure Mg (Fig. 5.45). The addition of β-TCP also enhanced microhardness, with a 17% improvement in the Mg-1.5 TCP composite. Corrosion resistance showed significant improvement due to faster pH stabilization

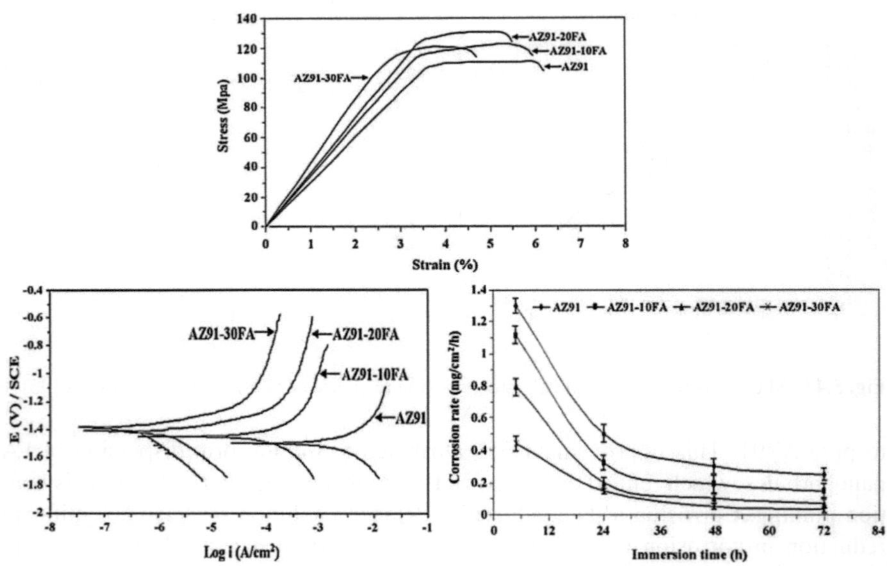

Fig. 5.44 Mechanical and corrosion properties of nano-FA reinforced AZ91 composites from Razavi et al. [97–99]

and the formation of a protective $Mg(OH)_2$ and Ca-P-based layer, resulting in a ninefold reduction in corrosion rates for Mg-1.5 TCP compared to pure Mg (Fig. 5.45). Additionally, damping properties were enhanced, with a 113% increase in damping loss rate and a 15.7% rise in damping capacity for the Mg-1.5 TCP composite. The study highlighted the role of β-TCP particles in promoting grain boundary pinning, particle-stimulated nucleation, and Hall-Petch strengthening mechanisms. Furthermore, the composites exhibited elastic modulus values (~45 GPa) closer to natural bone, mitigating stress-shielding effects and enhancing implant performance.

Table 5.10 summarizes the grain size, tensile and compressive response of biodegradable magnesium-based nanocomposites.

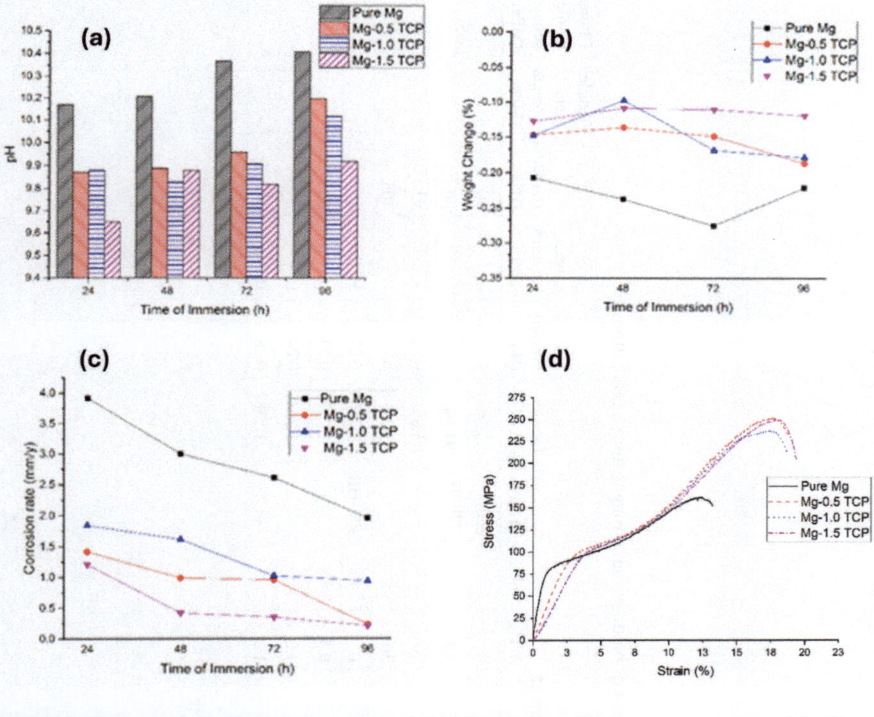

Fig. 5.45 Corrosion properties of nano-sized β-TCP reinforced Mg nanocomposites (**a**) pH vs. Time of immersion; (**b**) Weight change vs. immersion time, (**c**) Corrosion rate vs. immersion time, and (**d**) Compressive properties of β-TCP reinforced Mg nanocomposites [100]

Table 5.10 Mechanical properties of biodegradable magnesium nanocomposites

References	Matrix	Reinforcement type	Wt.%	Vol.%	Process	Density (g/cc)	Porosity (%)	Average grain size (μm)	Hardness (Hv)	Tensile yield strength (MPa)	Ultimate tensile strength (MPa)	Ductility (%)	Comp yield strength (MPa)	Ultimate comp strength (MPa)	Ductility (%)
Razavi et al. [97, 99]	AZ91	FA	0		Ball milling	92.7%	–	–	80	–	–	–	105.3	–	6.15
			10			94.1%	–	–	86	–	–	–	116.2	–	5.78
			20			95.8%	–	–	93	–	–	–	123.2.	–	5.32
			30			93.2%	–	–	105	–	–	–	112.4	–	4.51
Parande et al. [100]	Mg	TCP	0		PM-HE	1.7363	0.21	34	46 ± 3	–	–	–	77	156	15.8
			0.5			1.7371	0.23	18	52 ± 2	–	–	–	92	258	18.5
			1.0			1.7381	0.24	13	54 ± 3	–	–	–	96	223	17.2
			1.5			1.7387	0.28	10	54 ± 1	–	–	–	103	240	19.3

5.5 Metallic Nanoparticles

The incorporation of nanoscale metallic particles with limited or no solubility in Mg has also demonstrated notable improvements in the mechanical performance of magnesium. For the first time, Zhong et al. [101] investigated the effects of nano-Al particulates on Mg composites fabricated using a blend-press-sinter-extrusion powder metallurgy technique, with Al content varying from 0.25 to 1 vol.%. The composite containing 0.5 vol.% Al demonstrated the best mechanical performance, achieving a 62% increase in 0.2% yield strength (218 MPa), a 43% rise in ultimate tensile strength (271 MPa), a 35% improvement in failure strain, and a 109% enhancement in work of fracture compared to pure Mg. The hardness of the composite also increased, from 46 HV for pure Mg to 57 HV for the 0.5 vol.% Al composite (Fig. 5.46). Microstructural analysis revealed a uniform distribution of Al nanoparticles, grain refinement (grain size reduced from 23 μm in pure Mg to 16 μm in Mg/1 vol.% Al), and strong matrix–reinforcement bonding. Strengthening mechanisms, including Orowan looping, load transfer, grain boundary strengthening, and dislocation generation due to thermal expansion mismatch, were identified as key contributors to the increased strength. Fracture surface analysis further indicated a transition from brittle fracture in pure Mg to mixed-mode failure in the composite, attributed to the activation of non-basal slip systems.

In a similar study, Wong and Gupta [102] fabricated Mg/Cu nanocomposites by microwave-assisted powder metallurgy. Composites with 0.3, 0.6, and 1.0 vol.% Cu

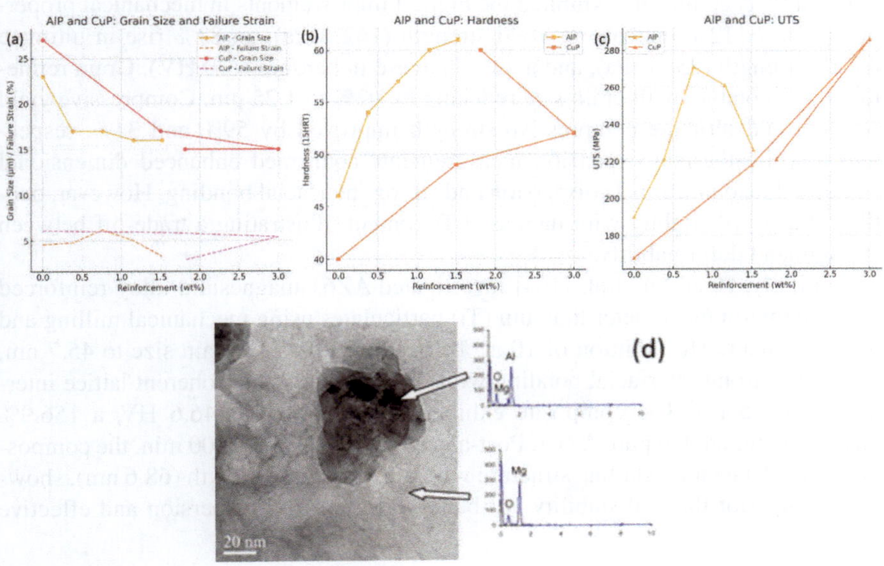

Fig. 5.46 (**a–c**) Properties of nano-Al/nano-Cu reinforced Mg nanocomposites. (**d**) TEM micrograph showing good interfacial integrity between Mg matrix and nano-Al particulates. The EDX spectrum verifies the presence of nano-aluminum particulate [101]

nanoparticulates were evaluated for their microstructural, thermal, and mechanical properties. Microstructural analysis revealed the presence of a continuous network of Cu nanoparticles and Mg_2Cu intermetallic phases along grain boundaries, promoting grain refinement from 27 μm for pure Mg to 15 μm for the 1.0 vol.% Cu composite. Porosity levels remained minimal, with the highest recorded porosity at 0.41% for the Mg/0.6Cu composite. Coefficient of thermal expansion (CTE) measurements showed marginal reductions due to the lower CTE of Cu and its constraint on matrix expansion. Mechanical characterization revealed significant enhancements, with the Mg/0.6Cu composite achieving the best performance. Elastic modulus increased by 40% (to 59.7 GPa), 0.2% yield strength improved by 82% (to 237 MPa), ultimate tensile strength rose by 51% (to 286 MPa), and work of fracture increased by 40% (to 16.5 MJ/m^3) compared to pure Mg. However, failure strain decreased slightly from 6.1% for pure Mg to 5.4% for Mg/0.6Cu, attributed to the higher rigidity and brittleness of the Cu phase. Similar to nano-Al added Mg composites, the strength increase was attributed to the strengthening mechanisms such as Orowan looping, thermal expansion mismatch-induced dislocations, grain boundary pinning, and effective load transfer from the matrix to the reinforcements.

Nano-sized titanium-reinforced magnesium composites have demonstrated significant potential for structural applications due to their enhanced mechanical properties, grain refinement, and thermal stability. Meenashisundaram and Gupta [103] synthesized magnesium nanocomposites reinforced with 0.58, 0.97, and 1.98 vol.% Ti particulates using disintegrated melt deposition followed by hot extrusion. The 1.98 vol.% Ti composite exhibited the highest improvements in mechanical properties, with a 112% increase in yield strength (162 MPa), an 81% rise in ultimate tensile strength (231 MPa), and a 43% increase in hardness (70 HV). Grain refinement was notable, with grain size reducing by 92% to 1.25 μm. Compressive yield strength and ultimate compressive strength improved by 59% and 34%, respectively, while thermal expansion measurements confirmed enhanced dimensional stability due to uniform Ti dispersion and strong interfacial bonding. However, ductility decreased slightly with increased Ti content, illustrating a trade-off between strength and deformability.

Similarly, Huan Yu et al. [104] investigated AZ61 magnesium alloy reinforced with submicron/nanometer titanium (Ti) particulates using mechanical milling and heat treatment. The addition of 10 at.% Ti reduced the Mg grain size to 45.7 nm, achieving strong interfacial bonding with coherent and semi-coherent lattice interfaces (Fig. 5.47). The composite exhibited a hardness of 146.6 HV, a 156.9% increase compared to pure AZ61. Post-annealing at 673 K for 600 min, the composite retained a nanocrystalline structure with minimal grain growth (68.6 nm), showcasing superior thermal stability attributed to uniform Ti dispersion and effective

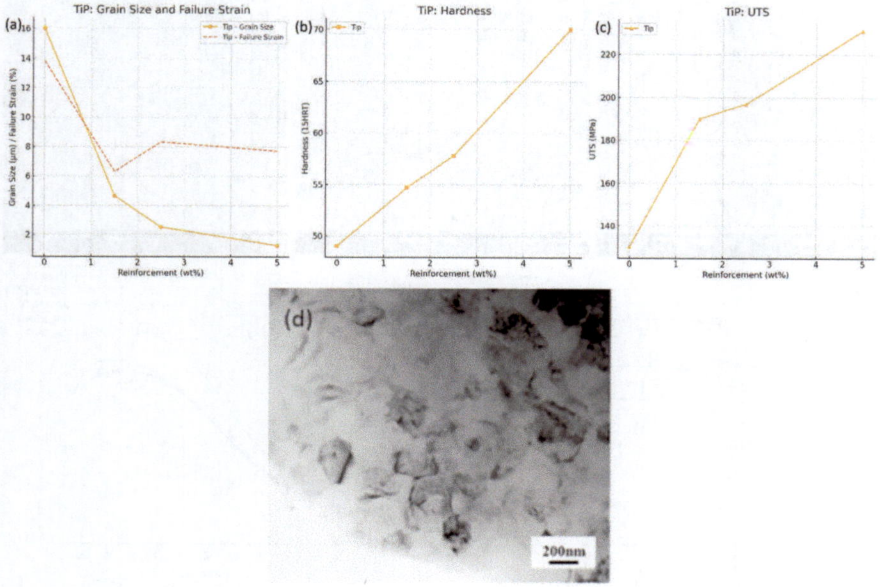

Fig. 5.47 (**a–c**) Properties of nano-Ti reinforced Mg composites. (**d**) High-resolution TEM image showing the distribution of nano-Ti particles in AZ61 alloy matrix [104]

grain boundary pinning. Strengthening mechanisms such as grain boundary pinning, Orowan strengthening, and solid solution strengthening were instrumental in enhancing strength and stability, making this composite suitable for high-performance applications.

Nano-sized NiTi (Nitinol) particles have also shown great potential as reinforcements for developing Mg composites with significantly enhanced properties for promising applications in structural and biomedical fields. Razzaghi et al. [105] synthesized Mg–3Zn–0.5Ag composites with 15, 30, and 45 wt.% NiTi using mechanical alloying followed by sintering. The optimal composite containing 15 wt.% NiTi achieved a 56% increase in compressive strength, a 7.6% rise in elongation, and a 46.8% improvement in microhardness compared to the matrix alloy. Grain size decreased by 16.7%, enhancing strength through Hall–Petch and Orowan mechanisms. However, corrosion resistance decreased with higher NiTi content due to galvanic interactions. In vitro biocompatibility tests confirmed that the 15 wt.% NiTi composite exhibited favorable cell adhesion and cytotoxicity results, making it suitable for orthopedic implants.

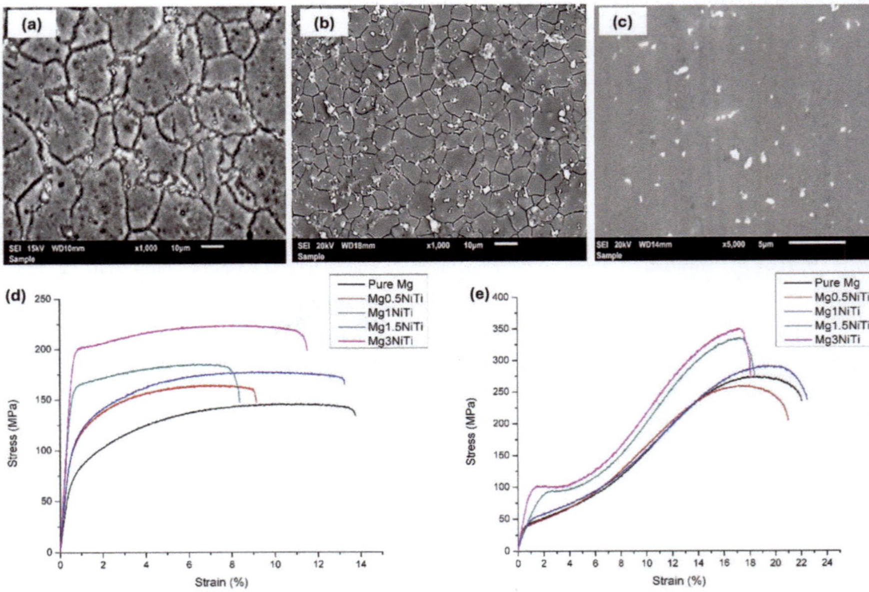

Fig. 5.48 SEM images showing the grain characteristics of (**a**) pure Mg, (**b**) Mg/3NiTi. (**c**) Distribution of NiTi particles in the Mg matrix. Stress–Strain curves corresponding to Mg/NiTi composites under (**d**) tension and (**e**) compression loading. From Parande et al. [99]

Parande et al. [106] fabricated pure Mg composites reinforced with 0.5, 1, 1.5, and 3 wt.% NiTi using disintegrated melt deposition (DMD) followed by hot extrusion. The 3 wt.% NiTi composite showed the best overall performance, achieving a 129% increase in tensile yield strength (193 MPa), a 46% improvement in ultimate tensile strength (217 MPa), and a 31% rise in hardness (75 HV) compared to pure Mg (Fig. 5.48). Grain size was refined from 40 to 9 μm, and the coefficient of thermal expansion decreased by 10% due to the low CTE of NiTi. Unlike Razzaghi's composites, these maintained moderate ductility and absorbed 35% more energy under tensile fracture although the biocompatibility is slightly inferior.

Table 5.11 summarizes the effects of metallic particles (Cu, Al and Ti) on grain size, tensile and compressive response of pure magnesium.

The overall snapshot of tensile and compressive response of magnesium-based nanocomposites is presented in Fig. 5.49.

Table 5.11 Effects of metallic nanoparticle addition on the microstructural and mechanical properties of Mg

References	Matrix	Reinforcement type	Wt.%	Vol.%	Process	Density (g/cc)	Porosity (%)	Average grain size (μm)	Hardness (Hv)	Tensile yield strength (MPa)	Ultimate tensile strength (MPa)	Ductility (%)	Comp yield strength (MPa)	Ultimate comp strength (MPa)	Ductility (%)
Wong et al. [102]	Mg	Cu	0	0	PM-MWS-HE	1.738	0.12	27 ± 7	40	116 ± 11	168 ± 10	6.1 ± 2.0	–	–	–
			0.3	1.5		1.758	0.19	17 ± 5	49	188 ± 13	218 ± 11	5.9 ± 1.1	–	–	–
			0.6	3.0		1.776	0.41	15 ± 4	52	237 ± 24	286 ± 8	5.4 ± 1.2	–	–	–
			1.0	4.9		1.809	0.13	15 ± 4	60	194 ± 17	221 ± 17	2.9 ± 0.4	–	–	–
Zhong et al. [101]	Mg	Al	0	0	PM-MWS-HE	1.731	0.5	23 ± 6	46 ± 3	134 ± 11	190 ± 10	4.6 ± 0.6	–	–	–
			0.25	0.38		1.738	0.2	17 ± 5	54 ± 1	181 ± 14	221 ± 15	4.8 ± 0.4	–	–	–
			0.5	0.76		1.746	0.4	17 ± 8	57 ± 1	218 ± 16	271 ± 11	6.2 ± 0.9	–	–	–
			0.75	1.16		1.751	0	16 ± 7	60 ± 1	202 ± 7	261 ± 10	5.0 ± 1.6	–	–	–
			1.0	1.52		1.755	0	16 ± 6	61 ± 1	185 ± 9	226 ± 12	3.3 ± 1.0	–	–	–
Meenashi-sundaram et al. [103]	Mg	Ti	0		DMD-HE	1.7375	0.08	16 ± 4	49 ± 2	73.5 ± 5.4	130.3 ± 4.4	13.82 ± 1.42	86 ± 1	326 ± 1	20.8 ± 1.7
			0.58			1.7540	0.04	4.65 ± 1.5	55 ± 1	134 ± 7	190 ± 7	6.3 ± 0.6	129 ± 2	431 ± 8	17.4 ± 0.3
			0.97			1.7651	0.02	2.5 ± 1.5	58 ± 1	135 ± 3	197 ± 8	8.3 ± 0.6	130 ± 8	413 ± 15	18.5 ± 0.6
			1.98			1.793	0.01	1.25 ± 0.25	70 ± 2	162 ± 5	231 ± 12	7.7 ± 0.1	120 ± 5	415 ± 4	17.7 ± 0.8
Parande et al. [106]	Mg	NiTi	0		DMD-HE	1.72	0.85	40 ± 9.4	57 ± 4.6	84 ± 6	148 ± 6	13 ± 0.9	46 ± 2	274 ± 4	23 ± 0.9
			0.5			1.73	0.48	17 ± 2.2	59 ± 3.3	106 ± 12	167 ± 14	9 ± 0.8	42 ± 4	265 ± 15	23 ± 0.3
			1			1.74	0.51	13 ± 1	72 ± 2.5	123 ± 2	176 ± 9	9 ± 4.2	52 ± 6	290 ± 4	23 ± 0.7
			1.5			1.75	0.50	10 ± 3.3	73 ± 3.9	163 ± 9	187 ± 8	9 ± 0.4	78 ± 5	337 ± 4	18 ± 0.4
			3			1.76	0.67	9 ± 1.8	75 ± 2.6	193 ± 17	217 ± 16	11 ± 0.2	94 ± 3	345 ± 5	18 ± 0.9

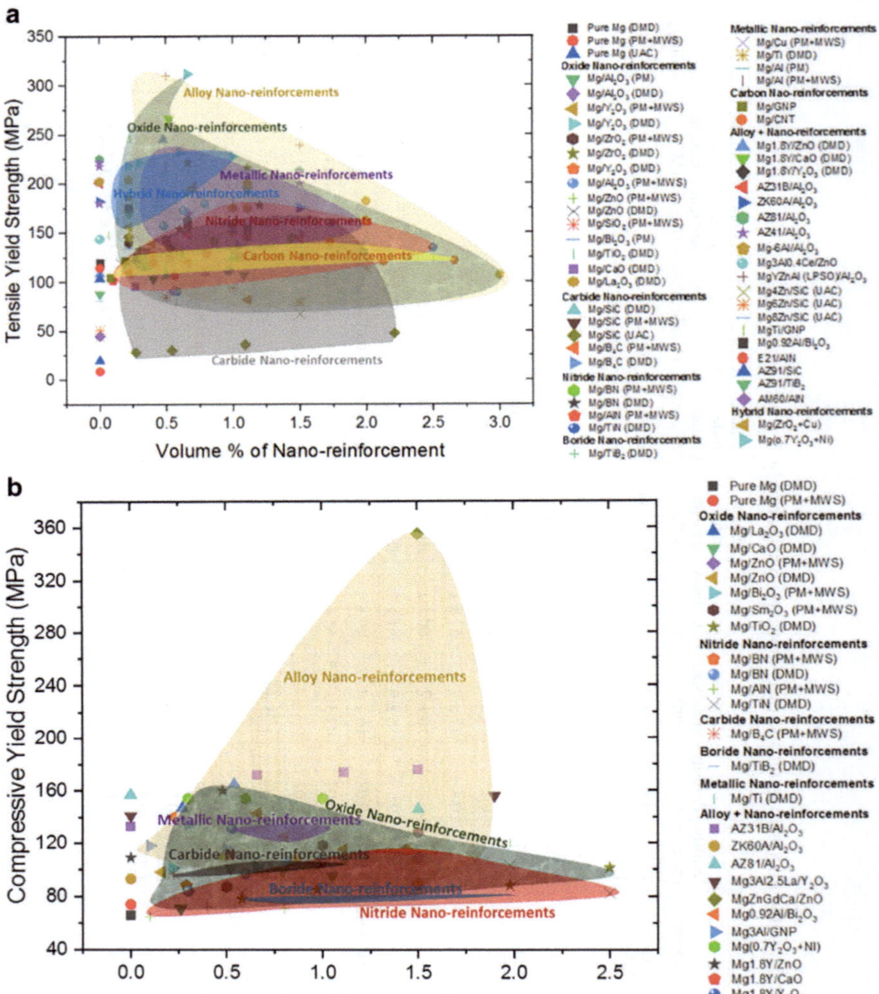

Fig. 5.49 (**a**) Tensile and (**b**) compressive properties of Mg nanocomposites

5.6 Summary

Nanoscale reinforcements have emerged as a transformative approach to enhance the properties of magnesium-based composites. These reinforcements, which include oxide, carbide, nitride, and boride ceramics, carbon-based materials like carbon nanotubes (CNTs) and graphene, calcium phosphate derivatives for biomedical applications, and metallic nanoparticles, contribute significantly to the mechanical, thermal, and biological performance of magnesium composites. The use of such reinforcements has shown great potential in applications ranging from structural and aerospace components to biomedical implants, offering tailored enhancements that address the inherent limitations of magnesium.

Ceramic reinforcements such as aluminum oxide (Al_2O_3), titanium carbide (TiC), titanium nitride (TiN), and boron carbide (B4C) are particularly effective in improving the strength, hardness, and wear resistance of magnesium composites. These nanoparticles contribute to grain refinement and provide resistance to dislocation motion, making them valuable for high-strength applications. However, their effectiveness largely depends on achieving a uniform dispersion within the magnesium matrix, which can be challenging due to the tendency of nano particles to agglomerate during processing.

Carbon-based reinforcements, including carbon nanotubes (CNTs) and graphene, have shown remarkable potential in enhancing not only the mechanical properties but also the thermal and electrical conductivity of magnesium composites. Due to their exceptional strength and large surface area, even small additions of CNTs or graphene significantly improve tensile strength and modulus. However, as with ceramic nanoparticles, the uniform dispersion and strong interfacial bonding between the matrix and these carbon-based reinforcements are critical for maximizing their beneficial effects. Addressing the challenges of agglomeration and poor bonding remains an area of ongoing research.

In the realm of biomedical applications, calcium phosphate derivatives such as hydroxyapatite (HA), fluorapatite (FA), and β-tricalcium phosphate (β-TCP) have emerged as promising reinforcements for magnesium composites. These compounds are biocompatible and bioactive, closely mimicking the mineral components of natural bone, making them ideal for use in biodegradable implants. The integration of calcium phosphate nanoparticles improves the bioactivity of magnesium implants, enhances bone cell adhesion, and promotes osteointegration. However, as with other reinforcements, controlling the particle dispersion and avoiding agglomeration are essential for maintaining both mechanical integrity and controlled biodegradation rates.

Metallic nanoparticles, such as copper (Cu), aluminum (Al), titanium (Ti), and nickel-titanium (NiTi) alloys, also play a crucial role in enhancing the mechanical properties of magnesium composites. These metallic particles improve strength, hardness, and wear resistance through mechanisms such as grain refinement and precipitation hardening. For instance, the addition of Cu or Al nanoparticles leads to significant improvements in yield strength and ultimate tensile strength. However, the presence of metallic reinforcements can also lead to the formation of brittle intermetallic phases, such as Mg_2Cu, which may reduce the composite's ductility. As with other reinforcement types, careful control over the amount and distribution of these metallic nanoparticles is critical in achieving the desired balance of strength and ductility.

Despite the considerable advancements in the field, several challenges remain. Uniform dispersion of nanoscale reinforcements within the magnesium matrix is one of the most significant hurdles. Agglomeration of nanoparticles, poor interfacial bonding, and the formation of undesirable phases can diminish the reinforcing effects and introduce weaknesses in the composite. Various fabrication methods, such as powder metallurgy, friction stir processing, and disintegrated melt deposition, have been developed to improve nanoparticle distribution and bonding, yet further optimization is required. Processing conditions, such as sintering temperatures and compaction pressures, also play a key role in determining the final properties of the composite.

Looking forward, the future of magnesium-based composites with nanoscale reinforcements is promising. Research is increasingly focused on developing multifunctional composites that combine high strength, improved wear resistance, and enhanced biocompatibility. In biomedical applications, for instance, the goal is to create biodegradable implants that not only possess the mechanical strength to support tissue regeneration but also degrade in a controlled manner, leaving behind no harmful residues. In structural applications, the development of lightweight magnesium composites with improved thermal and electrical properties opens up new possibilities in aerospace, automotive, and electronics industries.

In conclusion, nanoscale reinforcements, whether ceramic, carbon-based, calcium phosphate derivatives, or metallic—have proven to be effective in enhancing the performance of magnesium-based composites. Each type of reinforcement offers unique benefits, but challenges related to particle dispersion, interfacial bonding, and phase formation must be overcome to fully realize their potential. Continued advancements in fabrication techniques and deeper understanding of particle–matrix interactions will drive the next generation of magnesium composites, enabling their application in a broader range of high-performance and biomedical fields.

References

1. S.F. Hassan, *Creation of New Magnesium-Based Material Using Different Types of Rein Forcements* (National University of Singapore, 2006)
2. W.L.E. Wong, M. Gupta, Improving overall mechanical performance of magnesium using nano-alumina reinforcement and energy efficient microwave assisted processing route. Adv. Eng. Mater. (2007). https://doi.org/10.1002/adem.200700169
3. H.M.A. Zaid, A.R.N. Abed, H.S. Hasan, Improvement of mechanical properties of magnesium (Mg) matrix composites reinforced with nano alumina (Al_2O_3) particles. IOP Conf. Ser. Mater. Sci. Eng. **671**(1), 012162 (2020). https://doi.org/10.1088/1757-899X/671/1/012162
4. M. Shanthi, G.L.X. Xia, M. Gupta, Effect of Mg/nano-Al_2O_3 interaction time during stirring on microstructure and mechanical properties of Mg–Al_2O_3 composite. Mater. Sci. Technol. **27**(8), 1341–1346 (2011). https://doi.org/10.1179/026708310X12701095964568
5. S. Han, M.E. Alam, A.M.S. Hamouda, Q.B. Nguyen, M. Gupta, Enhancing mechanical properties of AZ31 magnesium alloy through simultaneous addition of aluminum and nano-Al_2O_3, in *IMECE2010, Volume 12: Processing and Engineering Applications of Novel Materials*, (ASME, 2010), pp. 145–152. https://doi.org/10.1115/IMECE2010-39901
6. M. Habibnejad-Korayem, R. Mahmudi, W.J. Poole, Enhanced properties of Mg-based nano-composites reinforced with Al_2O_3 nano-particles. Mater. Sci. Eng. A **519**(1), 198–203 (2009). https://doi.org/10.1016/j.msea.2009.05.001

7. N.Q. Bau, *Development of Nano-composites Based on Magnesium Alloys System AZ31B* (National University of Singapore, 2009)
8. M. Paramsothy, S.F. Hassan, N.Q. Bau, N. Srikanth, M. Gupta, Selective enhancement of tensile/compressive strength and ductility of AZ31 magnesium alloy via nano-Al_2O_3 reinforcement integration method alteration, in *Materials Science Forum*, (Trans Tech Publications, 2009), pp. 423–427
9. A. Ercetin, D.Y. Pimenov, Microstructure, mechanical, and corrosion behavior of Al_2O_3 reinforced Mg_2Zn matrix magnesium composites. Materials **14**(17), 4819 (2021). https://doi.org/10.3390/ma14174819
10. Y. Radi, R. Mahmudi, Effect of Al_2O_3 nano-particles on the microstructural stability of AZ31 Mg alloy after equal channel angular pressing. Mater. Sci. Eng. A **527**(10), 2764–2771 (2010). https://doi.org/10.1016/j.msea.2010.01.029
11. M. Paramsothy, J. Chan, R. Kwok, M. Gupta, Al_2O_3 nanoparticle addition to commercial magnesium alloys: multiple beneficial effects. Nanomaterials **2**(2), 147–162 (2012). https://doi.org/10.3390/nano2020147
12. D. Sameer Kumar, K.N.S. Suman, C. Tara Sasanka, K. Ravindra, P. Poddar, S.B. Venkata Siva, Microstructure, mechanical response and fractography of AZ91E/Al_2O_3 (p) nano composite fabricated by semi solid stir casting method. J. Magnes. Alloys **5**(1), 48–55 (2017). https://doi.org/10.1016/j.jma.2016.11.006
13. A. Gnanavelbabu, K.S. Surendran, S. Kumar, Influence of ultrasonication power on grain refinement, mechanical properties and wear behaviour of AZ91D/nano-Al_2O_3 composites. Mater. Res. Express **7**(1), 016544 (2020)
14. D. Ahmadkhaniha, M.H. Sohi, A. Salehi, R. Tahavvori, Formations of AZ91/Al_2O_3 nano-composite layer by friction stir processing. J. Magnes. Alloys **4**(4), 314–318 (2016)
15. X.H. Tan, W.K.H. Chee, J.K.W. Chan, R.W.O. Kwok, M. Gupta, Stretching the engineering strain of high strength LPSO quaternary Mg-Y-Zn-Al alloy via integration of nano-Al_2O_3. J. Mater. Sci. (2016). https://doi.org/10.1007/s10853-016-9742-9
16. M. Paramsothy, X.H. Tan, J. Chan, R. Kwok, M. Gupta, Al_2O_3 nanoparticle addition to concentrated magnesium alloy AZ81: enhanced ductility. J. Alloys Compd. (2012). https://doi.org/10.1016/j.jallcom.2012.08.020
17. M. Paramsothy, J. Chan, R. Kwok, M. Gupta, The effective reinforcement of magnesium alloy ZK60A using Al_2O_3 nanoparticles. J. Nanoparticle Res. **13**(10), 4855–4866 (2011). https://doi.org/10.1007/s11051-011-0464-2
18. M. Paramsothy, J. Chan, R. Kwok, M. Gupta, Nanoparticle interactions with the magnesium alloy matrix during physical deformation: tougher nanocomposites. Mater. Chem. Phys. **137**(2), 472–482 (2012). https://doi.org/10.1016/j.matchemphys.2012.09.036
19. M.K. Habibi, A.S. Hamouda, M. Gupta, Using hierarchical composite approach to improve mechanical response of Mg and Mg–Bi_2O_3 nano-composites. Mater. Des. **49**, 627–637 (2013)
20. S. Tekumalla, M. Gupta, K.H. Min, Using CaO nanoparticles to improve mechanical and ignition response of magnesium. Curr. Nanomater. (2018). https://doi.org/10.2174/2405461503666180502101957
21. S. Tekumalla et al., A strong and deformable in-situ magnesium nanocomposite igniting above 1000°C. Sci. Rep. (2018). https://doi.org/10.1038/s41598-018-25527-0
22. M. Johanes, M. Gupta, The promise of turning induced deformation process for synthesizing magnesium based materials with superior mechanical response. Technologies **9**(4), 69 (2021)
23. M. Johanes, M. Gupta, An investigation into the potential of turning induced deformation technique for developing porous magnesium and Mg-SiO_2 nanocomposite. Materials **16**(6), 2463 (2023)
24. G. Parande, V. Manakari, G.K. Meenashisundaram, M. Gupta, Enhancing the hardness/compression/damping response of magnesium by reinforcing with biocompatible silica nanoparticulates. Int. J. Mater. Res. **107**(12), 1091–1099 (2016)
25. C. Lee, J. Huang, P. Hsieh, Mg based nano-composites fabricated by friction stir processing. Scr. Mater. **54**(7), 1415–1420 (2006)

26. M.S. Kujur et al., Effect of samarium oxide nanoparticles on degradation and invitro biocompatibility of magnesium. Mater. Today Commun. **26**, 102171 (2021)
27. G.K. Meenashisundaram, M.H. Nai, A. Almajid, M. Gupta, Development of high performance Mg–TiO_2 nanocomposites targeting for biomedical/structural applications. Mater. Des. 1980–2015 **65**, 104–114 (2015)
28. G.K. Meenashisundaram, M.H. Nai, M. Gupta, Effects of primary processing techniques and significance of hall-petch strengthening on the mechanical response of magnesium matrix composites containing TiO_2 nanoparticulates. Nanomaterials **5**(3), 1256–1283 (2015)
29. G.K. Meenashisundaram, M.H. Nai, A. Almajid, K.A. Khalil, H.S. Abdo, M. Gupta, Effects of TiO_2 powder morphology on the mechanical response of pure magnesium: 1D nanofibers versus 0D nanoparticulates. J. Alloys Compd. **664**, 45–58 (2016)
30. M.S. Kujur, A. Mallick, V. Manakari, G. Parande, K.S. Tun, M. Gupta, Significantly enhancing the ignition/compression/damping response of monolithic magnesium by addition of Sm_2O_3 nanoparticles. Metals **7**(9), 357 (2017). https://doi.org/10.3390/met7090357
31. G. Parande, V. Manakari, G.K. Meenashisundaram, M. Gupta, Enhancing the tensile and ignition response of monolithic magnesium by reinforcing with silica nanoparticulates. J. Mater. Res. **32**(11), 2169–2178 (2017). https://doi.org/10.1557/jmr.2017.194
32. K.S. Tun, M. Gupta, Improving mechanical properties of magnesium using nano-yttria reinfor cement and microwave assisted powder metallurgy method. Compos. Sci. Technol. **67**(13), 2657–2664 (2007). https://doi.org/10.1016/j.compscitech.2007.03.006
33. K. Tun, M. Gupta, Effect of heating rate during hybrid microwave sintering on the tensile properties of magnesium and Mg/Y_2O_3 nanocomposite. J. Alloys Compd. **466**(1–2), 140–145 (2008)
34. K.S. Tun, M. Gupta, Effect of extrusion ratio on microstructure and mechanical properties of microwave-sintered magnesium and Mg/Y_2O_3 nanocomposite. J. Mater. Sci. **43**, 4503–4511 (2008). https://doi.org/10.1007/s10853-008-2649-3
35. S.F. Hassan, Effect of primary processing techniques on the microstructure and mechanical properties of nano-Y_2O_3 reinforced magnesium nanocomposites. Mater. Sci. Eng. A **528**(16), 5484–5490 (2011). https://doi.org/10.1016/j.msea.2011.03.063
36. A. Kumar, K. Tun, A. Kohadkar, M. Gupta, Improved compressive, damping and coefficient of thermal expansion response of Mg–3Al–2.5La alloy using Y_2O_3 nano reinforcement. Metals (2017). https://doi.org/10.3390/met7030104
37. K.S. Tun, P. Jayaramanavar, Q.B. Nguyen, J. Chan, R. Kwok, M. Gupta, Investigation into tensile and compressive responses of Mg–ZnO composites. Mater. Sci. Technol. **28**(5), 582–588 (2012). https://doi.org/10.1179/1743284711Y.0000000108
38. S. Sankaranarayanan, U. Pranav Nayak, R.K. Sabat, S. Suwas, A. Almajid, M. Gupta, Nano-ZnO particle addition to monolithic magnesium for enhanced tensile and compressive response. J. Alloys Compd. **615**, 211–219 (2014). https://doi.org/10.1016/j.jallcom.2014.06.163
39. Z. Zhang, D.L. Chen, Consideration of Orowan strengthening effect in particulate-reinforced metal matrix nanocomposites: a model for predicting their yield strength. Scr. Mater. **54**(7), 1321–1326 (2006). https://doi.org/10.1016/j.scriptamat.2005.12.017
40. S. Tekumalla, N. Farhan, T. Srivatsan, M. Gupta, Nano-ZnO particles' effect in improving the mechanical response of Mg-3Al-0.4Ce alloy. Metals **6**(11), 276 (2016). https://doi.org/10.3390/met6110276
41. Y. Chen, S. Tekumalla, Y.B. Guo, M. Gupta, Introducing Mg-4Zn-3Gd-1Ca/ZnO nanocomposite with compressive strengths matching/exceeding that of mild steel. Sci. Rep. **6**, 32395 (2016). https://doi.org/10.1038/srep32395
42. K. Qiao et al., Mg/ZrO_2 metal matrix nanocomposites fabricated by friction stir processing: microstructure, mechanical properties, and corrosion behavior. Front. Bioeng. Biotechnol. **9**, 605171 (2021). https://doi.org/10.3389/fbioe.2021.605171
43. H. Ferkel, B.L. Mordike, Magnesium strengthened by SiC nanoparticles. Mater. Sci. Eng. A **298**(1–2), 193–199 (2001). https://doi.org/10.1016/S0921-5093(00)01283-1

44. W.L.E. Wong, M. Gupta, Simultaneously improving strength and ductility of magnesium using nano-size SiC particulates and microwaves. Adv. Eng. Mater. **8**(8), 735–740 (2006). https://doi.org/10.1002/adem.200500209
45. G. Cao, H. Konishi, X. Li, Mechanical properties and microstructure of Mg/SiC nanocomposites fabricated by ultrasonic cavitation based nanomanufacturing. J. Manuf. Sci. Eng. Trans. ASME (2008). https://doi.org/10.1115/1.2823086
46. G. Cao, H. Choi, H. Konishi, S. Kou, R. Lakes, X. Li, Mg-6Zn/1.5%SiC nanocomposites fabricated by ultrasonic cavitation-based solidification processing. J. Mater. Sci. (2008). https://doi.org/10.1007/s10853-008-2785-9
47. M. De Cicco et al., Strong, ductile magnesium-zinc nanocomposites. Metall. Mater. Trans. A Phys. Metall. Mater. Sci. (2009). https://doi.org/10.1007/s11661-009-0013-0
48. Z.H. Wang, X.D. Wang, Y.X. Zhao, W.B. Du, SiC nanoparticles reinforced magnesium matrix composites fabricated by ultrasonic method. Trans. Nonferrous Met. Soc. China Engl. Ed. (2010). https://doi.org/10.1016/S1003-6326(10)60625-5
49. L. Shiying, G. Feipeng, Z. Qiongyuan, L. Wenzhen, Mechanical properties and microstrutures of nano-sized SiC particles reinforced AZ91D nanocomposites fabricated by high intensity ultrasonic assisted casting. Mater. Sci. Forum (2009). https://doi.org/10.4028/www.scientific.net/MSF.618-619.449
50. K.B. Nie, X.J. Wang, L. Xu, K. Wu, X.S. Hu, M.Y. Zheng, Influence of extrusion temperature and process parameter on microstruc tures and tensile properties of a particulate reinforced magnesium matrix nanocomposite. Mater. Des. (2012). https://doi.org/10.1016/j.matdes.2011.11.020
51. S. Sankaranarayanan, R.K. Sabat, S. Jayalakshmi, S. Suwas, M. Gupta, Effect of nanoscale boron carbide particle addition on the microstructural evolution and mechanical response of pure magnesium. Mater. Des. **56**, 428–436 (2014). https://doi.org/10.1016/j.matdes.2013.11.031
52. M.K. Habibi, A.S. Hamouda, M. Gupta, Hybridizing boron carbide (B4C) particles with aluminum (Al) to enhance the mechanical response of magnesium based nano-composites. J. Alloys Compd. **550**, 83–93 (2013). https://doi.org/10.1016/j.jallcom.2012.09.128
53. M. Paramsothy, M. Gupta, The opposing nanoscale and macroscale effects of selected nanoparticle addition to AZ91/ZK60A hybrid magnesium alloy. J. Nanoparticle Res. (2013). https://doi.org/10.1007/s11051-013-1938-1
54. S. Tekumalla, N.J. Yuan, M. Haghshenas, M. Gupta, Enhancing properties of aerospace alloy Elektron 21 using boron carbide nanoparticles as reinforcement. Appl. Sci. Switz. (2019). https://doi.org/10.3390/app9245470
55. M. Paramsothy, J. Chan, R. Kwok, M. Gupta, Adding TiC nanoparticles to magnesium alloy ZK60A for strength/ductility enhancement. J. Nanomater. (2011). https://doi.org/10.1155/2011/642980
56. G.K. Meenashisundaram, S. Seetharaman, M. Gupta, Enhancing overall tensile and compressive response of pure Mg using nano-TiB_2 particulates. Mater. Charact. **94**, 178–188 (2014). https://doi.org/10.1016/j.matchar.2014.05.021
57. H. Choi, Y. Sun, B.P. Slater, H. Konishi, X. Li, AZ91D/TiB_2 nanocomposites fabricated by solidification nanoprocessing. Adv. Eng. Mater. (2012). https://doi.org/10.1002/adem.201100313
58. M. Paramsothy, M. Gupta, *ZrB_2 nanoparticle induced nano-LPSO-grain and nano-LPSO-layer reinforced ultra-high strength Mg-RE alloy* (J. Mater. Sci, 2013). https://doi.org/10.1007/s10853-013-7647-4
59. S. Sankaranarayanan, M.K. Habibi, S. Jayalakshmi, K. Jia Ai, A. Almajid, M. Gupta, Nano-AlN particle reinforced Mg composites: microstructural and mechanical properties. Mater. Sci. Technol. **31**(9), 1122–1131 (2015). https://doi.org/10.1179/1743284714Y.0000000686
60. M.A. Thein, L. Lu, M.O. Lai, Mechanical properties of nanostructured Mg-5 wt%Al-x wt%AlN composite synthesized from Mg chips. Compos. Struct. (2006). https://doi.org/10.1016/j.compstruct.2006.04.059

61. G. Cao, H. Choi, J. Oportus, H. Konishi, X. Li, Study on tensile properties and microstructure of cast AZ91D/AlN nanocomposites. Mater. Sci. Eng. A (2008). https://doi.org/10.1016/j.msea.2008.04.070
62. H. Dieringa et al., Ultrasound assisted casting of an AM60 based metal matrix nanocomposite, its properties, and recyclability. Metals (2017). https://doi.org/10.3390/met7100388
63. L. Katsarou et al., Microstructure, mechanical properties and creep of magnesium alloy Elektron21 reinforced with AlN nanoparticles by ultrasound-assisted stirring. Mater. Sci. Eng. A (2016). https://doi.org/10.1016/j.msea.2016.02.042
64. S. Sankaranarayanan, R. Sabat, S. Jayalakshmi, S. Suwas, A. Almajid, M. Gupta, Mg/BN nanocomposites: nano-BN addition for enhanced room temperature tensile and compressive response. J. Compos. Mater. **49**(24), 3045–3055 (2015). https://doi.org/10.1177/0021998314559278
65. S. Seetharaman, J. Subramanian, K.S. Tun, A.S. Hamouda, M. Gupta, Synthesis and characterization of nano boron nitride reinforced magnesium composites produced by the microwave sintering method. Materials **6**(5), 1940–1955 (2013). https://doi.org/10.3390/ma6051940
66. M. Paramsothy, J. Chan, R. Kwok, M. Gupta, Enhanced mechanical response of hybrid alloy AZ31/AZ91 based on the addition of Si_3N_4 nanoparticles. Mater. Sci. Eng. A (2011). https://doi.org/10.1016/j.msea.2011.05.003
67. M. Paramsothy, J. Chan, R. Kwok, M. Gupta, Enhanced mechanical response of magnesium alloy ZK60A containing Si_3N_4 nanoparticles. Compos. Part Appl. Sci. Manuf. (2011). https://doi.org/10.1016/j.compositesa.2011.09.019
68. G.K. Meenashisundaram, M.H. Nai, A. Almajid, M. Gupta, Reinforcing low-volume fraction nano-tin particulates to monolithical, pure MG for enhanced tensile and compressive response. Materials (2016). https://doi.org/10.3390/ma9030134
69. G. Radhakrishnan, S. Mathialagan, S. Shankaranarayanan, J. Paulmar Pushparaj, Investigation on the machinability of TiN nano particulate reinforced magnesium matrix composite. Int. Conf. Process. Charact. Mater. **72**, 2281–2288 (2023). https://doi.org/10.1016/j.matpr.2022.09.217
70. H. Saravanan, D. Veeman, D. Srinivasan, M. Ravichandran, Influence of titanium nitride (TiN) particle addition on the mechanical, corrosion, and biocompatibility properties of ZK60A magnesium alloy composites synthesized via powder metallurgy (PM) technique. Ceram. Int. (2024). https://doi.org/10.1016/j.ceramint.2024.12.148
71. R. Vidu, M. Rahman, M. Mahmoudi, M. Enachescu, T.D. Poteca, I. Opris, *Nanostructures: a platform for brain repair and augmentation* (2014). https://doi.org/10.3389/fnsys.2014.00091
72. C.S. Goh, J. Wei, L.C. Lee, M. Gupta, Development of novel carbon nanotube reinforced magnesium nanocomposites using the powder metallurgy technique. Nanotechnology (2006). https://doi.org/10.1088/0957-4484/17/1/002
73. J. Hou, W. Du, G. Parande, M. Gupta, S. Li, Significantly enhancing the strength + ductility combination of Mg-9Al alloy using multi-walled carbon nanotubes. J. Alloys Compd. (2019). https://doi.org/10.1016/j.jallcom.2019.03.243
74. M. Zhou et al., The effects of carbon nanotubes on the mechanical and wear properties of AZ31 alloy. Materials (2017). https://doi.org/10.3390/ma10121385
75. M. Paramsothy, J. Chan, R. Kwok, M. Gupta, Addition of CNTs to enhance tensile/compressive response of magnesium alloy ZK60A. Compos. Part Appl. Sci. Manuf. (2011). https://doi.org/10.1016/j.compositesa.2010.11.001
76. M. Paramsothy, X.H. Tan, J. Chan, R. Kwok, M. Gupta, Carbon nanotube addition to concentrated magnesium alloy AZ81: enhanced ductility with occasional significant increase in strength. Mater. Des. (2013). https://doi.org/10.1016/j.matdes.2012.09.001
77. S.Y. Liu, F.P. Gao, Q.Y. Zhang, X. Zhu, W.Z. Li, Fabrication of carbon nanotubes reinforced AZ91D composites by ultrasonic processing. Trans. Nonferrous Met. Soc. China Engl. Ed. (2010). https://doi.org/10.1016/S1003-6326(09)60282-X
78. Q. Li, B. Tian, Compression behavior of magnesium/carbon nanotube composites. J. Mater. Res. (2013). https://doi.org/10.1557/jmr.2013.167

79. S.J. Huang, A. Abbas, B. Ballóková, Effect of CNT on microstructure, dry sliding wear and compressive mechanical properties of AZ61 magnesium alloy. J. Mater. Res. Technol. (2019). https://doi.org/10.1016/j.jmrt.2019.07.037
80. M.H. Nai, J. Wei, M. Gupta, Interface tailoring to enhance mechanical properties of carbon nanotube reinforced magnesium composites. Mater. Des. (2014). https://doi.org/10.1016/j.matdes.2014.04.011
81. L.Y. Chen et al., Novel nanoprocessing route for bulk graphene nanoplatelets reinforced metal matrix nanocomposites. Scr. Mater. (2012). https://doi.org/10.1016/j.scriptamat.2012.03.013
82. S.L. Xiang, M. Gupta, X.J. Wang, L.D. Wang, X.S. Hu, K. Wu, Enhanced overall strength and ductility of magnesium matrix composites by low content of graphene nanoplatelets. Compos. Part Appl. Sci. Manuf. **100**, 183–193 (2017). https://doi.org/10.1016/j.compositesa.2017.05.011
83. K. Munir, C. Wen, Y. Li, Graphene nanoplatelets-reinforced magnesium metal matrix nanocomposites with superior mechanical and corrosion performance for biomedical applications. J. Magnes. Alloys (2020). https://doi.org/10.1016/j.jma.2019.12.002
84. M. Rashad, F. Pan, H. Hu, M. Asif, S. Hussain, J. She, Enhanced tensile properties of magnesium composites reinforced with graphene nanoplatelets. Mater. Sci. Eng. A (2015). https://doi.org/10.1016/j.msea.2015.02.002
85. M. Rashad et al., Improved strength and ductility of magnesium with addition of aluminum and graphene nanoplatelets (Al+GNPs) using semi powder metallurgy method. J. Ind. Eng. Chem. (2015). https://doi.org/10.1016/j.jiec.2014.08.024
86. M. Rashad, F. Pan, M. Asif, A. Tang, Powder metallurgy of Mg–1%Al–1%Sn alloy reinforced with low content of graphene nanoplatelets (GNPs). J. Ind. Eng. Chem. **20**(6), 4250–4255 (2014). https://doi.org/10.1016/j.jiec.2014.01.028
87. M. Rashad, F.S. Pan, M. Asif, A. Ullah, Improved mechanical properties of magnesium-graphene composites with copper-graphene hybrids. Mater. Sci. Technol. (2015). https://doi.org/10.1179/1743284714Y.0000000726
88. L. Meng et al., Graphene nanoplatelets reinforced Mg matrix composite with enhanced mechanical properties by structure construction. Mater. Sci. Eng. A (2018). https://doi.org/10.1016/j.msea.2018.07.056
89. X. Du, W. Du, Z. Wang, K. Liu, S. Li, Ultra-high strengthening efficiency of graphene nanoplatelets reinforced magnesium matrix composites. Mater. Sci. Eng. A (2018). https://doi.org/10.1016/j.msea.2017.11.040
90. L. Chen et al., Development of AZ91D magnesium alloy-graphene nanoplatelets composites using thixomolding process. J. Alloys Compd. (2019). https://doi.org/10.1016/j.jallcom.2018.11.148
91. S. Ramezanzade, G.R. Ebrahimi, M.T. Parizi, H.R. Ezatpour, Microstructure and mechanical characterizations of graphene nanoplatelets-reinforced Mg–Sr–Ca alloy as a novel composite in structural and biomedical applications. J. Compos. Mater. (2020). https://doi.org/10.1177/0021998319867464
92. N. Eliaz, N. Metoki, Calcium phosphate bioceramics: a review of their history, structure, properties, coating technologies and biomedical applications. Materials (2017). https://doi.org/10.3390/ma10040334
93. M. Haghshenas, Mechanical characteristics of biodegradable magnesium matrix composites: a review. J. Magnes. Alloys (2017). https://doi.org/10.1016/j.jma.2017.05.001
94. K.A. Khalil, A new-developed nanostructured Mg/HAp nanocomposite by high frequency induction heat sintering process. IOP Conf. Ser. Mater. Sci. Eng. (2012). https://doi.org/10.1088/1757-899X/40/1/012031
95. M. Khodaei, F. Nejatidanesh, M.J. Shirani, S. Iyengar, H. Sina, O. Savabi, Magnesium/nano-hydroxyapatite composite for bone reconstruction: the effect of processing method. J. Bionic Eng. (2020). https://doi.org/10.1007/s42235-020-0007-6
96. G. Cao et al., Microstructure and properties of nano-hydroxyapatite reinforced WE43 a lloy fabricated by friction stir processing. Materials (2019). https://doi.org/10.3390/ma12182994

97. M. Razavi, M.H. Fathi, M. Meratian, Fabrication and characterization of magnesium-fluorapatite nanocomposite for biomedical applications. Mater. Charact. (2010). https://doi.org/10.1016/j.matchar.2010.09.008
98. M. Razavi, M.H. Fathi, M. Meratian, Bio-corrosion behavior of magnesium-fluorapatite nanocomposite for biomedical applications. Mater. Lett. (2010). https://doi.org/10.1016/j.matlet.2010.07.079
99. M. Razavi, M.H. Fathi, M. Meratian, Microstructure, mechanical properties and bio-corrosion evaluation of biodegradable AZ91-FA nanocomposites for biomedical applications. Mater. Sci. Eng. A (2010). https://doi.org/10.1016/j.msea.2010.07.063
100. G. Parande, V. Manakari, H. Gupta, M. Gupta, Magnesium-β-tricalcium phosphate composites as a potential orthopedic implant: a mechanical/damping/immersion perspective. Metals (2018). https://doi.org/10.3390/met8050343
101. X.L. Zhong, W.L.E. Wong, M. Gupta, Enhancing strength and ductility of magnesium by integrating it with aluminum nanoparticles. Acta Mater. **55**(18), 6338–6344 (2007). https://doi.org/10.1016/j.actamat.2007.07.039
102. W.L.E. Wong, M. Gupta, Development of Mg/Cu nanocomposites using microwave assisted rapid sin tering. Compos. Sci. Technol. **67**(7–8), 1541–1552 (2007). https://doi.org/10.1016/j.compscitech.2006.07.015
103. G.K. Meenashisundaram, M. Gupta, Low volume fraction nano-titanium particulates for improving the mechanical response of pure magnesium. J. Alloys Compd. (2014). https://doi.org/10.1016/j.jallcom.2013.12.157
104. H. Yu, Y. Sun, L. Hu, Z. Wan, H. Zhou, Microstructure and properties of mechanically milled AZ61 powders dispersed with submicron/nanometer Ti particulates. Mater. Charact. **127**, 272–278 (2017). https://doi.org/10.1016/j.matchar.2017.03.017
105. M. Razzaghi, M. Kasiri-Asgarani, H.R. Bakhsheshi-Rad, H. Ghayour, Microstructure, mechanical properties, and in-vitro biocompatibility of nano-NiTi reinforced Mg–3Zn-0.5Ag alloy: Prepared by mechanical alloying for implant applications. Compos. Part B Eng. **190**, 107947 (2020). https://doi.org/10.1016/j.compositesb.2020.107947
106. G. Parande, V. Manakari, S. Wakeel, M.S. Kujur, M. Gupta, Enhancing mechanical response of monolithic magnesium using nano-NiTi (Nitinol) particles. Metals (2018). https://doi.org/10.3390/met8121014

Chapter 6
Magnesium Composites Containing Hybrid Reinforcements

This chapter provides a detailed review of magnesium-based metal matrix composites (MMCs) enhanced with hybrid reinforcements, focusing on diverse reinforcement strategies, including fibers, particulates, and nanoscale additions. It examines their effects on key properties such as strength, stiffness, thermal stability, and creep resistance. Key studies explore the influence of reinforcement combinations, fabrication techniques (e.g., squeeze casting, powder metallurgy, friction stir processing), and microstructural factors on the overall performance of these composites. Challenges such as reduced ductility and corrosion resistance are also addressed, alongside innovative solutions like hierarchical reinforcements and in-situ synthesis. This chapter underscores the potential of hybrid magnesium composites for structural, automotive, and aerospace applications, providing valuable insights into fabrication methodologies, property optimization, and future research directions.

6.1 Introduction

Magnesium-based hybrid composites are an advanced class of materials that integrate multiple types of reinforcements—such as fibers, particulates, and nanoscale additions—into a magnesium matrix. By leveraging the unique advantages of each reinforcement type, these composites achieve synergistic improvements in mechanical, thermal, and functional properties, making them ideal for lightweight and high-performance applications. To address inherent challenges like reduced ductility and corrosion resistance, recent advancements in fabrication techniques—such as squeeze casting, powder metallurgy, and friction stir processing—play a critical role in optimizing microstructure and overall performance. Emerging strategies, including hierarchical reinforcements and in situ synthesis, further enhance their capabilities and broaden their potential applications.

S. Seetharaman et al., *Magnesium-based Syntactic, Metastable and Nano-composites*, https://doi.org/10.1007/978-3-031-96514-2_6

6.2 Magnesium-Based Hybrid Composites Containing Fibers

The research efforts on Mg-based metal matrix composites with hybrid reinforcements have evolved significantly over the years, beginning with the early work by Schröder and Kainer [1]. They explored liquid infiltration method to incorporate (10 vol.%) Al_2O_3 short saffil fibers and 15 vol.% SiC particles into a high-temperature Mg alloy, Mg-2.5Ag-2RE-0.6Zr (Fig. 6.1). The addition of hybrid reinforcement led to significant improvements in mechanical properties, thermal stability, and cost efficiency compared to traditional Mg composites. Microstructural analysis revealed uniform reinforcement dispersion, with fine precipitates along the fiber–matrix interface, contributing to a reduced grain size and enhanced strength. Hardness increased by 70%–85% compared to the unreinforced matrix alloy, with directional dependence in fiber orientation. Mechanical testing highlighted the hybrid composite's superior performance, with a 0.2% proof stress of 330 MPa and ultimate tensile strength of 240 MPa at room temperature, compared to the unreinforced matrix's 190 MPa and 150 MPa, respectively. The bending strength reached 560 MPa, a 51% improvement over the 370 MPa of the matrix alloy. Young's modulus rose from 45 to 77 GPa for the hybrid composite, enhancing stiffness. Additionally, the composites also exhibited isotropic properties, alongside reduced thermal expansion, and minimized chemical reactions during infiltration, due to the low silica content of the Saffil fibers.

Kumar et al. [2] investigated the thermal cycling behavior of AE42 Mg alloy composites reinforced with varying combinations of saffil fibers and SiC particulates in the longitudinal direction. The study revealed that adding SiC particulates significantly reduced the coefficient of thermal expansion (CTE), with a maximum reduction observed in composites containing 10 vol.% saffil fibers and 15 vol.% SiC. This reduction, attributed to complex interactions between the fibers and particulates, exceeded predictions based on the rule of mixtures. Additionally, thermal cycling introduced residual strain and hysteresis due to compressive plastic deformation at approximately 250 °C, which was most pronounced in the initial cycles. In the subsequent study [3], they examined these composites in the transverse direction, confirming similar trends in CTE reduction and residual strain behavior. The transverse CTE values were consistently lower than theoretical predictions due to the planar random orientation of the saffil fibers. The inclusion of SiC particulates further enhanced the thermal stability, with CTE values decreasing by up to 14.0×10^{-6} °C^{-1} in hybrids containing 10 vol.% saffil and 15 vol.% SiC (Table 6.1). These studies also highlighted that no significant degradation occurred at the matrix–reinforcement interface, even after thermal cycling, supporting the long-term stability of the composites for high-temperature applications.

Mondal and Kumar [4] examined the longitudinal creep response of AE42/Saffil-SiC hybrid composites, revealing that the composites exhibited a transition from normal to reverse creep behavior with increasing temperature and stress levels (Fig. 6.2). At 175 °C, normal creep was observed at all stresses, while at 300 °C, reverse creep occurred due to fiber breakage. The steady-state creep rate was lower

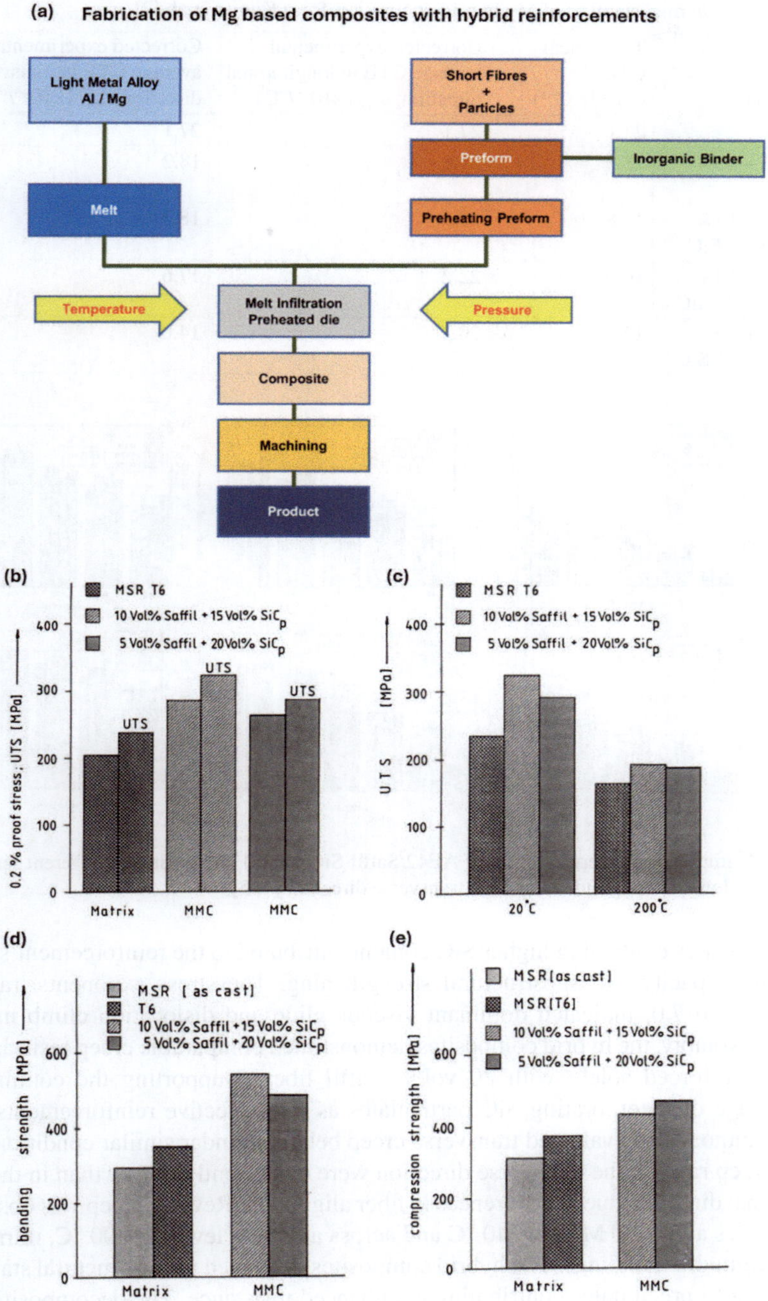

Fig. 6.1 Processing (**a**) and properties (**b–e**) of hybrid Mg composites as reported by Schröder and Kainer [1]

Table 6.1 Thermal stability of Mg hybrid composites from Kumar et al. [2]

Materials	Theoretical CTE ($\times 10^{-6}$/°C^{-1})	Corrected experimental average CTE in longitudinal direction, α_{mT} ($\times 10^{-6}$/°C)	Corrected experimental average CTE in transverse direction, α_{mL} ($\times 10^{-6}$/°C)
AE42	27.1	27.1	27.1
AE42 + 20% saffil	15.2	21.8	18.2
AE42 + 15% saffil+5% SiC	15.8	22.5	18.3
AE42 + 10% saffil+10% SiC	16.9	22.2	17.6
AE42 + 10% saffil+15% SiC	15.7	20.3	14.0

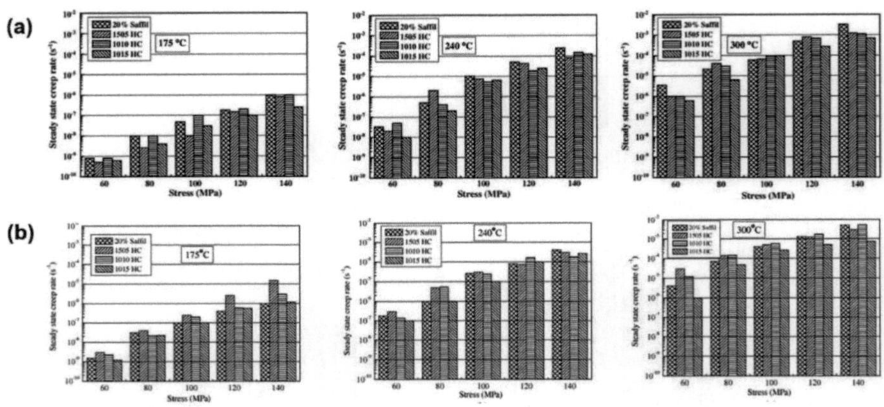

Fig. 6.2 Impression creep response of AE42/Saffil-SiC hybrid composites at different temperatures (**a**) along the longitudinal and (**b**) transverse direction [4, 5]

in composites containing higher SiC content, attributed to the reinforcement's load-bearing capacity and substructural strengthening. The stress exponents, ranging from 3.9 to 7.0, indicated dominant viscous glide and dislocation climb mechanisms. Notably, the hybrid composites demonstrated comparable creep resistance to those reinforced solely with 20 vol.% Saffil fibers, supporting the commercial advantage of incorporating SiC particulates as cost-effective reinforcements. The same authors also evaluated transverse creep behavior under similar conditions [5]. The creep rates in the transverse direction were consistently higher than in the longitudinal direction due to differences in fiber alignment. Reverse creep was observed at stresses above 80 MPa at 240 °C and across all stress levels at 300 °C, mirroring the longitudinal findings. The hybrid composites displayed microstructural stability, with Mg_2Si precipitates contributing to enhanced resistance. The decomposition of the Al_4RE phase into Al_2RE and additional precipitation during creep improved matrix stability at elevated temperatures. Despite directional differences, the hybrid composites maintained comparable performance to their longitudinal counterparts,

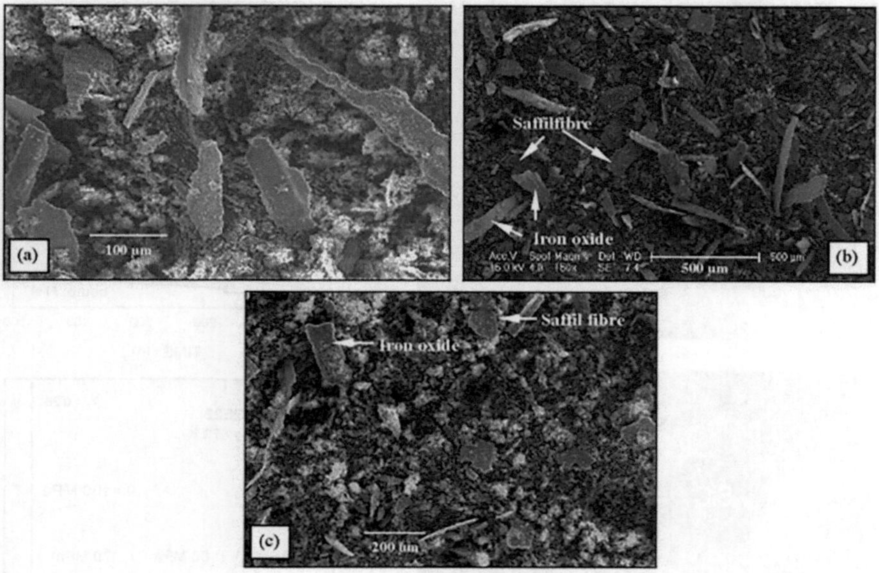

Fig. 6.3 SEM micrographs of the wear debris generated at 40 N load for the (**a**) AE42 alloy; (**b**) 20% Saffil composite and (**c**) 1015 HC. (Reported by Mondal et al. [6, 7])

reinforcing the effectiveness of combining Saffil fibers and SiC particulates for high-temperature applications.

Mondal and Kumar also investigated longitudinal and transverse wear behavior of the above-mentioned hybrid composites [6, 7]. Along the longitudinal direction, the composite containing 10 vol.% SSF and 15 vol.% SiC (1015 HC) exhibited the lowest wear rate with 34% lower wear rate than the 20% SSF composite [6]. The dominant wear mechanism in these composites was abrasion, accompanied by fiber fracture and iron-rich transfer layer formation at higher loads (Fig. 6.3). The wear resistance in transverse direction was consistently lower than along the longitudinal direction due to lower load-bearing capacity and increased fiber fracture and pull-out due to the fiber orientation relative to the counter-face [7]. Among the composites, the 1015 HC composite once again outperformed others, showing a 35% reduction in wear rate compared to the 20% SSF composite. The Microstructural analysis revealed intact SiC particles in the worn subsurface, further supporting their role in enhancing wear resistance through effective load transfer and preventing severe deformation.

Svoboda et al. [8] examined the microstructure and creep behavior of hybrid composites made from Mg alloys (AZ91 and QE22) reinforced with short carbon fibers and SiC particles using squeeze casting. The AZ91 hybrid composites demonstrated improved creep resistance compared to the unreinforced alloy (Fig. 6.4). In particular, the composite with 8 vol.% carbon fibers and 16 vol.% SiC particles exhibited the best performance, attributed to effective load transfer and stress

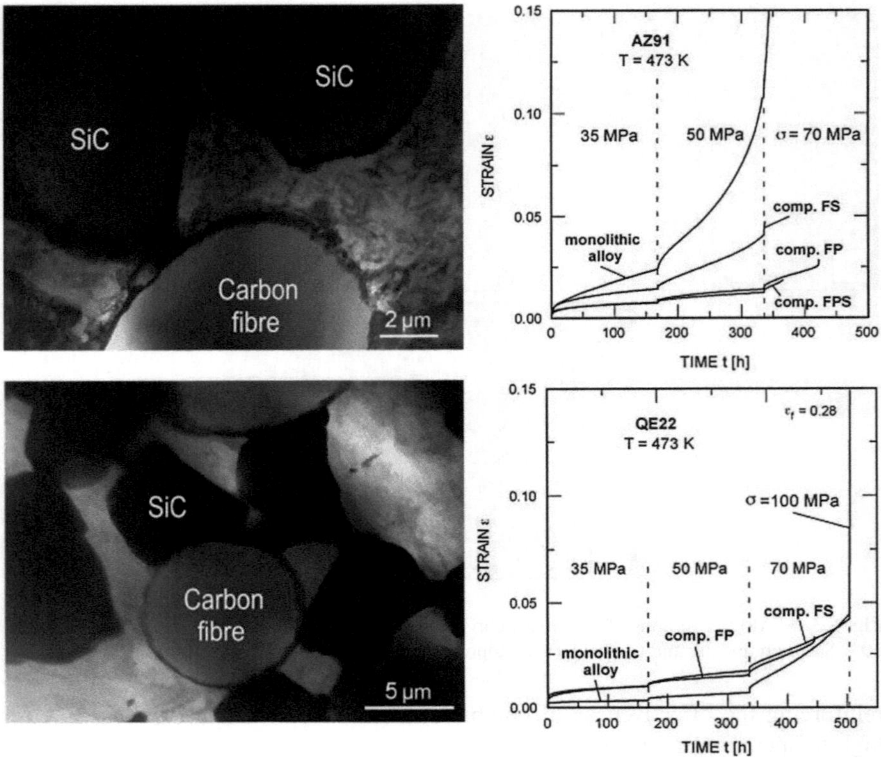

Fig. 6.4 TEM images and creep curves of SiC and carbon fiber reinforced AZ91 and QE22 composites [8]

redistribution within the matrix. The AZ91 composite containing 7 vol.% carbon fibers and 4 vol.% Si showed a moderate improvement in creep resistance due to the formation of Mg_2Si during infiltration. Microstructural analysis revealed the presence of $Mg_{17}Al_{12}$ precipitates, thicker reaction zones of MgO at fiber–matrix interfaces, and clusters of Mg_2Si particles, which contributed to the composite's creep strengthening. By contrast, the QE22-based hybrid composites displayed minimal improvement in creep resistance compared to the unreinforced matrix. This was attributed to weaker interfacial bonding between carbon fibers and the matrix, debonding at interfaces, and the formation of Mg_3(Ag, Nd) and Mg_2Si precipitates that did not effectively constrain plastic deformation. Fractographic analysis confirmed that interfacial debonding and fiber breakage were primary failure modes under creep conditions. The study concluded that the AZ91 hybrid composites benefitted more significantly from hybrid reinforcement than the QE22 counterparts, emphasizing the importance of selecting appropriate matrix–reinforcement combinations to optimize creep performance for high-temperature applications.

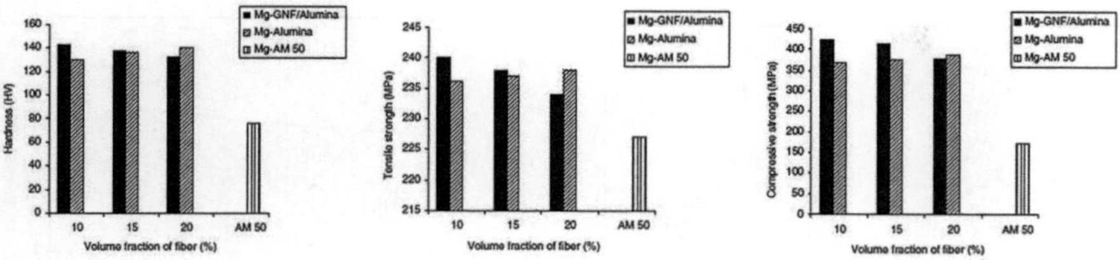

Fig. 6.5 Mechanical properties of Al_2O_3 short fibers and graphite nanofibers reinforced AM50 composites [9]

Babu et al. [9] investigated the mechanical properties of Mg-based hybrid composites reinforced with graphite nanofibers (GNFs) and alumina short fibers (Al_2O_3sf), fabricated using an infiltration method. The study revealed that these hybrid composites achieved optimal mechanical properties at a fiber volume fraction of 10%, with microhardness reaching 150 HV, tensile strength increasing to 240 MPa, and compressive strength improving to 390 MPa (Fig. 6.5). SEM analysis confirmed uniform dispersion of Al_2O_3sf within the Mg matrix and the presence of $Mg_{17}Al_{12}$ precipitates, which enhanced interfacial bonding and contributed to the mechanical strength. However, GNFs exhibited a tendency to agglomerate at higher volume fractions, leading to porosity and reduced mechanical properties when fiber content exceeded 10%. The strong interfacial bonding between GNFs, Al_2O_3sf, and the Mg matrix enabled effective load sharing, significantly improving the composites' performance compared to monolithic Mg or single-reinforced composites. In the complementary study [10], nanoindentation testing was used to evaluate the mechanical properties of aluminum hybrid composites reinforced with GNFs and Al_2O_3sf. The results showed that hardness and elastic modulus were highest near the GNFs/Al_2O_3sf interface, with values of 4.9 GPa and 95 GPa, respectively. The formation of Al_4C_3 phases near the interface contributed to these enhancements. Load-displacement profiles indicated that the GNF regions exhibited high elastic recovery due to their inherent stiffness, while the matrix regions demonstrated plastic deformation. Despite the improvements, localized yielding and nanopores at higher fiber contents were identified as factors limiting performance.

Zhang and Hu [11] studied Mg alloy AM60-based hybrid composites reinforced with Al_2O_3 short fibers and particles fabricated via a preform-squeeze casting method under 90 MPa pressure. The microstructural analysis showed uniform dispersion of fibers and particles without agglomeration, contributing to enhanced mechanical properties. The addition of 4 vol.% Al_2O_3 particles and 9 vol.% Al_2O_3 fibers reduced the matrix grain size by over 50%, with refined grains promoting improved mechanical performance (Fig. 6.6). Compared to unreinforced AM60 matrix, the hybrid composite achieved an 88% increase in yield strength (150 MPa), a 16% rise in ultimate tensile strength (199 MPa), and a 40% improvement in elastic modulus (59 GPa). However, elongation to failure decreased significantly to 1.5%, indicating a trade-off between strength and ductility. Hardness measurements

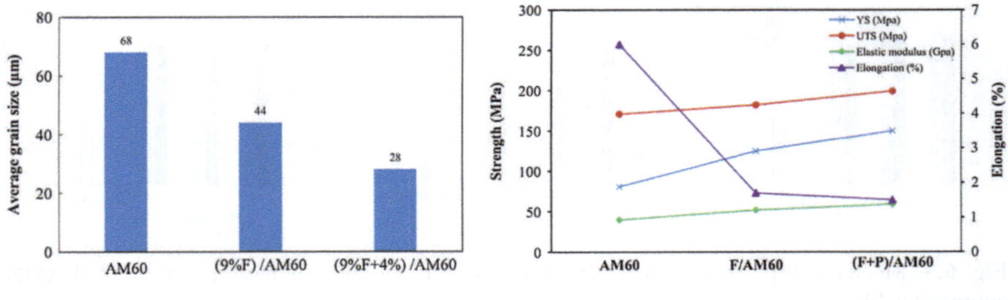

Fig. 6.6 Microstructure and mechanical properties of AM60 hybrid composites reported by Zhang et al. [11]

revealed a substantial increase, attributed to the hard ceramic reinforcements and their constraint on localized deformation. Fractographic analysis highlighted the brittle nature of the hybrid composites, with predominant failure modes including matrix cracking, interface debonding, and fiber breakage. The reinforcement combination leveraged the advantages of fibers for load transfer and particles for grain refinement, resulting in a composite with enhanced stiffness and strength.

Furthermore, Mondal et al. [12] investigated the corrosion behavior of AE42 Mg alloy-based hybrid composites reinforced with Saffil short fibers and SiC particulates, fabricated using direct squeeze casting. The study evaluated four reinforcement combinations, including: (i) 20 vol.% fibers, (ii) 15 vol.% fibers with 5 vol.% SiC (1505 HC), (iii) 10 vol.% fibres with 10 vol.% SiC (1010 HC), and (iv) 10 vol.% fibers with 15 vol.% SiC (1015 HC). Results indicated that the unreinforced AE42 alloy exhibited the best corrosion resistance with a rate of 1 mm/year due to the formation of a dense and stable quasi-passive film. Among the composites, the 20% SSF composite showed the lowest corrosion rate of 3 mm/year, attributed to its more uniform surface film compared to the hybrids. Hybrid composites exhibited higher corrosion rates, with the 1010 HC, 1505 HC, and 1015 HC composites showing rates of 3.4 mm/year, 4.3 mm/year, and 5 mm/year, respectively. The degradation of corrosion resistance in hybrid composites was primarily attributed to defects in the quasi-passive film caused by matrix–reinforcement interfacial discontinuities and increased impurity levels introduced during fabrication. Galvanic coupling between the matrix and reinforcements, as well as changes in grain boundary precipitation, were found to have minimal impact. Microscopic analysis revealed irregular and loosely adherent corrosion films on composite surfaces, further confirming the role of surface defects in the observed corrosion behavior. This study highlighted the trade-off between improved mechanical properties and reduced corrosion resistance in hybrid composites, emphasizing the need for optimized fabrication techniques to minimize surface defects and impurities for improved durability in structural applications.

6.3 Magnesium-Based Hybrid Composites Containing Various Particle Reinforcements

Zhou et al. [13] examined the properties of AZ61 alloy composites reinforced with 5 vol.% SiC particles and varying contents of CNTs. Here, the composites were developed using powder metallurgy method followed by hot extrusion. The AZ61-5SiC-0.5CNT composite exhibited ultra-high mechanical properties with a yield strength of 345 MPa, an ultimate tensile strength of 412 MPa, and a ductility of 8.0% (Fig. 6.7). These improvements were attributed to the synergistic effects of micron SiC particles and CNTs, which refined the grain size from 36.4 to 13.5 µm, induced heterogeneous nucleation, and enhanced interfacial bonding through in situ MgO formation. However, increasing CNT content beyond 0.5 vol.% led to clustering, resulting in reduced ductility and porosity. In a similar study, Zhou and colleagues developed AM60 Mg alloy hybrid composites reinforced with 3 vol.% nano-sized alumina (Al_2O_3) particles and 5 vol.% Al_2O_3 short fibers using preform-squeeze casting. The hybrid composites achieved significant grain refinement, reducing the matrix grain size from 68 to 20 µm. Mechanical properties were markedly enhanced, with a 73% increase in yield strength (140 MPa), a 26% improvement in ultimate tensile strength (216 MPa), and a restored ductility of 3.5% compared to the 6% elongation of the unreinforced AM60 alloy. The dual reinforcement leveraged the load-bearing capacity of fibers and grain refinement induced by nanoparticles, overcoming the traditional trade-off between strength and ductility. The same authors also developed AM60-based hybrid composites containing Al_2O_3 reinforcements in fiber and particle forms through a squeeze casting process [14]. In this study, while the combined inclusion of fibers and micron sized particles led to significant increases in ultimate tensile strength and yield strength, the ductility was reduced. However, when nano-sized Al_2O_3 particles were introduced, the ductility was restored. Similar results were seen when nano-sized AlN particles were used in place of Al_2O_3, demonstrating the potential of nano-reinforcements to improve both strength and ductility in hybrid composites.

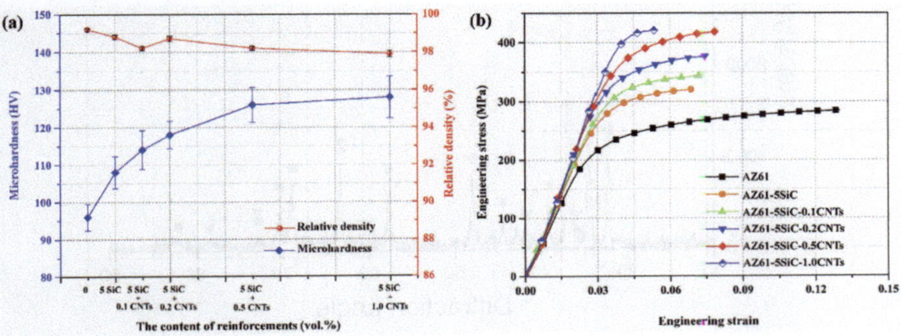

Fig. 6.7 Mechanical properties (**a–b**) of AZ61/SiC-CNT hybrid composites developed by Zhou et al. [13]

Recent literature has highlighted the in-situ synthesis of hybrid (TiB_2 + TiC) reinforcements through solid-state chemical reactions between Ti, B, and C phases. Ma et al. [15] fabricated AZ91 Mg alloy composites containing 5 wt.% (TiB_2 + TiC) particulates using a master alloy route with an Al-Ti-B_4C system via self-propagating high-temperature synthesis (SHS). The alloy was later re-melted and diluted using the stir casting method which resulted in a uniform distribution of fine particulates (0.8–2 μm) with strong matrix–reinforcement interfacial integrity. The chemical reactions involving Ti, Al, and B_4C phases were detailed as follows: ($3Ti + B_4C \rightarrow 2TiB_2 + TiC$; $\Delta G = -712{,}300 + 47.99$ T J/mol) and ($3Al + Ti \rightarrow Al_3Ti$; $\Delta G = -37{,}023 + 13.533$ T J/mol), with subsequent reactions forming TiB_2 and TiC. Mechanical tests revealed significant enhancements in hardness, which increased to 79 HB from 56 HB for unreinforced AZ91, and wear resistance, with volumetric wear rates decreasing by 57% under high loads. The hybrid reinforcements also contributed to improved thermal stability and dimensional stability during abrasive wear.

Xiuqing et al. [16] employed a remelting and dilution (RD) technique to synthesize AZ91 composites reinforced with 3.9 vol.% (TiB_2 + TiC) (Fig. 6.8). Microstructural analysis revealed a defect-free interface and uniform dispersion of particulates, resulting in enhanced tensile properties. Compared to T4 heat-treated AZ91, the composites achieved a 78% increase in 0.2% yield strength (95 MPa), a 28% rise in ultimate tensile strength (298 MPa), and a 17% improvement in elastic modulus (53 GPa). However, ductility decreased significantly from 6.0% to 2.4% due to the brittle nature of the ceramic reinforcements. Strengthening mechanisms

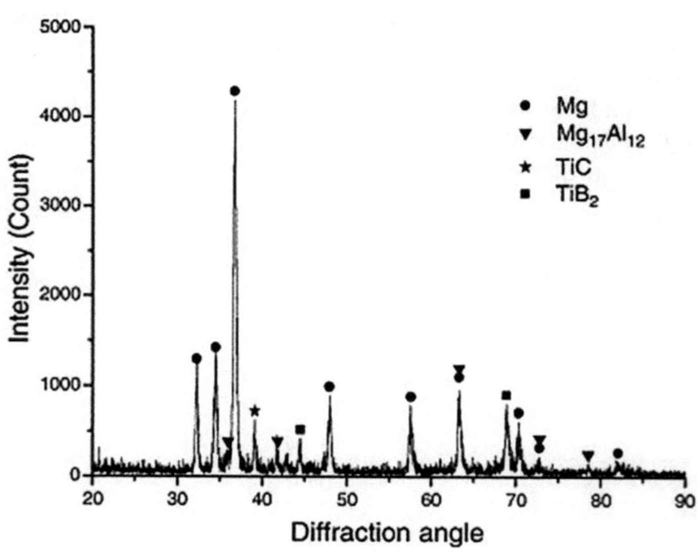

Fig. 6.8 XRD patterns showing TiB_2 and TiC phases in the composite material produced by Xiuqing et al. [16]

included grain refinement, Orowan looping, and improved load transfer between the matrix and particulates. Both studies underscored the effectiveness of in situ synthesized (TiB_2 + TiC) particulates in improving the mechanical and wear properties of Mg matrix composites. While Ma et al. [15] highlighted the wear resistance and thermal stability benefits for structural applications, Xiuqing et al. [16] focused on achieving superior tensile properties, albeit with reduced ductility.

Ugandhar et al. [17] studied Mg-based hybrid composites reinforced with submicron SiC and micron-sized Ti particulates, fabricated via the disintegrated melt deposition technique followed by hot extrusion. Two compositions, Mg/1.1Ti/2.8SiC and Mg/1.1Ti/8.8SiC, were evaluated for their microstructural, thermal, and mechanical properties. Microstructural analysis revealed uniform dispersion of SiC and Ti particles with strong matrix–reinforcement interfacial integrity, while the grain size decreased significantly with higher reinforcement content, from 21 μm in pure Mg to 15 μm in Mg/1.1Ti/8.8SiC. The hybrid composites showed enhanced thermal stability, with the coefficient of thermal expansion (CTE) decreasing from 28.02×10^{-6}/°C in pure Mg to 19.91×10^{-6}/°C in Mg/1.1Ti/8.8SiC, attributed to the low CTE of the reinforcements and their effective dispersion. Mechanical testing demonstrated a 38% increase in macrohardness and a 61% rise in microhardness for Mg/1.1Ti/8.8SiC compared to pure Mg. The 0.2% yield strength improved from 153 MPa in pure Mg to 217 MPa in Mg/1.1Ti/8.8SiC, while ultimate tensile strength rose to 233 MPa (Fig. 6.9). However, ductility decreased from 9.2% in pure Mg to 1.04% in the composite with higher reinforcement content. Strengthening mechanisms included load transfer from the matrix to the reinforcements, dislocation

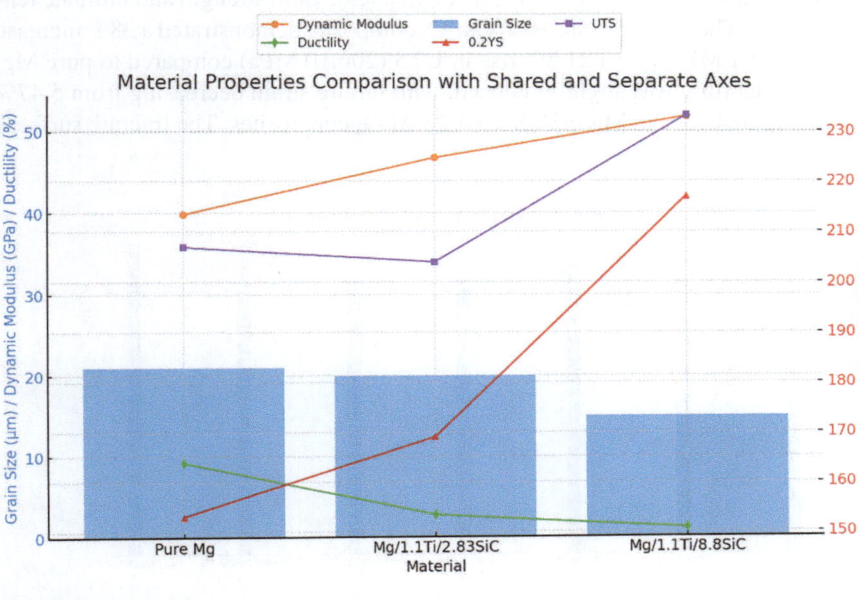

Fig. 6.9 Properties of Mg composites fabricated by Ugandhar et al. [17]

generation due to thermal mismatch, and grain boundary pinning. Ti reinforcements contributed significantly to strengthening through a load-sharing mechanism, as evidenced by fractured Ti particles under tensile loads.

Sinha et al. [18] further examined the scratch hardness and wear behavior of similar Mg/SiC and Mg/SiC/Ti hybrid composites (Fig. 6.10). The scratch hardness increased consistently with reinforcement content, reaching 969 MPa for Mg/1.1Ti/8.8SiC compared to 721 MPa for pure Mg under a 0.55 N load. SEM analysis revealed a transition in material removal modes, with Mg/SiC and Mg/SiC/Ti composites exhibiting a cutting mode of deformation, higher wear debris formation, and edge cracking compared to the ductile plowing observed in pure Mg. The study highlighted that while the addition of Ti marginally improved scratch resistance, the wear resistance remained influenced by brittle behavior and reinforcement content.

Thakur et al. [19, 20] investigated Mg-based hybrid composites containing nano-sized SiC and Al_2O_3 particles synthesized using a microwave-assisted powder metallurgy technique followed by hot extrusion. The addition of nano-sized reinforcements significantly enhanced the composites' mechanical and thermo-mechanical properties. Microstructural analysis showed a uniform dispersion of the reinforcements with minimal porosity, ensuring good interfacial bonding. Grain size reduced from 29.92 μm in pure Mg to 25.79 μm in Mg/0.3%SiC+0.7%Al_2O_3 composites, indicating effective grain refinement due to the reinforcement. The CTE decreased markedly due to hybrid reinforcements, with the lowest CTE value of 26.73×10^{-6}/°C observed in Mg/0.3%SiC+0.7%Al_2O_3 composites compared to 29.33×10^{-6}/°C for pure Mg. Mechanical testing revealed that the hybrid composites exhibited significant improvements in hardness, yield strength and ultimate tensile strength. The Mg/0.3%SiC+0.7%Al_2O_3 composite demonstrated a 38% increase in YS (164.91 MPa) and a 21.8% rise in UTS (206.10 MPa) compared to pure Mg. However, ductility was slightly reduced, with failure strain decreasing from 5.47% in pure Mg to 4.17% in Mg/0.3%SiC+0.7%Al_2O_3 composites. The fracture surfaces

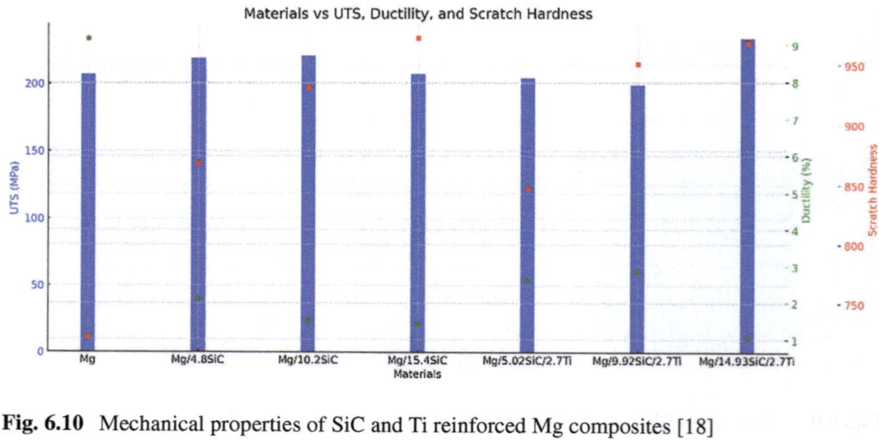

Fig. 6.10 Mechanical properties of SiC and Ti reinforced Mg composites [18]

of the composites revealed brittle failure modes with cleavage steps due to the inherent characteristics of the Mg matrix and the reinforcing particles.

Nguyen et al. [21–24] developed several hybrid composite formulations based on AZ31 alloy with nano-Al_2O_3 and either nickel or copper, by employing the DMD technique. Firstly, they investigated the properties of AZ31B composites with varying amounts of nano-Al_2O_3 and Ni [21]. Among the developed nanocomposites, the AZ31B–1.5Al_2O_3–1.5Ni composite demonstrated optimal performance, achieving a 25% increase in yield strength, a 13% rise in ultimate tensile strength, and a 62% improvement in hardness compared to the unreinforced matrix (Fig. 6.11). These enhancements were attributed to grain refinement, with a significant reduction in grain size, and uniform dispersion of reinforcements ensuring strong interfacial bonding. However, increasing the Ni content to 3.19 wt.% resulted in higher intermetallic content, leading to embrittlement that compromised ductility while further increasing strength. Similar increase in hardness and strength were also achieved when micrometer-sized Cu particles were added to AZ31B/1.5Al_2O_3 [22, 23]. The AZ31–4Cu–1.5Al_2O_3 composite displayed a remarkable 67% increase in YS (300 MPa), a 32% rise in UTS (350 MPa), and a 24% improvement in ductility (13.5%). The hybrid reinforcements reduced grain size to 0.3 μm and effectively balanced mechanical properties while further decreasing the CTE to

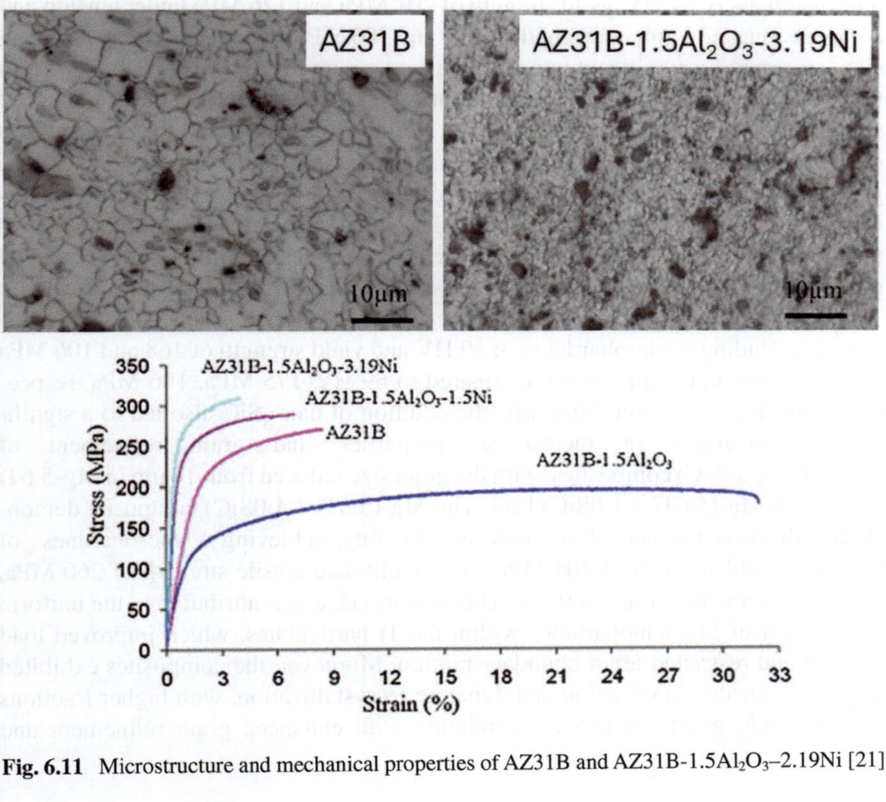

Fig. 6.11 Microstructure and mechanical properties of AZ31B and AZ31B-1.5Al_2O_3–2.19Ni [21]

22.15×10^{-6}/°C. Lastly, the amounts of both Cu and Al_2O_3 nanoparticles were modified and their effects in terms of microstructure and mechanical properties were studied [24]. The AZ31B–10Cu–3.3Al_2O_3 composite exhibited a 23% reduction in grain size (to 0.3 μm), alongside a 24% increase in YS (268 MPa), an 11% rise in UTS (294 MPa), and a 48% improvement in hardness (112 HV). The thermal expansion behavior also improved, with the CTE decreasing by 15%, enhancing the composite's stability at elevated temperatures.

Sankaranarayanan et al. [25–27] also conducted a series of studies exploring the effects of hybrid reinforcements on the microstructure and mechanical properties of Mg composites, employing the DMD technique followed by hot extrusion. One study evaluated the impact of mutually soluble and insoluble metallic reinforcements, namely Al and Ti, on Mg [27]. The addition of ball-milled Ti and Al powders to Mg resulted in a Mg/Ti-Al_3Ti composite with improved strength. However, direct addition of Al and Ti led to relatively lower strength and ductility, highlighting the critical role of Al_3Ti formation and the changes to reinforcement morphology due to ball milling. The Mg-$(3Al + 5.6Ti)_{BM}$ composite exhibited the highest performance, with a hardness of 93 HV, a UTS of 265 MPa, and exceptional thermal stability with the lowest coefficient of thermal expansion (26.13×10^{-6}/K). Another investigation examined the simultaneous addition of micron-sized Ti and nano-Cu particulates [25]. The Mg-$(Ti + Cu)_{BM}$ composite exhibited outstanding properties, including a microhardness of 91 HV, yield strength of 201 MPa and 126 MPa under tension and compression, and ultimate strength of 265 and 380 MPa under tension and compression, representing significant improvements over pure Mg. The ball milling process facilitated the formation of Ti_3Cu intermetallics, which contributed to enhanced strength while slightly reducing ductility. Further study indicated that the addition of 5.6 wt.% Ti particulates and 2.5 wt.% nanoscale Al_2O_3 significantly enhanced the properties of pure Mg [25]. While the direct addition of hybrid reinforcements improved strength at the expense of ductility, the pre-processed reinforcements led to enhanced ductility. The composite processed through ball milling (BM) displayed remarkable grain refinement, with grain sizes reduced from 21 μm in pure Mg to 7.64 μm. The BM-processed composites achieved superior mechanical performance, including a microhardness of 69 HV and yield strength of 168 and 106 MPa under tension and compression, compared to 69 HV, 175 MPa, 106 MPa, respectively, for direct addition. Similarly, the addition of nano-SiC also led to a significant enhancement in mechanical properties and grain refinement of Mg-(5.6Ti + x-SiC) composites, with the grain size reduced from 10 μm in Mg-5.6Ti to 6 μm in Mg-(5.6Ti + 1.0SiC) [28]. The Mg-(5.6Ti + 1.0SiC) composite demonstrated the best balance of strength and ductility, achieving a microhardness of 83 HV, a yield strength of 204 MPa, and an ultimate tensile strength of 260 MPa, alongside a fracture strain of 9.2%. This performance was attributed to the uniform distribution of SiC nanoparticles within the Ti particulates, which improved load transfer and restricted grain boundary motion. Moreover, the composites exhibited improved tensile twin fraction and dynamic recrystallization, with higher fractions of high-angle grain boundaries correlating with enhanced grain refinement and mechanical properties. In a parallel study involving Mg-(5.6Ti + x-B_4C) composites, nano-B_4C particles further refined the microstructure, reducing the grain size to 6 μm in Mg-(5.6Ti + 2.5B_4C) [29, 30]. This composite exhibited the highest mechanical properties, with a microhardness of 92 HV, YS of 215 MPa, and a UTS of 260 MPa (Fig. 6.12). The presence of Ti(B, C) intermetallic phases at the Mg/Ti

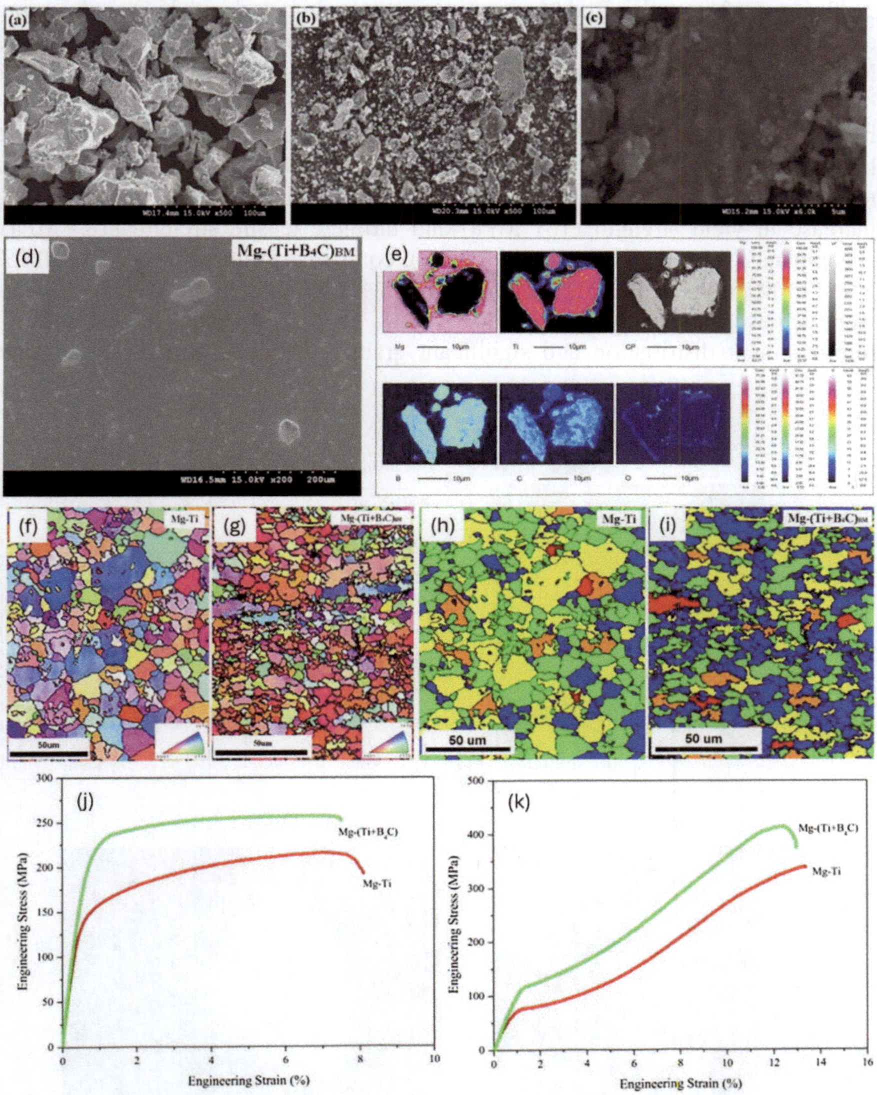

Fig. 6.12 SEM micrographs showing the microstructure of Ti- (**a**) and ball-milled (Ti + B_4C) powder (**b–c**), microstructure (**d–i**) and mechanical properties of Mg/Ti-B_4C hybrid nanocomposite under tension (**j**) and compression (**k**) loads [29]

interface, coupled with dynamic recrystallization and effective load transfer mechanisms, contributed to these improvements. Interestingly, the composites retained good ductility despite the strength enhancements, highlighting the benefits of hybrid reinforcement. Texture evolution studies revealed that nano-B_4C additions spread the basal texture and promoted recrystallization, further supporting ductility and mechanical performance.

Tun et al. [31] synthesized Mg nanocomposites reinforced with Y_2O_3 and varying amounts of Ni (0.3–1.0 vol.%) using PM and microwave sintering (Fig. 6.13). The Mg/(0.7Y_2O_3 + 0.6Ni) composite demonstrated the best performance, with a 35% increase in YS (205 MPa), a 23% increase in UTS (260 MPa), and improved ductility (6.8%). Hassan et al. [32] investigated the high-temperature properties of these hybrid composites containing nano-Y_2O_3 and nickel particles and found improved tensile properties up to 100 °C. The Mg/(Y_2O_3 + Ni) composite exhibited a notable increase in yield strength (161 MPa) and ultimate tensile strength (187 MPa), accompanied by enhanced ductility (7.2%), compared to unreinforced Mg (0.2% YS: 125 MPa, UTS: 143 MPa, elongation: 4.3%). Similar strength enhancements were achieved under compression loads [33]. The micrographs showed uniform reinforcement distribution and significant grain refinement, with the grain size

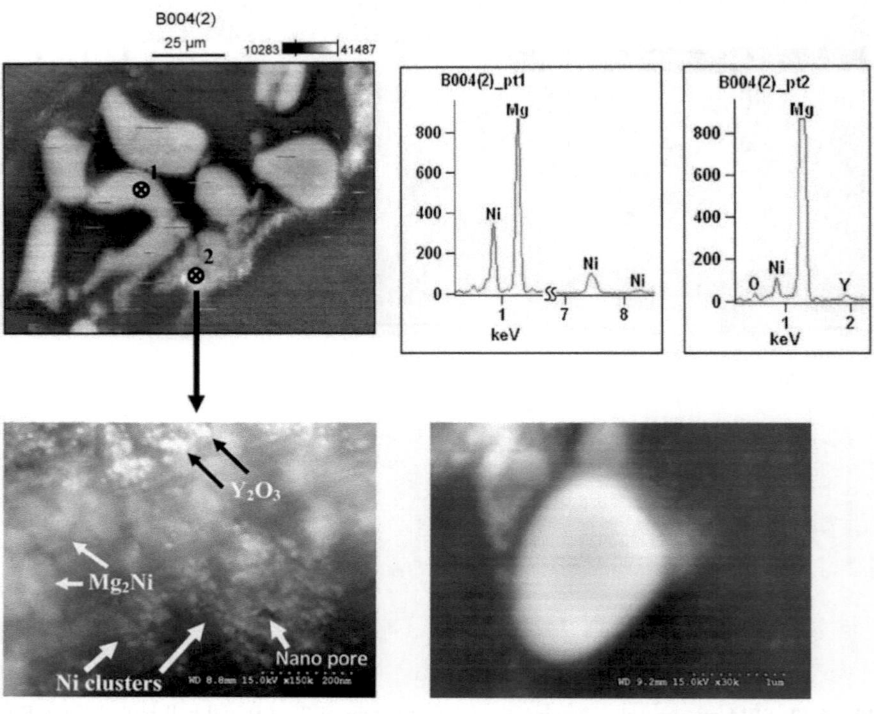

Fig. 6.13 Micrographs showing the presence of nickel, yttria, and Mg_2Ni intermetallics in Mg matrix [31]

reduced to 5 μm. The mechanical improvements were linked to the synergistic effects of Y_2O_3-induced dislocation generation and Ni-induced intermetallic formation (Mg_2Ni), which enhanced the load-bearing capacity and thermal stability of the composite. Additionally, the hybrid reinforcement enabled superior grain boundary pinning, contributing to improved thermal stability. However, as the testing temperature exceeded 150 °C, the strength benefits diminished due to thermal softening effects. The addition of Ni and TiO_2 also enhanced hardness and tensile properties [34], with similar results reported for copper and other metallic particles in Mg-based nanocomposites.

Tun et al. [35] also studied Mg composites reinforced with hybrid nano-ZrO_2 and metallic Cu reinforcements, synthesized using microwave-assisted PM. The Mg/(ZrO_2 + Cu) composite exhibited remarkable mechanical improvements, with a grain size reduction to 9 μm, microhardness increase to 50 HV (20% improvement), and tensile strength enhancements (0.2% YS: 180 MPa, UTS: 245 MPa) compared to pure Mg (Fig. 6.14). The presence of Cu facilitated the formation of Mg_2Cu intermetallics, contributing to grain refinement and strengthening through load transfer,

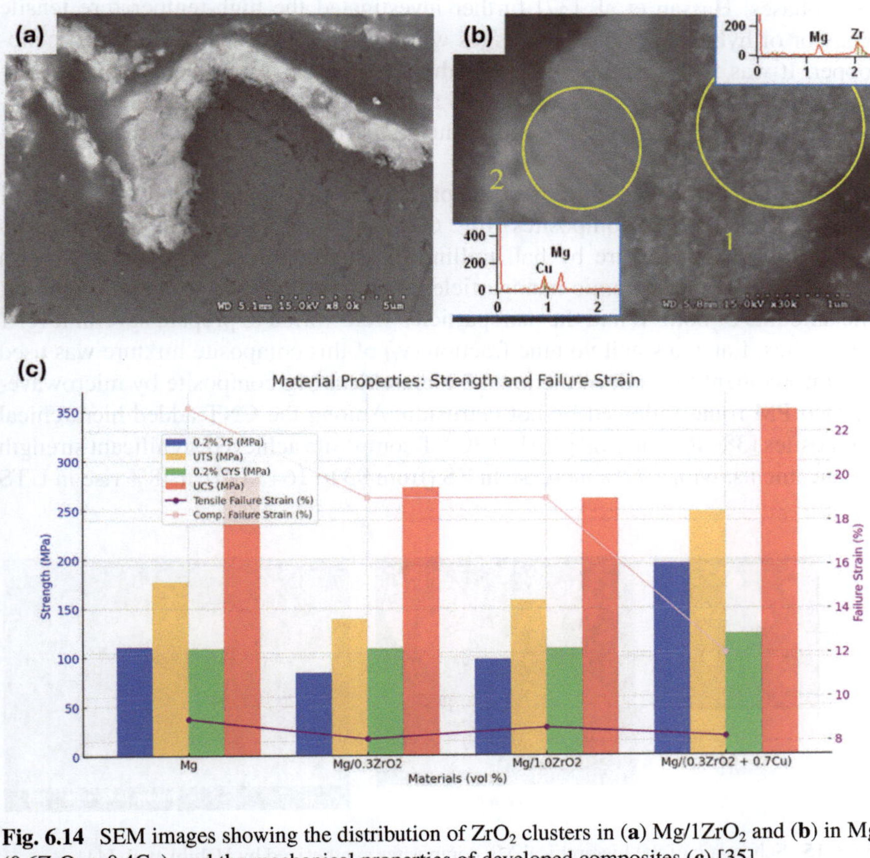

Fig. 6.14 SEM images showing the distribution of ZrO_2 clusters in (**a**) Mg/1ZrO_2 and (**b**) in Mg/(0.6ZrO_2 + 0.4Cu), and the mechanical properties of developed composites (**c**) [35]

Orowan, and Hall-Petch effects. The hybrid composite also demonstrated excellent compressive properties, with significant increases in ultimate compressive strength and work of fracture. However, the inclusion of Cu led to reduced ductility under compressive loading, attributed to the brittleness of the intermetallic phases.

Another study [36] explored the effects of varying Cu content (0.3–1.0 vol.%) on the microstructure and mechanical properties of 0.7 vol.% Al_2O_3 reinforced Mg nanocomposites. The Mg/(0.7Al_2O_3 + 0.6Cu) composite displayed the most balanced performance, achieving a significant grain size reduction to 6 μm, a 25% increase in microhardness (to 51 HV), and marked improvements in yield strength (175 MPa) and ultimate tensile strength (230 MPa). The addition of nano-Al_2O_3 enhanced grain boundary pinning, while Cu facilitated the formation of Mg_2Cu intermetallic phases, providing effective load transfer and increased dislocation density. Under compressive loading, the Mg/(0.7Al_2O_3 + 0.6Cu) composite achieved an ultimate compressive strength of 340 MPa, a 22% improvement compared to unreinforced Mg. Fracture surface analysis revealed ductile failure modes with microvoid coalescence, indicating improved toughness. However, at higher Cu content, ductility decreased slightly due to the brittleness introduced by excess intermetallic phases. Hassan et al. [37] further investigated the high-temperature tensile behavior of hybrid Mg nanocomposites with 0.7 vol.% yttria and 0.3 vol.% nanocopper. It was found in this study that the nano-copper reinforcement effectively maintained the strengthening effect even at temperatures up to 100 °C, but its efficacy diminished at higher temperatures due to thermal softening and reduced resistance to dislocation motion.

Habibi et al. [38–45] extensively explored the development and properties of Mg-based hierarchical composites (Fig. 6.15). In these studies, they developed a level 1 composite mixture by ball milling the submicron-sized metallic Al with either CNT or other ceramic nanoparticles such as Al_2O_3, Bi_2O_3 or B_4C particles. The amounts of both Al and the nanoparticles were varied to prepare several level 1 composites. Later, a small volume fraction (v_f) of this composite mixture was used as reinforcement to produce the level 2 hierarchical Mg composite by microwave-assisted PM route followed by hot extrusion. Among the CNT added hierarchical composites [38–40], the Mg/1.0Al–0.3CNT composite achieved significant strength enhancements, with a 72% increase in YS (from 95 to 164 MPa), a 48% rise in UTS

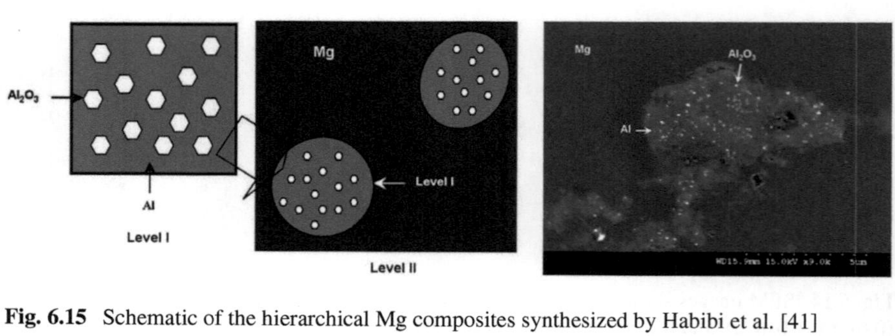

Fig. 6.15 Schematic of the hierarchical Mg composites synthesized by Habibi et al. [41]

(from 150 to 222 MPa), and an 80% improvement in compressive strength and the improvements were attributed to the uniform dispersion of Al-coated CNTs, which eliminated porosity and provided strong interfacial bonding. Similarly, among the nano-Al_2O_3 added composites, Mg/0.972Al–0.66Al_2O_3 with a refined grain structure demonstrated remarkable improvements in mechanical properties, including a 96% increase in yield strength, an 80% rise in ultimate tensile strength, and a 42% enhancement in ductility compared to unreinforced Mg [41–43]. For Bi_2O_3-reinforced composites [44], the Mg/0.92Al–0.22Bi_2O_3 hybrid exhibited a 27% increase in yield strength (138 MPa), along with improved thermal stability while the B_4C-reinforced Mg/0.92Al–0.66B_4C composite [45] demonstrated a superior tensile and compressive performance, achieving a 45% increase in ultimate tensile strength (from 165 to 240 MPa) and a 31% improvement in compressive strength.

Rashad et al. [46–48] conducted several studies focusing on Mg composites reinforced with a variety of hybrid- and nano-reinforcements, showcasing significant advancements in microstructural refinement, mechanical properties, and thermal stability. In one study, they explored Mg composites reinforced with hybrid Al_2O_3-SiC particulates synthesized using powder metallurgy. The Mg-3Al-1Zn-(1.5Al_2O_3–1.0SiC) composite demonstrated notable enhancements, including a 65.5% increase in compressive yield strength and a 7.5% rise in ultimate compressive strength compared to the base alloy. Microstructural analysis revealed a reduction in intermetallic phase size to 1–4 μm, contributing to improved hardness and overall mechanical strength. In another study, they investigated Mg composites reinforced with copper-graphene nanoplatelets (Cu-GNPs) hybrids [47] (Fig. 6.16). The Mg-1Cu-0.54GNP composite exhibited remarkable improvements in tensile and compressive behavior, with a 58% increase in ultimate tensile strength (UTS), a 61% rise in compressive strain, and significant grain refinement to 11 μm. The inclusion of Cu facilitated the formation of Mg_2Cu intermetallics, enhancing load transfer, while GNPs provided elasticity and improved fracture toughness. They also examined Ti-Al reinforced Mg composites fabricated via semi-powder metallurgy [48]. The Mg-10Ti-1Al composite achieved superior mechanical performance, including a 45% increase in UTS (238 MPa) and an enhancement in ductility to 21.2%. The addition of Ti refined grains to 10 μm, while Al improved matrix compatibility and reduced intermetallic formation, enabling a balance between strength and ductility.

Aatthisugan et al. [49] investigated the mechanical and wear behavior of AZ91D Mg matrix hybrid composites reinforced with boron carbide and graphite particles, manufactured via stir casting. The hybrid composite containing 1.5% B_4C and 1.5% Gr displayed superior performance compared to the unreinforced AZ91D alloy and AZ91D-B_4C composite. Microstructural analysis revealed uniform distribution of B_4C and Gr particles within the matrix, contributing to enhanced properties. The density of the hybrid composite increased to 1.8249 g/cm^3 compared to 1.8091 g/cm^3 for pure AZ91D, while porosity decreased to 1.11%. Mechanical testing showed that the hardness of the hybrid composite (22.5 BHN) was slightly lower than that of the AZ91D-B_4C composite (27.1 BHN) but higher than pure AZ91D (20.5 BHN). Ultimate tensile strength (UTS) was recorded at 86.3 MPa for the hybrid composite,

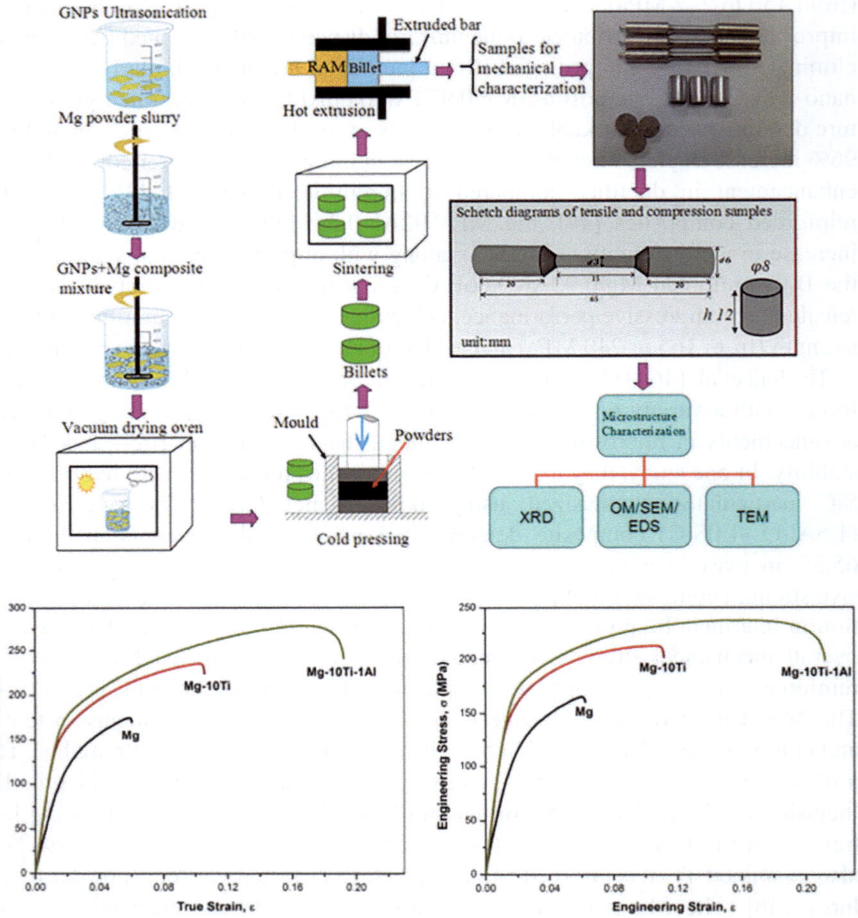

Fig. 6.16 Processing and properties of Mg hybrid composites from Rashad et al. [48]

compared to 99.2 MPa for the AZ91D-B_4C composite and 62.1 MPa for AZ91D alloy. The presence of Gr, acting as a solid lubricant, significantly enhanced wear resistance by reducing metal-to-metal contact and friction coefficient, although it slightly reduced hardness due to its softer nature. The hybrid composite demonstrated reduced wear loss under dry sliding conditions, with Gr contributing to lower surface adhesion and thermal softening (Fig. 6.17).

In another study, Arokiasamy et al. [50, 51] explored the application and effectiveness of friction stir processing in enhancing the mechanical properties of Mg-based hybrid metal matrix composites. The research aimed to improve surface hardness, grain refinement, tensile strength, and wear resistance of composites reinforced with silicon carbide and Al_2O_3. Utilizing stir casting as the manufacturing technique, the studies integrated SiC and Al_2O_3 as reinforcing agents to create

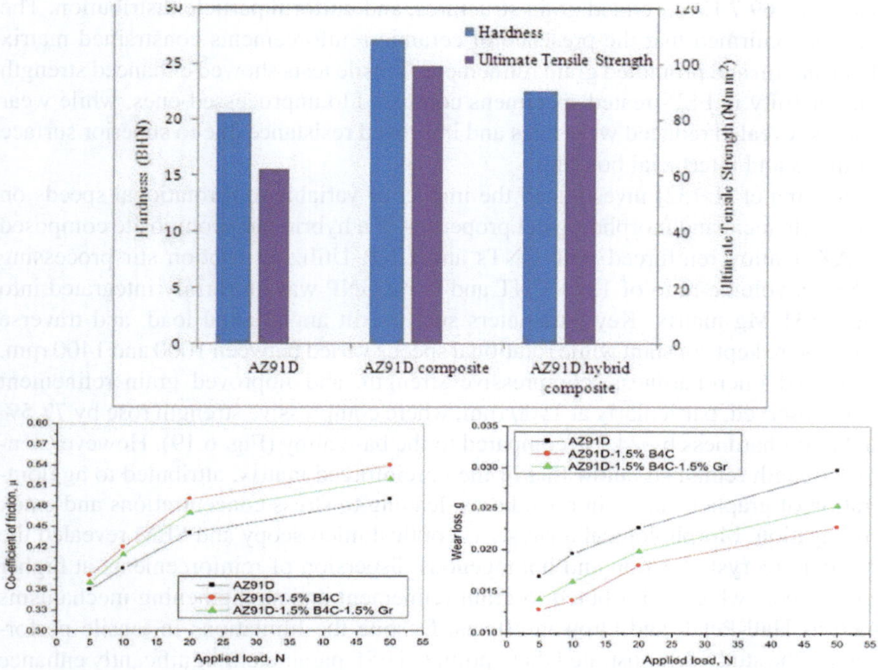

Fig. 6.17 Mechanical and tribological properties of AZ91D/SiC-Gr hybrid composites [49]

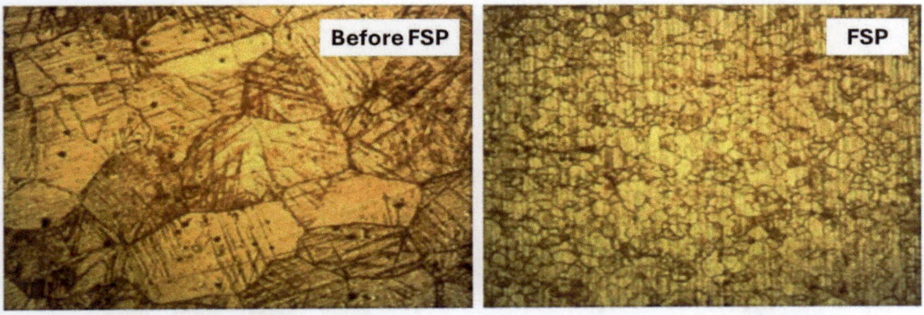

Fig. 6.18 Microstructure of FSP processed hybrid composite compared to the base composite [51]

hybrid composites, achieving a significant refinement in grain size, reducing it from approximately 82–84 μm in base materials to as fine as 7 μm in optimally processed samples (Fig. 6.18). FSP demonstrated its capability as a solid-state technique to enhance material properties through mechanical stirring, dynamic recrystallization, and grain structure homogenization. Experimental setups employed varying tool geometries, rotational speeds, and traverse speeds, revealing that optimum parameters, such as 540 rpm rotation and 30 mm/min traverse speed, yielded maximum hardness (69.7 HV), refined grain structures, and uniform particle distribution. The studies confirmed that the presence of ceramic reinforcements constrained matrix deformation and promoted grain refinement. Tensile tests showed enhanced strength and ductility in FSP-treated specimens compared to unprocessed ones, while wear studies revealed reduced wear rates and improved resistance due to superior surface hardness and interfacial bonding.

Sharma et al. [52] investigated the impact of variable tool rotational speeds on the mechanical and morphological properties of a hybrid nanocomposite composed of AZ31 alloy reinforced with CNTs and GNP. Utilizing friction stir processing (FSP), a volume ratio of 1.6% CNT and 0.3% GNP was uniformly integrated into the AZ31 Mg matrix. Key parameters such as tilt angle, axial load, and traverse speed were kept constant while rotational speeds varied between 1000 and 1400 rpm. Enhanced microhardness, compressive strength, and improved grain refinement were observed, particularly at 1400 rpm, where compressive strength rose by 77.5% and microhardness by 24.5% compared to the base alloy (Fig. 6.19). However, tensile strength remained below that of the unreinforced matrix, attributed to agglomeration of graphene at grain boundaries, leading to stress concentrations and crack propagation. Morphological analysis via optical microscopy and SEM revealed the dynamic recrystallization and homogenous dispersion of reinforcements at higher tool speeds, which contributed to grain refinement and strengthening mechanisms such as Hall-Petch and Orowan effects. Despite the limitations in tensile performance, the study demonstrated that optimized FSP parameters significantly enhance the composite's compressive properties, indicating its potential in lightweight, high-strength applications.

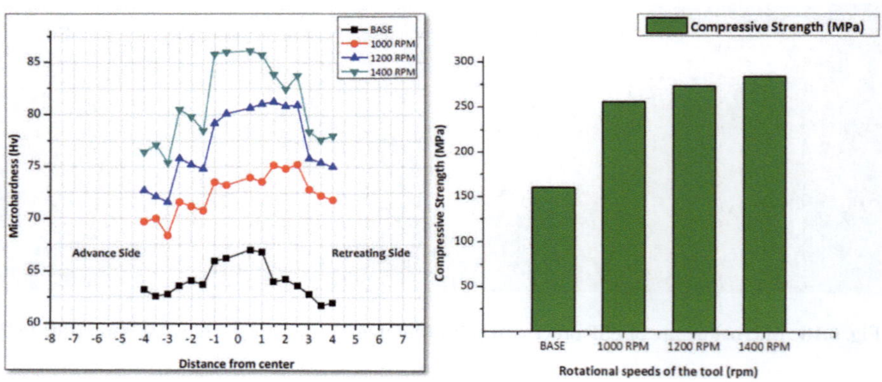

Fig. 6.19 Microhardness and compressive properties of FSP processed AZ31/CNT-GNP hybrid composites [52]

Zhou et al. [53] utilized ultrasonic cavitation-assisted semi-solid stirring to fabricate AZ91-based hybrid composites containing varying amounts of CNTs and SiC nanoparticles. By optimizing the hybrid reinforcement ratio to 7:3 (CNTs:SiC) and a total reinforcement content of 1.0 mass percent, the researchers observed significant grain refinement and uniform nanoparticle distribution. The hybrid composites demonstrated enhanced mechanical properties, with tensile strength and elongation increased by 45% and 50%, respectively, compared to the unreinforced AZ91 alloy (Fig. 6.20). The microstructural analysis revealed that the addition of hybrid reinforcements created heterogeneous nucleation sites, reducing the grain size and improving interfacial bonding. The study attributed these enhancements to several strengthening mechanisms: Orowan strengthening from particle dispersion, grain refinement via the Hall-Petch effect, mismatch in CTE between the matrix and reinforcements, and effective load transfer enabled by the hybrid reinforcement's

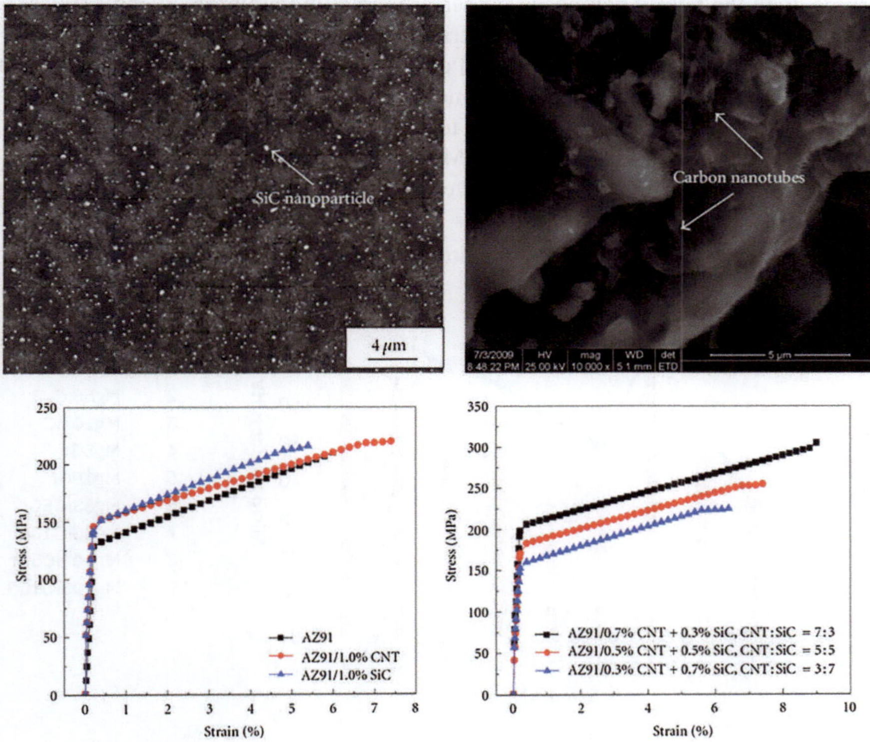

Fig. 6.20 Microstructure and mechanical properties of hybrid composites fabricated by Zhou et al. [53]

design. Fracture analysis indicated a transition from brittle failure in the base alloy to quasi-cleavage in the composites, highlighting improved ductility. In a similar study, the FSP processed AZ31 alloy composites reinforced nano-Al_2O_3 particles and CNTs were investigated [54]. They varied the reinforcement content and found that the one with 0.1% n-Al_2O_3 + 0.2% CNTs exhibited the best combination of properties including higher hardness (1.4 times greater than the unreinforced AZ31 alloy), and lower wear rates at higher loads (above 1.95 MPa).

Turan et al. [55] synthesized AZ91 Mg alloy composites reinforced with 0.3 wt% CNTs and a hybrid of 0.15 wt% CNT and 0.15 wt% GNPs via semi-powder metallurgy. The AZ91-CNT-GNP composite showed a 21% increase in hardness, reaching 88.3 HV, compared to 72.9 HV for unreinforced AZ91. The wear rate decreased by approximately 50% for the hybrid composite under a 40 N load, owing to the formation of a protective carbonaceous film during sliding. At a sliding speed of 250 mm/s and a 40 N load, wear mechanisms included oxidative and adhesive processes, with the hybrid composite outperforming both unreinforced and MWCNT-only composites in reducing wear and friction coefficients.

In another study, Prakash et al. [56] developed Mg composites reinforced with SiC and graphite (Gr) through powder metallurgy. The study revealed that the addition of 10 wt% SiC improved hardness to 80 HV, compared to 26 HV for pure Mg, while incorporating 5 wt% Gr into the Mg-10SiC composite slightly reduced hardness to 77 HV but significantly enhanced wear resistance (Fig. 6.21). Among the

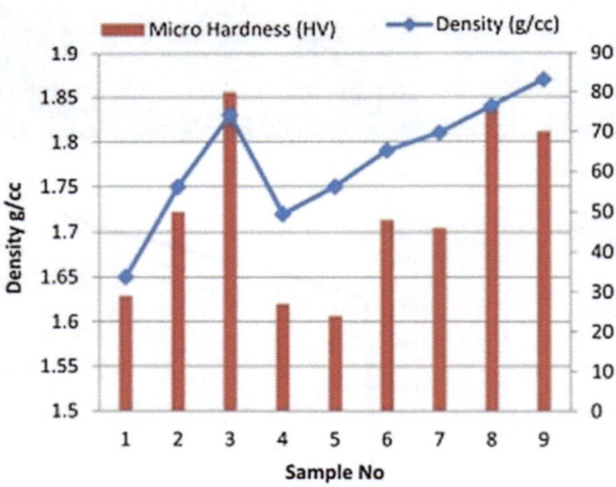

Sample	Material
1	Mg
2	Mg5SiC
3	Mg10SiC
4	Mg5Gr
5	Mg10Gr
6	Mg5SiC5Gr
7	Mg5SiC10Gr
8	Mg10SiC5Gr
9	Mg10SiC10G

Fig. 6.21 Hardness of different Mg composites fabricated by Prakash et al. [56]

composites, Mg-10SiC-5Gr exhibited the best wear performance due to the balance of hardness and lubrication, showing a 50% reduction in wear loss compared to pure Mg. SEM analysis indicated reduced abrasive wear and a shift to delamination mechanisms, though higher Gr content (10 wt%) caused increased brittleness and localized stress.

Sahoo and Panigrahi [57] investigated the impact of in situ TiC-TiB_2 reinforcement and aging treatments on the tension-compression yield asymmetry of AZ91 alloy and their composites. It examined four material states: untreated alloy, solution-treated alloy, peak-aged alloy, and composites with both reinforcements and precipitates. Results indicated that tension-compression asymmetry, initially 1.30, reduced to 1.04 after aging treatment in the composite due to reduced twin formation. This improvement was attributed to the combined presence of TiC-TiB_2 reinforcements and $Mg_{17}Al_{12}$ precipitates, which hindered twin nucleation and growth through back stress mechanisms. The aging treatment also enhanced yield strength and tensile strength of composites by 33% and 65%, respectively, relative to the untreated alloy, showcasing a synergy between reinforcement and precipitate structures in optimizing mechanical properties and reducing yield asymmetry. Microstructural analysis confirmed that reinforcements and precipitates altered twin interactions, demonstrating their critical role in deformation behavior (Fig. 6.22).

Table 6.2 summarizes the properties of magnesium-based hybrid composites containing various particle reinforcements.

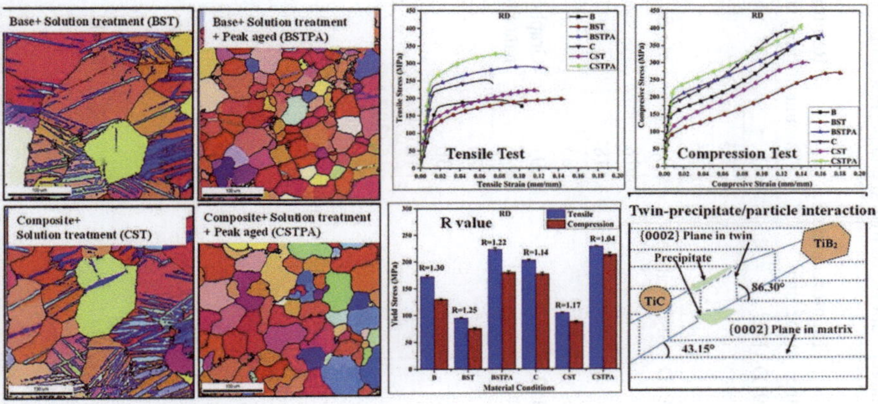

Fig. 6.22 Microstructure and property highlights from Sahoo and Panigrahi [57]

Table 6.2 Properties of magnesium-based hybrid composites containing various particle reinforcements

References	Matrix	Reinforcement 1, type, size	Amount	Reinforcement 2, type, size	Amount	Reinforcement 3, type, size	Amount	Process	Density	Porosity
Aatthisugan et al. [49]	AZ91D	B_4C	0	Gr	0			Stir casting	1.8091	1.42
			1.5		0				1.8197	1.34
			1.5		1.5				1.8249	1.11
Asano et al. [58]	AZ91D	Al_2O_3 Fibers	0	Mg_2Si particles	0			Infiltration	–	–
			18		0					
			18		12				–	–
			33		0					
Babu et al. [9]	AM50	Al_2O_3	0	Graphene nanofiber	0			Infiltration		
			10		0					
			20		0					
			30		0					
			10 (30%)		10 (70%)					
			15 (30%)		15 (70%)					
			20 (30%)		20 (70%)					
Hassan et al. [37]	Mg	Y_2O_3 (30–50 nm)	0.7	Cu (25 nm)	0			PM-MWS-HE	1.757	0.35
					0.3				1.775	0.45
Hassan et al. [59]	Mg	Y_2O_3	0	Cu	0				1.738	0.13
			0.7		0					
					0.3				1.775	0.45
					0.6				1.792	0.77
					1.0				1.825	0.62
Habibi et al. [39]	Mg	Al (7–15 μm)	0	CNT	0			PM-MWS-HE	1.7379	0.12
			0.5		0.18				1.7346	0.53
			1.0		0.18				1.7359	0.63
			1.5		0.18				1.7370	0.74

Habibi et al. [43]	Mg	Al (7–15 μm)	0	Al_2O_3 (50 nm)	0			PM-MWS-HE	1.7379	0.12
			0.972		0.22				1.7531	0.06
			0.972		0.66				1.7633	0.05
			0.972		1.11				1.7695	0.27
Habibi et al. [40]	Mg	Al (7–15 μm)	–	CNT	0			PM-MWS-HE	1.7379	0.12
			1 wt%		0.09 wt%				1.7395	0.40
			1 wt%		0.18 wt%				1.7359	0.63
			1 wt%		0.30 wt%				1.7369	0.60
			1 wt%		0.50 wt%				1.7311	0.98
Habibi et al. [45]	Mg	Al (7–15 μm)	0	B_4C (50 nm)	0			PM-MWS-HE		
			0		0.22					
			0		0.66					
			0		1.11					
			0.92		0.66					
Habibi et al. [44]	Mg	Al (7–15 μm)	0	Bi_2O_3 (90–210 nm)	0			PM-MWS-HE		
			0		0.22					
			0		0.66					
			0		0.11					
			0.92		0.22					
			0.92		0.66					
			0.92		1.11					

(continued)

Table 6.2 (continued)

References	Matrix	Reinforcement 1, type, size	Amount	Reinforcement 2, type, size	Amount	Reinforcement 3, type, size	Amount	Process	Density	Porosity
Habibi et al. [41]	Mg	Al (7–15 μm)	0	Al_2O_3 (50 nm)	0			PM-MWS-HE	1.7379	0.12
			0.647		0.66				1.7599	0.06
			0.972		0.66				1.7633	0.05
			1.298		0.66				1.7638	0.20
			1.95		0.66				1.7648	0.34
Habibi et al. [42]	Mg	Al (7–15 μm)	0	Al_2O_3				PM-MWS-HE	1.7379	0.12
			0.972		0.66				1.7631	0.06
			0.972		0.66				1.7630	0.07
			0.972		0.05				1.7633	0.05
Hassan et al. [32]	Mg	Y_2O_3 (30–50 nm)	0	Ni (20 nm)	0				1.738	0.13
			0.7		0.6				1.802	0.21
Ma et al. [15]	AZ91	TiB_2	5.5 (50%)	TiC	5.5 (50%)			Ball milling and melt processing		
Nguyen et al. [21]	AZ31B	Al_2O_3 (50 nm)	–	Ni (29 ± 19 μm	–			DMD	1.7689	0.06
			1.5		–				1.8086	0.07
			1.5		1.5				1.9077	0.12
			1.5		3.19				2.0279	0.14
Nguyen et al. [23]	AZ31B	Al_2O_3 (50 nm)	0	Cu (50 μm)	0			DMD	1·7681	0·11
			1.5		0				1·7681	0·11
			0		2				1·9095	0·14
			1.5		2				1·9432	0·13
			1.5		4				2·0522	0·15
			1.5		4				2·0850	0·14

Nguyen et al. [24]	AZ31B	Al_2O_3 (50 nm)	0	Cu (50 μm)				DMD	1.7679	0.12
			0		10 wt%				1.9196	0.24
			1.5		10 wt%				1.9601	0.16
			0		18 wt%				2.0625	0.29
			1.5		18 wt%				2.1093	0.21
Rashad et al. [46]	AZ31	Al_2O_3 (75 μm)	0	SiC (40 nm)	0			PM		
			1.5		0.2					
			1.5		0.5					
			1.5		1.0					
			0		0					
			1.5		0.2					
			1.5		0.5					
			1.5		1.0					
Rashad et al. [47]	Mg	Cu (40–75 μm)	0	GNP (5–15 nm)	0			Semi-solid powder metallurgy	1.7385	
			1		0.18				1.7493	
					0.36				1.7468	
					0.54				1.7480	
Rashad et al. [48]	Mg	Ti (3 μm)	0	Al (25 μm)	0			PM		
			10 wt%		0					
			10 wt%		1 wt%					

(continued)

Table 6.2 (continued)

References	Matrix	Reinforcement 1, type, size	Amount	Reinforcement 2, type, size	Amount	Reinforce-ment 3, type, size	Amount	Process	Density	Porosity
Sankaranarayanan et al. [28]	Mg	Ti (<140 μm)	0	SiC (50 nm)	0			DMD		
			5.6 wt%		0					
			5.6 wt%		0.5 wt%					
			5.6 wt%		1.0 wt%					
			5.6 wt%		2.0 wt%					
Sankaranarayanan et al. [30]	Mg	Ti (<140 μm)	5.6 wt%	B_4C (50 nm)	0					
			5.6 wt%		0.5 wt%					
			5.6 wt%		1.5 wt%					
			5.6 wt%		2.5 wt%					
Sankaranarayanan et al. [27]	Mg	Ti (<140 μm)	0	Al (15 μm)	0				1.7397	0.02
			0		3 wt%				1.7587	0.01
			5.6 wt%		0				1.8002	0.09
			5.6 wt%		3 wt%				1.8213	0.04
			5.6 wt%		3 wt%				1.8218	0.02
Sankaranarayanan et al. [25]	Mg	Ti (<140 μm)	0	Al_2O_3 (50 nm)	0				1.7129	0.02
			5.6 wt%		2.5 wt%				1.6755	0.09
			5.6 wt%		2.5 wt%				1.7826	0.04
Sankaranarayanan et al. [26]	Mg	Ti (<140 μm)	0	Cu	0				1.7397	0.02
			5.6 wt%		0				1.8002	0.09
			0		3 wt%				1.9681	0.04
			5.6 wt%		3 wt%				2.1089	0.16
			5.6 wt%		3 wt%				2.1096	0.09

Sinha et al. [18]	Mg	Ti (20 µm)	0	SiC (0.6 µm)	0				1.7380	
			0		4.8				1.7698	
			0		10.2				1.7931	
			0		15.4				1.8349	
			2.7		5.02				1.7971	
			2.7		9.92				1.818	
			2.7		14.93				1.879	
Prakash et al. [56]	Mg	SiC (35–40 µm)	0	Gr (20–25 µm)	0			PM	1.65	
			5		0				1.75	
			10		0				1.82	
			0		5				1.73	
			0		10				1.75	
			5		5				1.79	
			5		10				1.81	
			10		5				1.84	
			10		10				1.87	
Thakur et al. [20]	Mg	SiC (50 nm)	0	Al_2O_3 (50 nm)	0			PM-MWS-HE	1.7345	
			1		0				1.7414	
			0.5		0.5				1.7410	
			0.3		0.7				1.7407	

(continued)

Table 6.2 (continued)

References	Matrix	Reinforcement 1, type, size	Amount	Reinforcement 2, type, size	Amount	Reinforcement 3, type, size	Amount	Process	Density	Porosity
Thakur et al. [19]	Mg	CNT (40–70 nm)	0	SiC (50 nm)	0			PM-MWS-HE	1.737	
			0.3		0.7				1.742	
			0.5		0.5				1.740	
			0.7		0.3				1.739	
			1		0				1.736	
Tun et al. [36]	Mg	Al_2O_3 (50 nm)	–	Cu (25 nm)	–			PM-MWS-HE	1.736	0.2
			1		0.1				1.766	0.2
			1		0.3				1.776	0.4
			1		0.6				1.793	0.7
			1		0.9				1.821	0.3
Tun et al. [35]	Mg	ZrO_2 (51 to 65 nm)	0	Cu (25 nm)	0			PM-MWS-HE		
			0.3		0					
			0.6		0					
			1.0		0					
			1.0		0					
			0.3		0.7					
			0.6		0.4					
Tun et al. [31, 33]	Mg	Y_2O_3 (30–50 nm)	–	Ni (20 nm)	–			PM-MWS-HE	1.738	0.13
			0.7		–				1.757	0.35
			0.7		0.3				1.778	0.34
			0.7		0.6				1.802	0.21
			0.7		1.0				1.829	0.30

Tun et al. [60]	Mg	Y_2O_3 (30–50 nm)	0	Cu (25 nm)	0			PM-MWS-HE	1.738	0.13
			0.7		0.3				1.775	0.45
			0.7		0.6				1.792	0.77
			0.7		1.0				1.825	0.62
Turan et al. [55]	AZ91	CNT (8–18 nm)	0	GNP	0			PM		
			0.3		0					
			0.3		0.15					
Ugandhar et al. [17]	Mg	SiC (0.6 μm)	0	Ti (20 μm)				PM-MWS-HE	1.738	0.12
			2.8		1.1				1.797	0.69
			8.8		1.1				1.879	1.03
Xiuqing et al. [16]	AZ91	–	0	–	0			Remelting and dilution	1.81	
	Mg	TiB_2		TiC					1.93	
Zhang et al. [11]	AM60	Al_2O_3	0	Al_2O_3 short fibers	0			Perform-squeeze casting		
			9		0					
			9		2					
			9		4					
Zhou et al. [14, 61]	AM60	Al_2O_3 fiber	0	Al_2O_3 (5 μm)	0	Al_2O_3 (100 nm)	0	Squeeze casting		
			5		0		0			
			5		3		0			
			5		0		3			
Zhou et al. [13]	AZ61	CNT (30–50 nm)	0	SiC (8 μm)	0			PM-HE		
			0		5					
			0.1		5					
			0.2		5					
			0.5		5					
			1.0		5					

(continued)

Table 6.2 (continued)

References	Grain size	Hardness (HV/BHN*)	Tensile yield strength	Ultimate tensile strength	Tensile failure strain	Compressive yield strength	Ultimate compressive strength	Compressive failure strain	Remarks
Aatthisugan et al. [49]		20.5*		62.1					
		27.1*		99.2					
		22.5*		86.3					
Asano et al. [58]	–	–	203						
			238						
	–	–	250 (at 293 K)						
			270						
Babu et al. [9]		76		226			155		
		128		237			350		
		136		238			370		
		140		240			400		
		142		240			420		
		134		238			400		
		130		234			380		
Hassan et al. [37]	18		157	244	9.1				
	9		215	270	11.1				
Hassan et al. [59]	20		134	193	6.9				
			157	244	8.6				
	9		215	270	11.1				
	9		179	231	11.1				
	8		148	200	10.2				

Habibi et al. [39]	19 ± 4	40 ± 2	93 ± 01	153 ± 07	7.9 ± 3.4				
	10 ± 3	50 ± 4	116 ± 11	186 ± 12	10.9 ± 3.5				
	8 ± 3	58 ± 3	128 ± 12	208 ± 08	11.2 ± 2.9				
	8 ± 4	60 ± 4	156 ± 13	223 ± 12	7.0 ± 0.5				
Habibi et al. [43]	19 ± 4		93 ± 01	153 ± 07	7.9 ± 3.4	91 ± 03	239 ± 11	19.8 ± 1.7	
	9 ± 2		160 ± 07	225 ± 05	9.5 ± 0.4	151 ± 10	430 ± 11	13.1 ± 1.5	
	8 ± 4		182 ± 10	276 ± 08	11.2 ± 1.5	164 ± 10	429 ± 05	10.2 ± 1.4	
	8 ± 3		172 ± 12	236 ± 19	4.0 ± 0.3	120 ± 11	421 ± 04	13.2 ± 0.9	
Habibi et al. [40]	19 ± 4	40 ± 2	93 ± 01	153 ± 07	7.9 ± 3.4	91 ± 3	239 ± 11	19.8 ± 1.7	
	10 ± 2	55 ± 4	148 ± 02	206 ± 04	9.1 ± 0.6	135 ± 04	337 ± 14	10.1 ± 1.5	
	9 ± 3	58 ± 3	128 ± 12	208 ± 08	11.2 ± 2.9	126 ± 05	421 ± 13	12.5 ± 0.4	
	7 ± 2	60 ± 2	160 ± 08	227 ± 14	8.6 ± 0.4	148 ± 05	424 ± 12	12.4 ± 1.7	
	7 ± 3	58 ± 2	168 ± 02	220 ± 06	4.6 ± 0.5	144 ± 10	401 ± 15	10.9 ± 1.5	
Habibi et al. [45]	20 ± 3		109 ± 09	161 ± 16	8.9 ± 1.1	93 ± 02	246 ± 23	19.4 ± 0.8	
	11 ± 2		110 ± 15	159 ± 14	9.9 ± 0.6	97 ± 04	337 ± 14	10.0 ± 1.6	
	10 ± 3		120 ± 05	164 ± 06	10.0 ± 0.3	100 ± 05	335 ± 11	11.8 ± 1.8	
	10 ± 2		82 ± 11	119 ± 17	5.5 ± 1.2	105 ± 03	331 ± 10	13.3 ± 1.4	
	8 ± 3		130 ± 09	238 ± 15	7.0 ± 0.9	129 ± 05	356 ± 33	9.8 ± 1.5	
Habibi et al. [44]	20 ± 3		109 ± 09	161 ± 16	8.9 ± 1.1	93 ± 02	246 ± 23	19.4 ± 0.8	
	10 ± 2		135 ± 12	207 ± 13	7.5 ± 0.3	101 ± 09	341 ± 08	13.9 ± 0.7	
	11 ± 3		135 ± 13	201 ± 15	6.0 ± 0.5	102 ± 11	387 ± 17	12.0 ± 1.5	
	10 ± 2		139 ± 13	205 ± 18	4.9 ± 0.6	108 ± 03	411 ± 06	13.6 ± 1.4	
	7 ± 2		172 ± 09	250 ± 06	10.0 ± 1.0	140 ± 18	446 ± 10	13.1 ± 1.5	
	10 ± 3		158 ± 13	226 ± 09	6.2 ± 0.7	143 ± 13	430 ± 20	9.0 ± 2.0	
	11 ± 4		154 ± 09	216 ± 11	6.1 ± 0.5	115 ± 06	420 ± 15	8.5 ± 3.0	

(continued)

Table 6.2 (continued)

References	Grain size	Hardness (HV/BHN*)	Tensile yield strength	Ultimate tensile strength	Tensile failure strain	Compressive yield strength	Ultimate compressive strength	Compressive failure strain	Remarks
Habibi et al. [41]	19 ± 4	40 ± 2	93 ± 01	153 ± 07	7.9 ± 3.4				
	8 ± 4	59 ± 4	164 ± 18	254 ± 23	9.5 ± 1.6				
	8 ± 4	60 ± 3	182 ± 10	276 ± 08	11.2 ± 1.5				
	7 ± 4	62 ± 4	159 ± 11	255 ± 15	7.1 ± 0.6				
	7 ± 3	62 ± 3	148 ± 10	233 ± 07	6.1 ± 1.2				
Habibi et al. [42]	19	40	93	153	7.9	91	235	20.3	
	10	58	154	212	6.0	162	418	9.5	Al_2O_3 (1 μm)
	9	57	158	223	7.4	158	418	9.3	Al_2O_3 (0.3 μm)
	8	60	182	276	11.2	164	429	10.2	Al_2O_3 (0.05 μm)
Hassan et al. [32]	20		134	193	6.9				
	6		232	272	9.5				
Ma et al. [15]		56							
		79							
Nguyen et al. [21]	1.57 ± 0.36	63 ± 1	201 ± 7	270 ± 6	5.6 ± 1.4				
	1.06 ± 0.08	86 ± 3	144 ± 9	214 ± 16	29.5 ± 1.9				
	1.22 ± 0.23	93 ± 4	227 ± 4	298 ± 5	5.7 ± 0.7				
	1.24 ± 0.35	102 ± 4	251 ± 8	306 ± 6	2.5 ± 0.4				
Nguyen et al. [23]	6.2		180 ± 4	265 ± 7	10·9 ± 2·4				
	1.1		219 ± 4	295 ± 8	15 ± 2·3				
	0.5		238 ± 8	298 ± 6	4·5 ± 0·5				
	0.3		250 ± 11	301 ± 11	13·5 ± 2·1				
	0.5		265 ± 8	308 ± 9	1·9 ± 0·4				
	0.3		300 ± 12	350 ± 14	8·5 ± 1·6				

Nguyen et al. [24]	5.8 ± 0.36	63 ± 1	201 ± 7	270 ± 6	5.6 ± 1.4				
	0.5 ± 0.05	95 ± 3	240 ± 3	302 ± 2	3.8 ± 0.5				
	0.3 ± 0.08	103 ± 3	241 ± 8	313 ± 9	5.6 ± 0.5				
	0.5 ± 0.06	105 ± 3	268 ± 4	310 ± 5	1.1 ± 0.4				
	0.3 ± 0.07	112 ± 2	294 ± 12	344 ± 7	4.8 ± 0.4				
Rashad et al. [46]		55	166 ± 3.8	269 ± 3.1	16.90 ± 1.6	144 ± 3.6	374 ± 5.7	12.92 ± 1.6	As-extruded
		68	198 ± 4.1	293 ± 5.0	10.58 ± 2.0	178 ± 5.1	315 ± 4.4	08.44 ± 2.0	As-extruded
		80	208 ± 3.8	306 ± 3.8	07.54 ± 1.8	206 ± 4.2	345 ± 5.0	08.47 ± 1.3	As-extruded
		90	230 ± 2.5	322 ± 2.5	04.32 ± 1.3	238 ± 2.5	402 ± 2.6	07.67 ± 1.1	As-extruded
		67	175 ± 2.8	282 ± 3.8	17.42 ± 1.2				Heat treated
		71	174 ± 3.0	275 ± 5.9	16.50 ± 1.8				Heat treated
		67	188 ± 2.6	305 ± 5.3	13.51 ± 2.0				Heat treated
		75	196 ± 5.7	311 ± 4.8	14.75 ± 2.1				Heat treated
Rashad et al. [47]	29 ± 1.5	40 ± 3	104 ± 4	164 ± 5	6.2 ± 1.8	123 ± 5	264 ± 6	9 ± 2.5	
	18 ± 1.0	44 ± 2	160 ± 6	240 ± 2	10.4 ± 2.1	140 ± 4	335 ± 8	10.3 ± 1.6	
	13 ± 1.2	46 ± 3	184 ± 3	252 ± 3	12.2 ± 1.3	143 ± 6	338 ± 5	11.7 ± 2	
	11 ± 1.0	56.7 ± 1	226 ± 5	260 ± 5	4.8 ± 2.5	166 ± 3	420 ± 6	14.5 ± 1.4	
Rashad et al. [48]	29 ± 3.5	~40	~104	~164	~6.2				
	20 ± 4.1	~46	~147	~212	~11.1				
	10 ± 3.2	~55	~163	~238	~21.2				
Sankaranarayanan et al. [28]		48 ± 1	120 ± 9	169 ± 11	6.2 ± 0.7	65 ± 4	248 ± 8	19.2 ± 1.1	
		71 ± 2	158 ± 6	226 ± 6	8.0 ± 1.5	77 ± 3	347 ± 5	13.5 ± 0.8	
		74 ± 4	167 ± 4	228 ± 5	9.6 ± 0.6	101 ± 7	358 ± 9	14.5 ± 1.2	
		83 ± 3	204 ± 7	260 ± 3	9.2 ± 1.1	117 ± 6	391 ± 7	13.1 ± 1.1	
		86 ± 5	192 ± 4	250 ± 3	4.6 ± 0.8	121 ± 6	379 ± 10	10.7 ± 0.9	

(continued)

Table 6.2 (continued)

References	Grain size	Hardness (HV/BHN*)	Tensile yield strength	Ultimate tensile strength	Tensile failure strain	Compressive yield strength	Ultimate compressive strength	Compressive failure strain	Remarks
Sankaranarayanan et al. [30]	18 ± 7	71 ± 2	158 ± 6	226 ± 6	8.0 ± 1.5	77 ± 3	347 ± 5	13.5 ± 0.8	
	17 ± 9	70 ± 4	156 ± 9	228 ± 12	11.7 ± 0.4	77 ± 7	372 ± 3	15.8 ± 0.5	
	9 ± 6	87 ± 5	180 ± 5	238 ± 6	9.8 ± 0.7	96 ± 7	395 ± 5	15.6 ± 1.3	
	6 ± 4	92 ± 7	215 ± 9	260 ± 8	8.1 ± 0.3	118 ± 5	419 ± 7	13.0 ± 0.5	
Sankaranarayanan et al. [27]	21 ± 3	48 ± 1	125 ± 9	169 ± 11	6.2 ± 0.7	74 ± 3	273 ± 11	22.7 ± 4.9	
	9 ± 3	65 ± 2	161 ± 15	250 ± 20	7.5 ± 1.1	127 ± 2	437 ± 5	21.7 ± 4.3	
	15 ± 4	71 ± 2	158 ± 6	226 ± 6	8.0 ± 1.5	85 ± 3	360 ± 5	13.6 ± 1.2	
	10 ± 5	78 ± 3	167 ± 5	236 ± 5	4.8 ± 0.5	104 ± 1	378 ± 13	12.6 ± 1.3	
	8 ± 2	93 ± 3	194 ± 2	265 ± 2	4.8 ± 0.6	139 ± 6	431 ± 14	12.9 ± 1.6	Ball milling
Sankaranarayanan et al. [25]	21.30 ± 2.60	125 ± 9	169 ± 11	6.2 ± 0.7	87 ± 4	240 ± 9	19.2 ± 0.7		
	14.88 ± 1.55	175 ± 4	227 ± 10	3.3 ± 0.2	106 ± 0	424 ± 15	14.4 ± 1.2		
	7.64 ± 1.47	168 ± 8	214 ± 8	6.8 ± 0.8	106 ± 3	349 ± 9	15.2 ± 1.7		Ball milling
Sankaranarayanan et al. [26]	21 ± 3	48 ± 1	125 ± 9	169 ± 11	6.2 ± 0.7	74 ± 3	273 ± 11	22.7 ± 4.9	
	15 ± 4	71 ± 2	158 ± 6	226 ± 6	8.0 ± 1.5	85 ± 3	360 ± 5	13.6 ± 1.2	
	11 ± 2	82 ± 4	182 ± 4	220 ± 4	8.9 ± 0.9	102 ± 8	386 ± 4	21.1 ± 1.7	
	9 ± 3	86 ± 2	196 ± 9	227 ± 4	5.7 ± 1.6	116 ± 5	342 ± 3	16.1 ± 1.4	
	7 ± 3	91 ± 3	201 ± 7	265 ± 11	7.5 ± 0.8	126 ± 8	380 ± 6	19.1 ± 2.9	
Sinha et al. [18]		47 ± 1	153 ± 8	207 ± 4	9.2 ± 1.0				
		58 ± 1	182 ± 2	219 ± 2	2.1 ± 0.9				
		59 ± 1	171 ± 3	221 ± 14	1.5 ± 0.2				
		63 ± 1	155 ± 1	207 ± 9	1.4 ± 0.1				
		48 ± 2	169 ± 8	204 ± 17	2.67 ± 0.67				
		52 ± 1	161 ± 9	199 ± 16	2.87 ± 0.95				
		65 ± 1	217 ± 2	233 ± 6	1.04 ± 0.05				

Prakash et al. [56]		30							
		50							
		80							
		27							
		21							
		43							
		42							
		80							
		72							
Thakur et al. [20]	29.92 ± 1.91	39.08 ± 0.8	119.39 ± 7.63	169.22 ± 4.4	5.47 ± 1.57				
	28.71 ± 1.52	43.08 ± 0.7	131.14 ± 12.3	182.40 ± 9.4	5.01 ± 0.46				
	26.27 ± 1.56	46.12 ± 0.9	155.66 ± 6.99	197.38 ± 1.83	4.58 ± 2.08				
	25.79 ± 1.87	47.96 ± 0.5	164.91 ± 1.43	206.10 ± 5.28	4.17 ± 1.78				
Thakur et al. [19]		41 ± 1	111.9 ± 7.7	155.8 ± 2.1	5.9 ± 1.2				
		46 ± 1	152.9 ± 4.1	195.4 ± 4.7	3.3 ± 0.7				
		45 ± 1	152.1 ± 1.2	188.5 ± 2.7	2.3 ± 0.6				
		44 ± 1	139.5 ± 6.5	182.9 ± 7.5	2.1 ± 0.5				
		43 ± 1	117 ± 6.2	153.8 ± 2.8	1.5 ± 0.3				
Tun et al. [36]	21 ± 3	40 ± 1	121 ± 5	179 ± 6	11.4 ± 1.1	103 ± 4	263 ± 5	22 ± 2	40 ± 3
	17 ± 5	49 ± 2	169 ± 11	205 ± 11	2.8 ± 0.4	115 ± 4	265 ± 13	10 ± 1	24 ± 1
	13 ± 4	56 ± 2	190 ± 13	225 ± 13	4.1 ± 0.6	119 ± 8	349 ± 7	10 ± 4	35 ± 2
	10 ± 3	68 ± 2	184 ± 7	224 ± 8	9.5 ± 1.2	136 ± 8	397 ± 7	13 ± 1	46 ± 4
	7 ± 3	59 ± 2	202 ± 7	232 ± 7	4.1 ± 0.3	156 ± 13	368 ± 7	9 ± 4	47 ± 7

(continued)

Table 6.2 (continued)

References	Grain size	Hardness (HV/BHN*)	Tensile yield strength	Ultimate tensile strength	Tensile failure strain	Compressive yield strength	Ultimate compressive strength	Compressive failure strain	Remarks
Tun et al. [35]	25 ± 7	42.0 ± 1.6	111 ± 7.8	177 ± 10	9.0 ± 2.2	109 ± 4	284 ± 11	23 ± 3	
	24 ± 7	40.0 ± 1.0	84.8 ± 8.0	139 ± 7.5	8.1 ± 1.6	109 ± 6	273 ± 13	19 ± 1	
	29 ± 3	41.6 ± 2.1	117 ± 11	182 ± 14	9.4 ± 2.7				
	25 ± 4	42.1 ± 1.9	97.8 ± 6.3	158 ± 12	8.6 ± 2.2	109 ± 5	262 ± 18	19 ± 4	
	23 ± 6	39.9 ± 1.4	122 ± 7.7	188 ± 5.9	10 ± 1.3				
	9 ± 2	47.6 ± 1.0	196 ± 16	249 ± 7.5	8.2 ± 1.1	124 ± 7	352 ± 18	12 ± 3	
	11 ± 3	50.2 ± 0.9	139 ± 22	193 ± 21	11.4 ± 2.9				
Tun et al. [31, 33]	20 ± 3	37 ± 2	134 ± 7	193 ± 1	6.9 ± 2.5	70 ± 7	280 ± 9	28.1 ± 2.9	
	18 ± 3	45 ± 2	157 ± 10	244 ± 1	9.1 ± 0.6				
	9 ± 3	54 ± 4	221 ± 7	262 ± 6	9.0 ± 0.9	154 ± 5	402 ± 3	17.2 ± 0.2	
	6 ± 2	60 ± 4	232 ± 8	272 ± 2	9.5 ± 0.9	154 ± 9	394 ± 4	16.3 ± 0.4	
	5 ± 2	63 ± 4	228 ± 8	271 ± 6	5.5 ± 0.7	154 ± 3	406 ± 8	16.0 ± 0.4	
Tun et al. [60]	20	37				70	280	26.9	
	9	53				114	388	13.1	
	9	57				117	396	12.8	
	8	61				156	411	12.5	
Turan et al. [55]		72.9 ± 1.4							
		82.7 ± 1.6							
		88.3 ± 2.1							
Ugandhar et al. [17]	21	41	158	207	9.2				
	20	48	169	204	2.67				
	15	66	217	233	1.04				
Xiuqing et al. [16]			82 ± 3	233 ± 0	6.0 ± 0.5				
			95 ± 2	298 ± 2	2.4 ± 0.4				

Zhang et al. [11]	68	60 HRC	80	160	6				
	44	85 HRC	120	180	1.8				
		90 HRC							
	28	98 HRC	148	200	1.4				
Zhou et al. [14, 61]			81 ± 6	171 ± 8	6.0 ± 1.3				
			120 ± 5	189 ± 12	2.2 ± 1.7				
			142 ± 7	192 ± 15	1.6 ± 1.1				
			140 ± 14	216 ± 5	3.5 ± 1.2				
Zhou et al. [13]	36.4	96 ± 3.5	208 ± 3.5	285 ± 4.3	13.2 ± 0.4				
	27.5	108 ± 4.3	235 ± 4.2	318 ± 5.5	7.0 ± 0.2				
	18.7	114 ± 5.2	262 ± 3.7	345 ± 3.2	7.4 ± 0.4				
	15.9	118 ± 3.7	292 ± 4.1	376 ± 3.9	7.6 ± 0.5				
	14.0	126 ± 4.6	345 ± 4.6	412 ± 5.1	8.0 ± 0.3				
	13.5	128 ± 5.6	368 ± 5.5	420 ± 4.0	4.7 ± 0.6				

Values marked with * represent hardness readings in the Brinell scale

6.4 Summary

The development of magnesium-based metal matrix composites (MMCs) with hybrid reinforcements has led to significant advancements in mechanical properties, thermal stability, wear resistance, and cost efficiency. By combining fibers, particulates, and nanoscale reinforcements, researchers have achieved notable improvements in tensile strength, yield strength, hardness, and stiffness. These enhancements are primarily attributed to effective load transfer, grain refinement, and strong interfacial bonding. Hybrid composites also exhibit superior thermal stability, with reduced coefficients of thermal expansion and enhanced resistance to thermal cycling, making them suitable for high-temperature applications. Wear resistance has been significantly improved through synergistic reinforcement effects, which provide superior load-bearing capacity and minimize deformation. Additionally, creep resistance is enhanced by mechanisms such as dislocation pinning and substructural strengthening, with fiber alignment and reinforcement composition playing critical roles. However, challenges remain, as hybrid reinforcements often reduce ductility and corrosion resistance due to brittleness and interfacial defects. Innovations such as in situ-synthesized reinforcements, hierarchical composites, and advanced processing techniques like friction stir processing and microwave-assisted powder metallurgy have expanded the design possibilities, enabling tailored properties. Future research aims to optimize reinforcement distribution, control interfacial reactions, and improve both ductility and corrosion resistance. These composites hold immense potential for structural, automotive, and aerospace applications, where high-strength, lightweight, and thermally stable materials are in demand.

References

1. J. Schröder, K.U. Kainer, Magnesium-base hybrid composites prepared by liquid infiltration. Mater. Sci. Eng. A **135**, 33–36 (1991). https://doi.org/10.1016/0921-5093(91)90532-R
2. S. Kumar, H. Dieringa, K.U. Kainer, Effect of particulate content on the thermal cycling behaviour of the magnesium alloy based hybrid composites. Compos. Part Appl. Sci. Manuf. **36**(3), 321–325 (2005). https://doi.org/10.1016/j.compositesa.2004.07.005
3. S. Kumar, H. Dieringa, K.U. Kainer, Thermal cycling behaviour of the magnesium alloy based hybrid composit es in the transverse direction. Mater. Sci. Eng. A **454–455**, 367–370 (2007). https://doi.org/10.1016/j.msea.2006.11.041
4. A.K. Mondal, S. Kumar, Impression creep behaviour of magnesium alloy-based hybrid composites in the longitudinal direction. Compos. Sci. Technol. **68**(15–16), 3251–3258 (2008). https://doi.org/10.1016/j.compscitech.2008.08.007
5. A.K. Mondal, S. Kumar, Impression creep behaviour of magnesium alloy-based hybrid composites in the transverse direction. Compos. Sci. Technol. **69**(10), 1592–1598 (2009). https://doi.org/10.1016/j.compscitech.2009.02.038
6. A.K. Mondal, S. Kumar, Dry sliding wear behaviour of magnesium alloy based hybrid composites in the longitudinal direction. Wear **267**(1–4), 458–466 (2009). https://doi.org/10.1016/j.wear.2008.12.036

7. A.K. Mondal, S. Kumar, Dry sliding wear behaviour of magnesium alloy based hybrid composites in transverse direction. Mater. Sci. Forum **783–786**, 1530–1535 (2014). https://doi.org/10.4028/www.scientific.net/msf.783-786.1530
8. M. Svoboda, M. Pahutová, K. Kuchařová, V. Sklenička, K.U. Kainer, Microstructure and creep behaviour of magnesium hybrid composites. Mater. Sci. Eng. A **462**(1–2), 220–224 (2007). https://doi.org/10.1016/j.msea.2006.02.466
9. J.S.S. Babu, K.P. Nair, G. Unnikrishnan, C.G. Kang, H.H. Kim, Fabrication and properties of magnesium (AM50)-based hybrid composites with graphite nanofiber and alumina short fiber. J. Compos. Mater. **44**(8), 971–987 (2010). https://doi.org/10.1177/0021998309349548
10. J.S.S. Babu, C.G. Kang, Nanoindentation behaviour of aluminium based hybrid composites with graphite nanofiber/alumina short fiber. Mater. Des. **31**(10), 4881–4885 (2010). https://doi.org/10.1016/j.matdes.2010.05.029
11. X. Zhang, Q. Zhang, H. Hu, Tensile behaviour and microstructure of magnesium AM60-based hybrid composite containing Al_2O_3 fibres and particles. Mater. Sci. Eng. A **607**, 269–276 (2014). https://doi.org/10.1016/j.msea.2014.03.069
12. A.K. Mondal, C. Blawert, S. Kumar, Corrosion behaviour of creep-resistant AE42 magnesium alloy-based hybr id composites developed for powertrain applications. Mater. Corros. **66**(10), 1150–1158 (2015). https://doi.org/10.1002/maco.201408071
13. M.Y. Zhou et al., Achieving ultra-high strength and good ductility in AZ61 alloy composites containing hybrid micron SiC and carbon nanotubes reinforcements. Mater. Sci. Eng. A **768**, 138447 (2019). https://doi.org/10.1016/j.msea.2019.138447
14. J. Zhou, X. Zhang, L. Fang, H. Hu, Processing and properties of as-cast magnesium AM60-based composite co ntaining alumina nano particles and micron fibres, in *Magnesium Technology*, (Springer, 2017), pp. 573–578. https://doi.org/10.1007/978-3-319-52392-7_79
15. B.X. Ma, H.Y. Wang, Y. Wang, Q.C. Jiang, Fabrication of (TiB_2—TiC)p/AZ91 magnesium matrix hybrid composite. J. Mater. Sci. **40**, 4501–4504 (2005). https://doi.org/10.1007/s10853-005-0886-2
16. Z. Xiuqing, W. Haowei, L. Lihua, T. Xinying, M. Naiheng, The mechanical properties of magnesium matrix composites reinforced wi th (TiB_2 + TiC) ceramic particulates. Mater. Lett. **59**(17), 2105–2109 (2005). https://doi.org/10.1016/j.matlet.2005.02.020
17. S. Ugandhar, M. Gupta, S.K. Sinha, Effect of hybrid metallic and ceramic reinforcements on the properties of pure magnesium. Solid State Phenom. **111**, 79–82 (2006). https://doi.org/10.4028/www.scientific.net/SSP.111.79
18. S.K. Sinha, S.U. Reddy, M. Gupta, Scratch hardness and mechanical property correlation for Mg/SiC and Mg/SiC/Ti metal-matrix composites. Tribol. Int. **39**(2), 184–189 (2006). https://doi.org/10.1016/j.triboint.2005.04.017
19. S.K. Thakur, G.T. Kwee, M. Gupta, Development and characterization of magnesium composites containing na no-sized silicon carbide and carbon nanotubes as hybrid reinforcements. J. Mater. Sci. **42**, 10040–10046 (2007). https://doi.org/10.1007/s10853-007-2004-0
20. S.K. Thakur, K. Balasubramanian, M. Gupta, Microwave synthesis and characterization of magnesium based composites containing nanosized SiC and hybrid (SiC +Al_2O_3) reinforcements. J. Eng. Mater. Technol. Trans. SME **129**(2), 194–199 (2007). https://doi.org/10.1115/1.2400279
21. Q.B. Nguyen, K.S. Tun, J. Chan, R. Kwok, J.V.M. Kuma, M. Gupta, Enhancing strength and hardness of AZ31B through simultaneous addition of nickel and nano-Al_2O_3 particulates. Mater. Sci. Eng. A **528**(3), 888–894 (2011). https://doi.org/10.1016/j.msea.2010.10.021
22. Q.B. Nguyen, K.S. Tun, M. Gupta, Effect of nano-alumina and copper micron size particulates on microstr ucture and mechanical properties of magnesium alloy AZ31, in *Magnesium Technology*, (Springer, 2012), pp. 187–189. https://doi.org/10.1007/978-3-319-48203-3_35
23. Q.B. Nguyen et al., Simultaneous effect of nano-Al_2O_3 and micrometre Cu particulates on microstructure and mechanical properties of magnesium alloy AZ31. Mater. Sci. Technol. **28**(2), 227–233 (2012). https://doi.org/10.1179/1743284711Y.0000000023

24. Q.B. Nguyen, M. Gupta, Enhancing mechanical response of AZ31B using Cu + nano-Al_2O_3 addition. Mater. Sci. Eng. A **527**(6), 1411–1416 (2010). https://doi.org/10.1016/j.msea.2009.11.002
25. S. Sankaranarayanan, S. Jayalakshmi, M. Gupta, Effect of ball milling the hybrid reinforcements on the microstructure and mechanical properties of Mg-(Ti + n-Al_2O_3) composites. J. Alloys Compd. **509**(26), 7229–7237 (2011). https://doi.org/10.1016/j.jallcom.2011.04.083
26. S. Seetharaman, J. Subramanian, M. Gupta, A.S. Hamouda, Influence of micron-Ti and Nano-Cu additions on the microstructure and mechanical properties of pure magnesium. Metals **2**(3), 274–291 (2012). https://doi.org/10.3390/met2030274
27. S. Sankaranarayanan, S. Jayalakshmi, M. Gupta, Effect of addition of mutually soluble and insoluble metallic elements on the microstructure, tensile and compressive properties of pure mag nesium. Mater. Sci. Eng. A **530**(1), 149–160 (2011). https://doi.org/10.1016/j.msea.2011.09.066
28. S. Sankaranarayanan, R.K. Sabat, S. Jayalakshmi, S. Suwas, M. Gupta, Effect of hybridizing micron-sized Ti with nano-sized SiC on the micro structural evolution and mechanical response of Mg-5.6Ti composite. J. Alloys Compd. **575**, 207–217 (2013). https://doi.org/10.1016/j.jallcom.2013.04.095
29. S. Sankaranarayanan, S. Jayalakshmi, M. Gupta, Hybridizing micro-Ti with nano-B_4C particulates to improve the microstructural and mechanical characteristics of Mg–Ti composite. J. Magnes. Alloys **2**(1), 13–19 (2014). https://doi.org/10.1016/j.jma.2014.03.001
30. S. Sankaranarayanan, R.K. Sabat, S. Jayalakshmi, S. Suwas, M. Gupta, Microstructural evolution and mechanical properties of Mg composites c ontaining nano-B_4C hybridized micro-Ti particulates. Mater. Chem. Phys. **143**(3), 1178–1190 (2014). https://doi.org/10.1016/j.matchemphys.2013.11.019
31. K.S. Tun, M. Gupta, Development of magnesium/(yttria + nickel) hybrid nanocomposites using hybrid microwave sintering: Microstructure and tensile properties. J. Alloys Compd. **487**(1–2), 76–82 (2009). https://doi.org/10.1016/j.jallcom.2009.07.117
32. S.F. Hassan, K.S. Tun, F. Patel, N. Al-Aqeeli, M. Gupta, Microwave sintered magnesium nanocomposite: Hybrid (Y_2O_3+Ni) nano-size reinforcement and ensile properties. Adv. Compos. Lett. **25**(4), 103–107 (2016). https://doi.org/10.1177/096369351602500404
33. K.S. Tun, M. Gupta, Compressive deformation behavior of Mg and Mg/(Y_2O_3+Ni) nanocomposites. Mater. Sci. Eng. A **527**(21–22), 5550–5556 (2010). https://doi.org/10.1016/j.msea.2010.05.025
34. O. Olalekan, *Development of Magnesium Based Hybrid Nanocomposite* (King Fahd University of Petroleum and Minerals (Saudi Arabia), 2016)
35. K.S. Tun, W.L.E. Wong, Q.B. Nguyen, M. Gupta, Tensile and compressive responses of ceramic and metallic nanoparticle reinforced Mg composites. Materials **6**(5), 1826–1839 (2013). https://doi.org/10.3390/ma6051826
36. K.S. Tun et al., Enhancing tensile and compressive strengths of magnesium using nanosiz e (Al_2O_3 + Cu) hybrid reinforcements. J. Compos. Mater. **46**(15), 1879–1887 (2012). https://doi.org/10.1177/0021998311427767
37. S. Fida Hassan, K.S. Tun, F. Patel, M. Gupta, Effect of copper nano particles on high temperature tensile behavior o f Mg-Y_2O_3 nanocomposite. Met. Mater. Int. **21**, 588–592 (2015). https://doi.org/10.1007/s12540-015-4188-1
38. M.K. Habibi, M. Paramsothy, A.M.S. Hamouda, M. Gupta, Tensile strength and ductility improvement of magnesium by using ball milled Al-CNT particles as reinforcement, in *Materials Science and Technology Conference and Exhibition 2010*, (Curran Associates, 2010)
39. M.K. Habibi, M. Paramsothy, A.M.S. Hamouda, M. Gupta, Using integrated hybrid (Al+CNT) reinforcement to simultaneously enhance strength and ductility of magnesium. Compos. Sci. Technol. **71**(5), 734–741 (2011). https://doi.org/10.1016/j.compscitech.2011.01.021
40. M.K. Habibi, A.M.S. Hamouda, M. Gupta, Enhancing tensile and compressive strength of magnesium using ball mil led Al+CNT reinforcement. Compos. Sci. Technol. **72**(2), 290–298 (2012). https://doi.org/10.1016/j.compscitech.2011.11.015

41. M.K. Habibi, S.P. Joshi, M. Gupta, Hierarchical magnesium nano-composites for enhanced mechanical response. Acta Mater. **58**(18), 6104–6114 (2010). https://doi.org/10.1016/j.actamat.2010.07.028
42. M.K. Habibi, S.P. Joshi, M. Gupta, Development of hierarchical magnesium composites using hybrid microwave sintering. J. Microw. Power Electromagn. Energy **45**(3), 112–120 (2011). https://doi.org/10.1080/08327823.2011.11689805
43. M.K. Habibi, M. Qian, A.S. Hamouda, M. Gupta, Differentiating the mechanical response of hierarchical magnesium nano-composites as a function of temperature. Mater. Des. **42**, 102–110 (2012). https://doi.org/10.1016/j.matdes.2012.05.037
44. M.K. Habibi, A.S. Hamouda, M. Gupta, Using hierarchical composite approach to improve mechanical response o f Mg and Mg-Bi_2O_3nano-composites. Mater. Des. **49**, 627–637 (2013). https://doi.org/10.1016/j.matdes.2013.02.028
45. M.K. Habibi, A.S. Hamouda, M. Gupta, Hybridizing boron carbide (B_4C) particles with aluminum (Al) to enhanc e the mechanical response of magnesium based nano-composites. J. Alloys Compd. **550**, 83–93 (2013). https://doi.org/10.1016/j.jallcom.2012.09.128
46. M. Rashad, F. Pan, W. Guo, H. Lin, M. Asif, M. Irfan, Effect of alumina and silicon carbide hybrid reinforcements on tensile, compressive and microhardness behavior of Mg-3Al-1Zn alloy. Mater. Charact. **106**, 382–389 (2015). https://doi.org/10.1016/j.matchar.2015.06.033
47. M. Rashad, F.S. Pan, M. Asif, A. Ullah, Improved mechanical properties of magnesium-graphene composites with c opper-graphene hybrids. Mater. Sci. Technol. **31**, 1452–1461 (2015). https://doi.org/10.1179/1743284714Y.0000000726
48. M. Rashad, F. Pan, M. Asif, J. She, A. Ullah, Improved mechanical proprieties of 'magnesium based composites' with t itanium-aluminum hybrids. J. Magnes. Alloys **3**(1), 1–9 (2015). https://doi.org/10.1016/j.jma.2014.12.010
49. I. Aatthisugan, A. Razal Rose, D. Selwyn Jebadurai, Mechanical and wear behaviour of AZ91D magnesium matrix hybrid composite reinforced with boron carbide and graphite. J. Magnes. Alloys **5**(1), 20–25 (2017). https://doi.org/10.1016/j.jma.2016.12.004
50. S. Arokiasamy, B. Anand Ronald, Experimental investigations on the enhancement of mechanical properties of magnesium-based hybrid metal matrix composites through friction stir processing. Int. J. Adv. Manuf. Technol. **93**, 493–503 (2017). https://doi.org/10.1007/s00170-017-0221-5
51. S. Arokiasamy and B. Anand Ronald, "Enhanced properties of magnesium based metal matrix composites via friction stir processing," in Materials Today: Proceedings, 5(2), 6934-6939. 2018. doi: https://doi.org/10.1016/j.matpr.2017.11.355
52. S. Sharma, A. Handa, S.S. Singh, D. Verma, Influence of tool rotation speeds on mechanical and morphological properties of friction stir processed nano hybrid composite of MWCNT-graphene-AZ31 magnesium. J. Magnes. Alloys **7**(3), 487–500 (2019). https://doi.org/10.1016/j.jma.2019.07.001
53. X. Zhou, D. Su, C. Wu, L. Liu, Tensile mechanical properties and strengthening mechanism of hybrid carbon nanotube and silicon carbide nanoparticle-reinforced magnesium alloy composites. J. Nanomater. **2012**, 851862 (2012). https://doi.org/10.1155/2012/851862
54. D. Lu, Y. Jiang, R. Zhou, Wear performance of nano-Al_2O_3 particles and CNTs reinforced magnesium matrix composites by friction stir processing. Wear **305**(1–2), 286–290 (2013). https://doi.org/10.1016/j.wear.2012.11.079
55. M.E. Turan, H. Zengin, Y. Sun, Dry sliding wear behavior of (MWCNT + GNPs) reinforced AZ91 magnesium matrix hybrid composites. Met. Mater. Int. **26**, 541–550 (2019). https://doi.org/10.1007/s12540-019-00338-8
56. K. Soorya Prakash, P. Balasundar, S. Nagaraja, P.M. Gopal, V. Kavimani, Mechanical and wear behaviour of Mg–SiC–Gr hybrid composites. J. Magnes. Alloys **4**(3), 197–206 (2016). https://doi.org/10.1016/j.jma.2016.08.001
57. B.N. Sahoo, S.K. Panigrahi, A study on the combined effect of in-situ (TiC-TiB_2) reinforcement and aging treatment on the yield asymmetry of magnesium matrix composite. J. Alloys Compd. **737**, 575–589 (2018). https://doi.org/10.1016/j.jallcom.2017.12.027

58. K. Asano, H. Yoneda, High temperature properties of AZ91D magnesium alloy composite reinfor ced with short alumina fiber and Mg_2Si particle. Mater. Trans. **49**(7), 1688–1693 (2008). https://doi.org/10.2320/matertrans.MER2008092
59. S. Fida Hassan, N. Al-Aqeeli, Z.M. Gasem, K.S. Tun, M. Gupta, Magnesium nanocomposite: Increasing copperisation effect on high temperature tensile properties. Powder Metall. **59**(1), 160209042533005 (2016). https://doi.org/10.1080/00325899.2015.1109816
60. K.S. Tun, M. Gupta, Role of microstructure and texture on compressive strength improvement of Mg/(Y_2O_3 + Cu) hybrid nanocomposites. J. Compos. Mater. **44**(25), 3033–3050 (2010). https://doi.org/10.1177/0021998310369591
61. J. Zhou, L. Ren, X. Geng, L. Fang, H. Hu, As-cast magnesium AM60-based hybrid nanocomposite containing alumina f ibres and nanoparticles: Microstructure and tensile behavior. Mater. Sci. Eng. A **740–741**, 305–314 (2019). https://doi.org/10.1016/j.msea.2017.10.070

Chapter 7
Magnesium-Based Composites Containing Novel Reinforcements

7.1 Introduction

Magnesium-based composites with novel reinforcements represent a cutting-edge approach to developing lightweight, high-performance materials for structural and industrial applications. By incorporating innovative reinforcements such as hollow alumina spheres, glass micro-balloons, fly ash cenospheres, and metallic glasses, these composites achieve unique combinations of mechanical, thermal, and corrosion-resistant properties. Advanced fabrication techniques, including sub-atmospheric pressure infiltration, microwave-assisted sintering, and powder metallurgy, further enhance their performance. These materials hold significant promise for aerospace, automotive, and biomedical applications, paving the way for sustainable and efficient engineering solutions.

7.2 Magnesium-Based Syntactic Foam Composites

Syntactic magnesium composites reinforced with hollow reinforcements have demonstrated remarkable potential for lightweight structural applications, owing to their unique properties and this section reviews the recent advances in magnesium-based syntactic foam composites focusing on the reinforcement types, the fabrication methods, density, mechanical properties, corrosion attributes, and potential applications. Techniques such as sub-atmospheric pressure infiltration, hot press sintering, and melt stir casting are discussed, highlighting their influence on material properties. Key parameters like sphere size, wall thickness-to-diameter ratio, and interfacial modifications are analyzed for their impact on compressive strength, stiffness, and energy absorption. The document concludes with insights into future directions, focusing on interfacial engineering and process optimization to enhance performance and broaden application versatility.

S. Seetharaman et al., *Magnesium-based Syntactic, Metastable and Nano-composites*, https://doi.org/10.1007/978-3-031-96514-2_7

A pivotal study described in Hartmann et al. [1] showcased the fabrication of syntactic foams incorporating thin-walled hollow alumina spheres within various magnesium alloy matrices, such as cp-Mg, AM20, AM50, and AZ91. These foams displayed a uniform closed-cell structure with high sphere volume fractions, approximately 63% for random packing and 74% for hexagonal close-packed arrays. The results of mechanical testing revealed superior specific stiffness and strength compared to aluminum foams, particularly under axial and bending loads (Fig. 7.1). Notably, the authors also reported that the compressive strength of the syntactic composites was largely influenced by the sphere diameter-to-wall thickness ratio, thereby highlighting the fact the energy absorption capacity of the composites can be tailored for potential energy-dissipative applications.

Newsome et al. [2] used a sub-atmospheric pressure infiltration method to produce AZ91D-Al_2O_3 syntactic foams which were then investigated for quasi-static compressive properties. Here, they studied influence of the size and wall thickness-to-diameter ratio of the hollow spheres on mechanical properties of syntactic composites. From the experimental results, they found that the smaller spheres (0.106–0.212 mm) with higher t/D ratios delivered superior performance, including higher energy absorption per unit weight and peak strengths (Fig. 7.2). Furthermore, the stress–strain analyses also revealed a distinct peak stress followed by a plateau phase and densification, affirming the high energy absorption of the composite during compression.

Li et al. [3] employed a hot press sintering method to fabricate closed-cell Mg matrix syntactic foams reinforced with hollow Al_2O_3 particles. They employed a

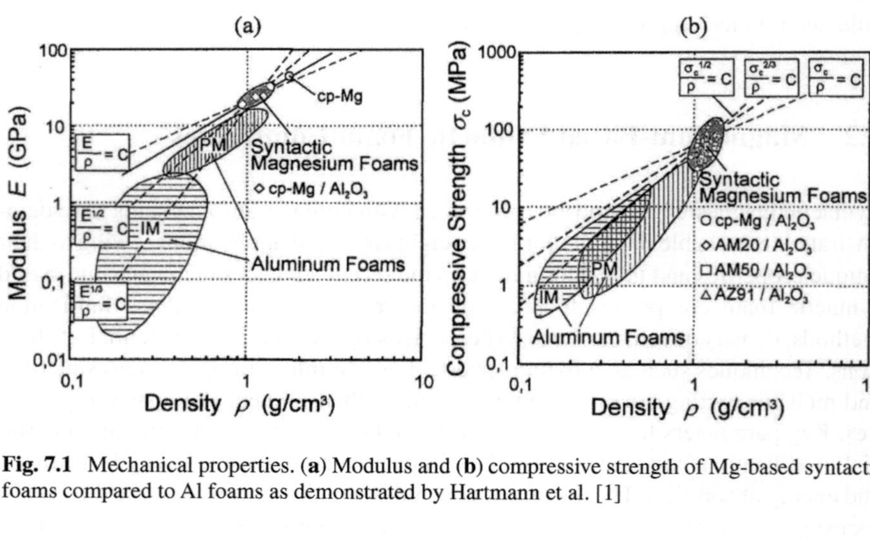

Fig. 7.1 Mechanical properties. (**a**) Modulus and (**b**) compressive strength of Mg-based syntactic foams compared to Al foams as demonstrated by Hartmann et al. [1]

range of sintering temperatures and found 723 K as the optimal sintering temperature to achieve a maximum compressive strength of 162 and 167.87 MPa under quasi-static and dynamic loading. The enhanced mechanical strengths were attributed to the formation of $MgAl_2O_4$ phases at the interface. On the other hand, excessive sintering temperatures led to brittleness and degraded performance (Fig. 7.3).

An innovative method involving gravity infiltration was presented by Qu et al. [4] to produce Mg matrix syntactic foams with a high porosity of more than 41.64 vol% and low density (1.17 g/cm^3). In this study, the hollow Al_2O_3 spheres were coated with MgO using $MgCl_2$ solution and high-temperature calcination which suppressed the interfacial reactions and improved the mechanical performance (Fig. 7.4). An optimal $MgCl_2$ concentration of 0.15 g/mL yielded a plateau stress of 55.96 MPa, and an energy absorption capacity of 39.27 MJ/m^3. In particular, the maximum specific energy absorption of 26.06 kJ/kg achieved in this study matched or exceeded those foams produced by other methods, thereby demonstrating scalability and industrial potential.

Squeeze casting produced AZ91 foams reinforced with hollow alumina microspheres were examined for machining characteristics [5]. Cutting experiments

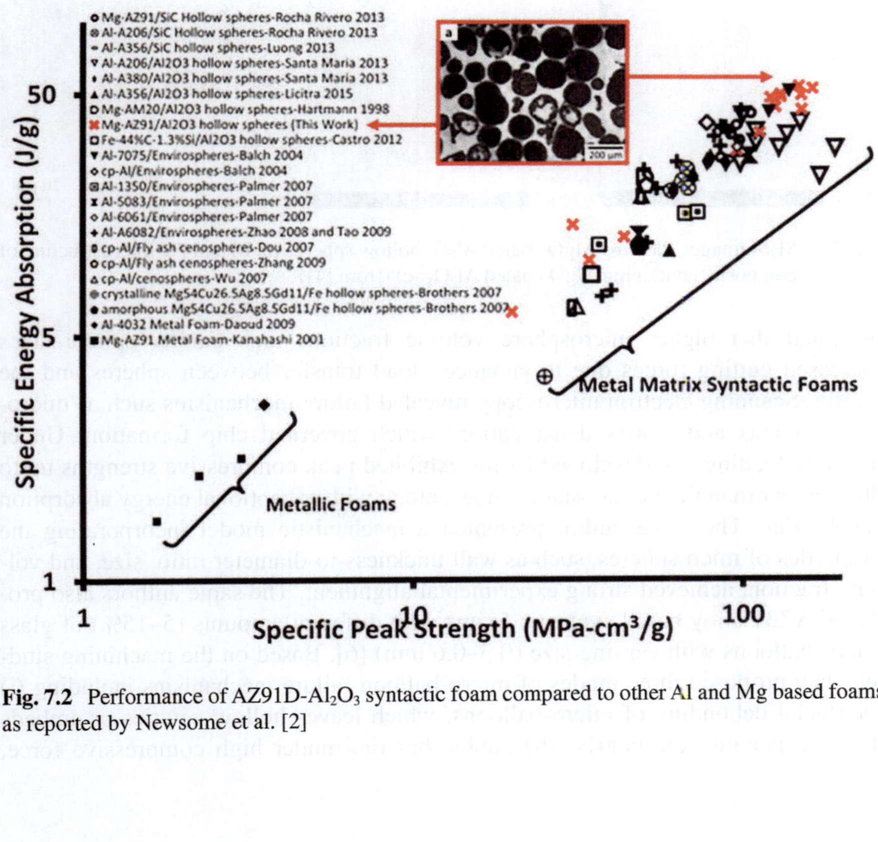

Fig. 7.2 Performance of AZ91D-Al_2O_3 syntactic foam compared to other Al and Mg based foams as reported by Newsome et al. [2]

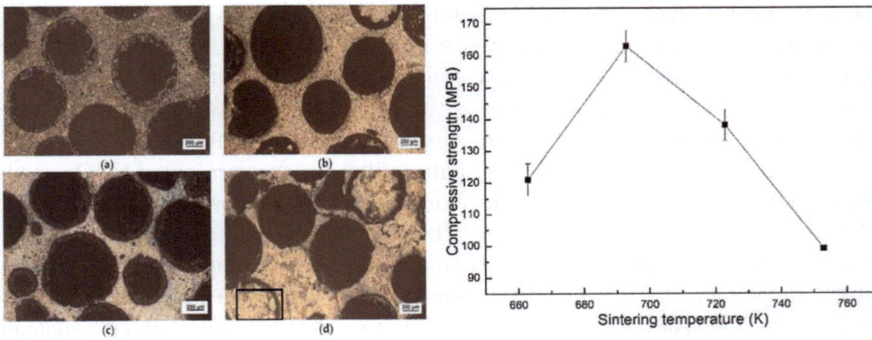

Fig. 7.3 Microstructure of Al_2O_{3hs}/AZ91D prepared at different temperatures: (**a**) 663 K, (**b**) 693 K, (**c**) 723 K, and (**d**) 753 K. (**e**) Quasi-static compressive strength of the Al_2O_{3hs}/AZ91D syntactic foams from Li et al. [3]

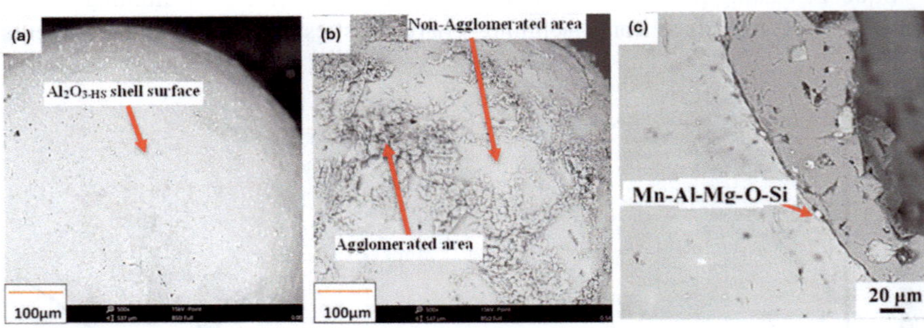

Fig. 7.4 SEM images showing MgO coated Al_2O_3 hollow spheres (**a, b**) and the microstructure of syntactic composite containing MgO coated Al_2O_3 (**c**) (from [4])

indicated that higher microsphere volume fractions and smaller sphere sizes increased cutting forces due to enhanced load transfer between spheres and the matrix. Scanning electron microscopy revealed failure mechanisms such as microsphere bursts and matrix densification, which governed chip formation. Under dynamic loading, the developed foams exhibited peak compressive strengths up to 30% higher than their quasi-static values, along with exceptional energy absorption capabilities. The same author presented a mechanistic model incorporating the properties of microspheres, such as wall thickness-to-diameter ratio, size, and volume fraction, achieved strong experimental alignment. The same authors also produced AZ31 alloy based syntactic foams with different amounts (5–15%) of glass micro-balloons with varying size (0.3–0.6 mm) [6]. Based on the machining studies, they proposed three modes of micro-balloon failure mechanisms including (i) interfacial debonding of micro-balloons, which leaves hollow cavities or embeds them deeper into the matrix; (ii) bubble bursting under high compressive force,

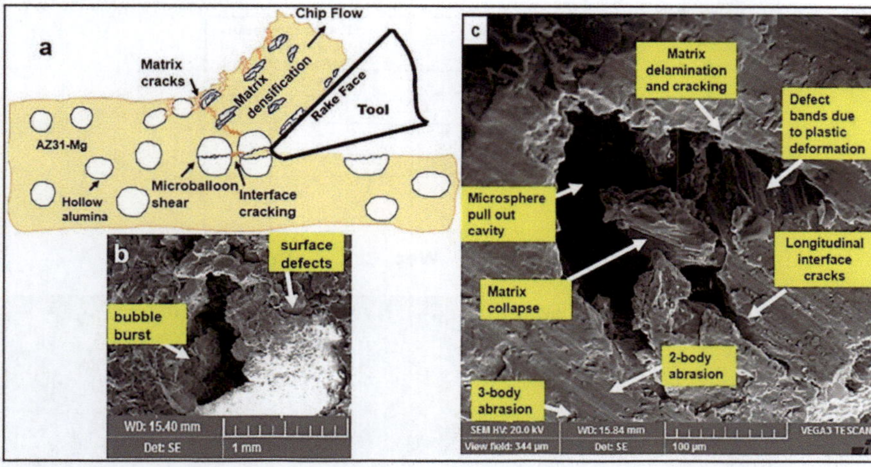

Fig. 7.5 Ceramic micro-balloon failure mechanisms constituting shear failure during cutting AZ31 metal syntactic foams. (**a**) representative chip formation (**b**) fractured bubble (**c**) key deformation mechanisms [6]

which transfers the load to the matrix, collapses cavities, and leads to densification through shear deformation; and (iii) shear failure at higher bubble diameter-to-feed ratios (Davg/h), where weaker bubbles are cut by the tool, causing cracks to propagate (Fig. 7.5).

In the following study, the high-speed machining trials revealed that finer micro-balloons and higher volume fractions increased cutting forces due to enhanced load transfer. The novel force model proposed by the authors characterized the interplay between matrix deformation and micro-balloon behavior, identifying bubble burst and fracture as key mechanisms (Equations). Similarly, the effectiveness of cryogenic machining was analyzed and the results showed a significant improvement in surface finish compared to wet and dry machining methods (Fig. 7.6) [7].

Magnesium-based syntactic foams incorporating hollow fly ash cenospheres were synthesized using the disintegrated melt deposition (DMD) technique [8]. Here, the uniformly dispersed cenospheres were found to enhance the tensile and compressive strengths of the composites while maintaining the densities as comparable to plastics (~1.8 g/cm^3) (Fig. 7.7). A similar approach was adopted by Sankaranarayanan et al. [9], where the fly ash cenospheres were introduced into the Mg matrix through a microwave sintering assisted powder metallurgy method. Here, the inclusion of up to 15 wt% cenospheres reduced the composite density by 18% with intact microstructures characterized by MgO and Mg_2Si intermetallic phases. Although tensile ductility decreased slightly, the composites retained their specific mechanical properties, underscoring their potential for lightweight applications while effectively utilizing industrial waste.

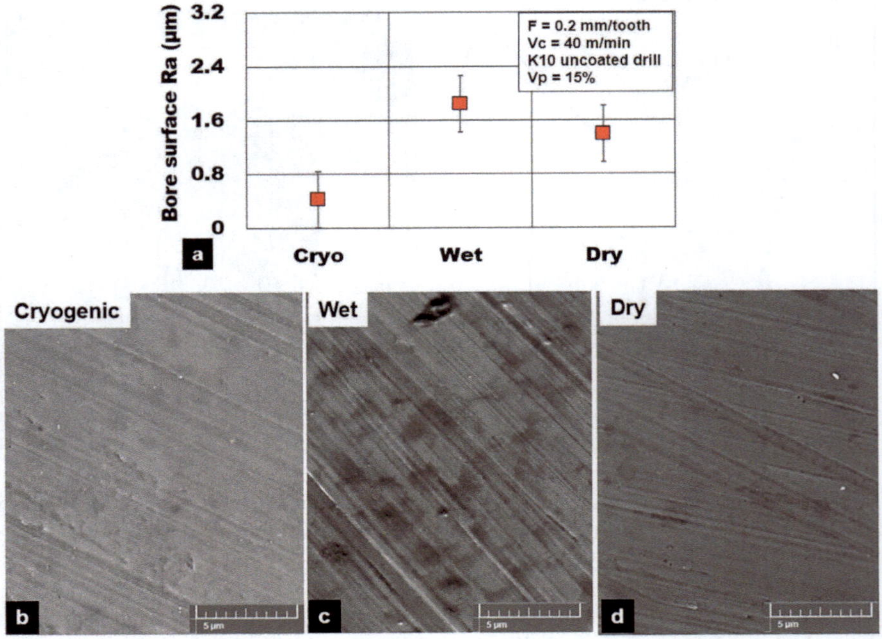

Fig. 7.6 Effectiveness of cryogenic machining on surface finish [7]

Huang et al. [10] used the melt stir casting method to produce AZ91D based syntactic composites reinforced with fly ash cenosphere particles. On investigating the microstructure and compressive response, they found significant improvements to the compressive strength and the energy absorption capabilities of the syntactic composites due to the formation of Mg_2Si and MgO phases at the interfaces (Fig. 7.8). The authors also highlighted the fact processing conditions, that controlled temperatures and holding times, refined the microstructure and maximized the mechanical performance.

In another study, AZ61 alloy based foam composites were produced using fly ash microspheres and microwave-assisted rapid sintering [11]. Here, the incorporation of up to 40 vol% fly ash microspheres reduced the density of syntactic composites by 23%. The developed composites were also tested for corrosion performance in 3 wt% NaCl solution and the results revealed an enhanced corrosion resistance due to the reduced matrix exposure during rapid sintering (Fig. 7.9). In a follow up study, the authors also developed a floatable structure with an innovative design of hybridized open- and closed-cells using AZ61 alloy syntactic foams [12]. In this study, a density as low as 0.79 g/cm^3 was achieved through the incorporation of hollow fly ash spheres and carbamide granules. With respect to the load bearing attributes, a compressive strength of 16 MPa was achieved as the rapid microwave sintering preserved the structural

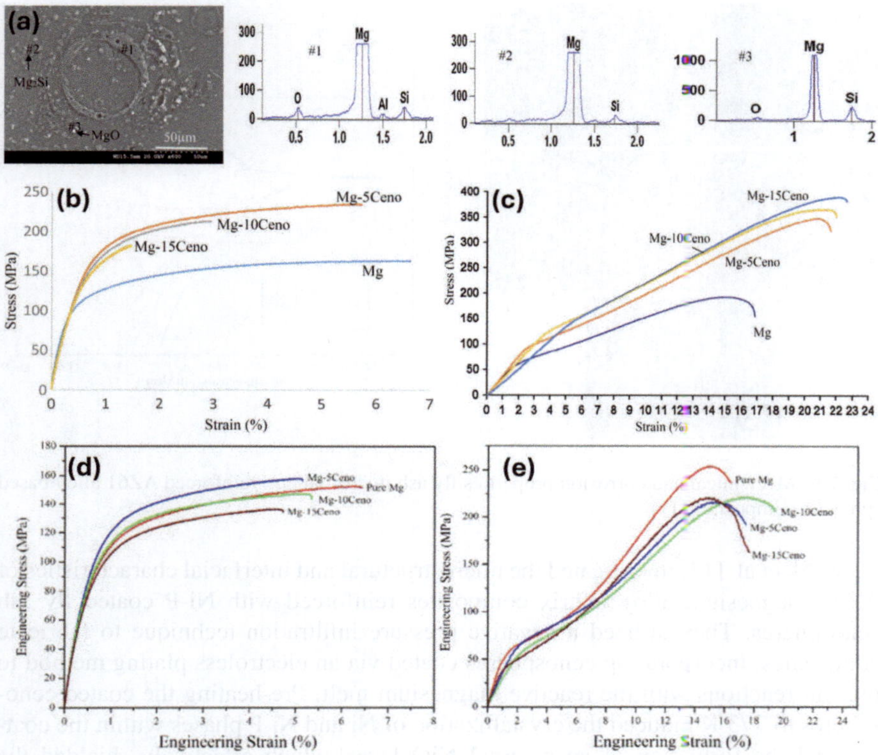

Fig. 7.7 Microstructure and mechanical properties of flyash cenosphere reinforced Mg composite produced by DMD (**a–c**) and microwave-assisted powder metallurgy (**d, e**) methods [8]

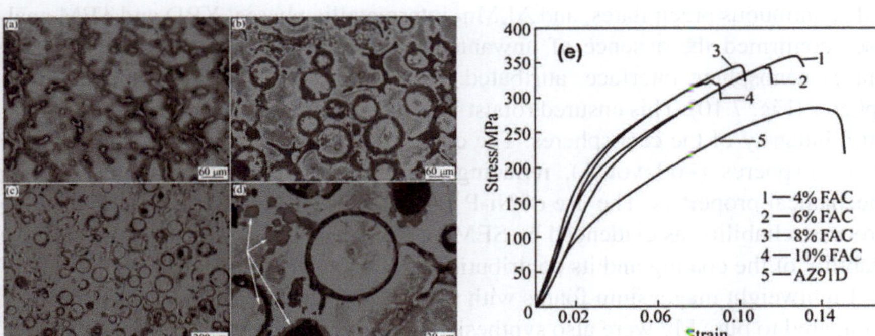

Fig. 7.8 Microstructure and mechanical properties of AZ91D [10]

integrity of the hollow spheres. Based on these properties, the authors proposed these materials as a potential alternative to polymers for buoyant applications and chemical release systems.

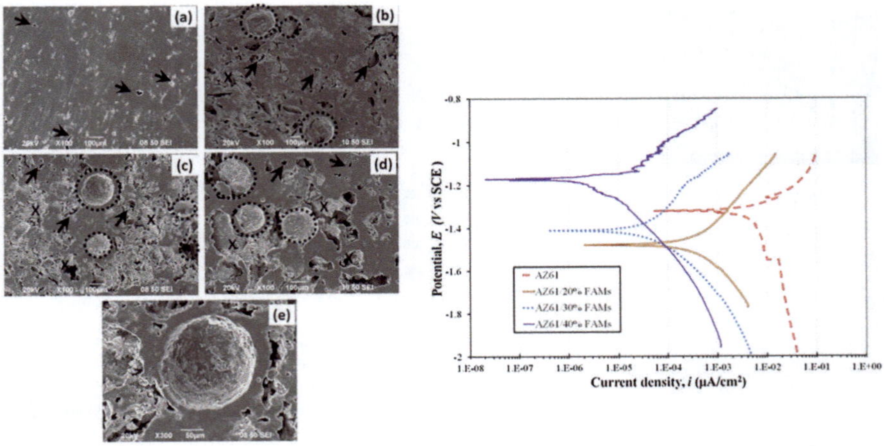

Fig. 7.9 Mechanical and corrosion properties fly ash micro-balloon reinforced AZ61 alloy based syntactic composites [11]

Malik et al. [13] investigated the microstructural and interfacial characteristics of AZ91 magnesium alloy matrix composites reinforced with Ni-P coated fly ash cenospheres. They utilized a negative pressure infiltration technique to fabricate composites, incorporating cenospheres coated via an electroless plating method to prevent reactions with the reactive magnesium melt. Pre-heating the coated cenospheres to 773 K induced the crystallization of Ni and Ni_3P phases within the coating and the formation of an external NiO layer, which effectively shielded the cenospheres from chemical degradation during processing. Microstructural analyses revealed a uniform distribution of cenospheres, which remained intact and unfilled by the matrix alloy. The matrix comprised α-Mg, an $\alpha + \gamma$ divorced eutectic, γ discontinuous precipitates, and Al_8Mn_5 intermetallic phases. XRD and TEM analyses confirmed the absence of unwanted reaction products such as MgO at the matrix-cenosphere interface, attributed to the protective NiO layer on the cenospheres (Fig. 7.10). This ensured robust interfacial bonding and preserved the structural integrity of the cenospheres. The composite achieved a high-volume fraction of cenospheres (~60 vol.%), reducing its density while maintaining excellent mechanical properties. The use of Ni-P coatings proved instrumental in enhancing process reliability, as evidenced by SEM and EDX analyses, which highlighted the stability of the coating and its contribution to strong interfacial bonding.

Lightweight magnesium foams with significant density reductions of up to 26% compared to pure Mg were also synthesized using hybrid microwave sintering [14]. Despite a decrease in yield strength with increased GMB content, the ultimate compressive strength, fracture strain, and energy absorption capabilities improved markedly. The Mg-5 wt.% GMB composite exhibited an ultimate compressive strength of 321 MPa, a 26% increase over pure magnesium. Meanwhile, the Mg-20 wt.% GMB composite achieved a 39% higher fracture strain and 65%

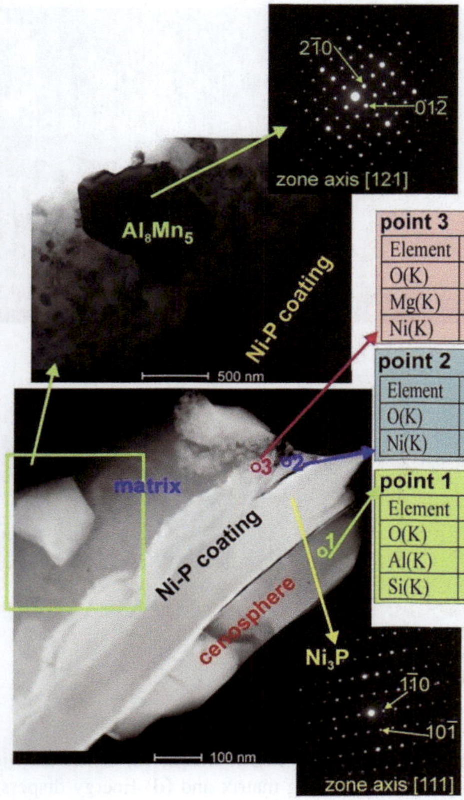

Fig. 7.10 TEM images (dark and bright field) showing the interfacial microstructure of the composite with EDAX results from designated points and diffraction patterns obtained from observed structural constituent [13]

greater energy absorption capability, showcasing these foams' suitability for energy-intensive applications. In a parallel study, syntactic foams based on pure Mg and glass micro-balloons with densities ranging from 1.47 to 1.67 g/cm^3 were fabricated using the disintegrated melt deposition technique (Fig. 7.11) [15]. The developed composites exhibited remarkable performance with 25 wt.% GMB addition resulting in a CYS and UCS of 161 and 232 MPa. Thus, the presence of glass micro-balloons enhanced the energy absorption capability by approximately 200%. Interestingly, the failure strain was also found to be as high as 37.7%.

In a related study, the wear behavior of Mg-based composite foams containing GMBs was examined [16]. Here, the incorporation of 25 wt.% GMB into Mg reduced the wear rate by approximately 2.5 times and lowered the coefficient of friction by 13% compared to pure Mg (Fig. 7.12). This improvement was attributed to the cushioning effect of hollow micro-balloons, which minimized surface scratching and delamination wear.

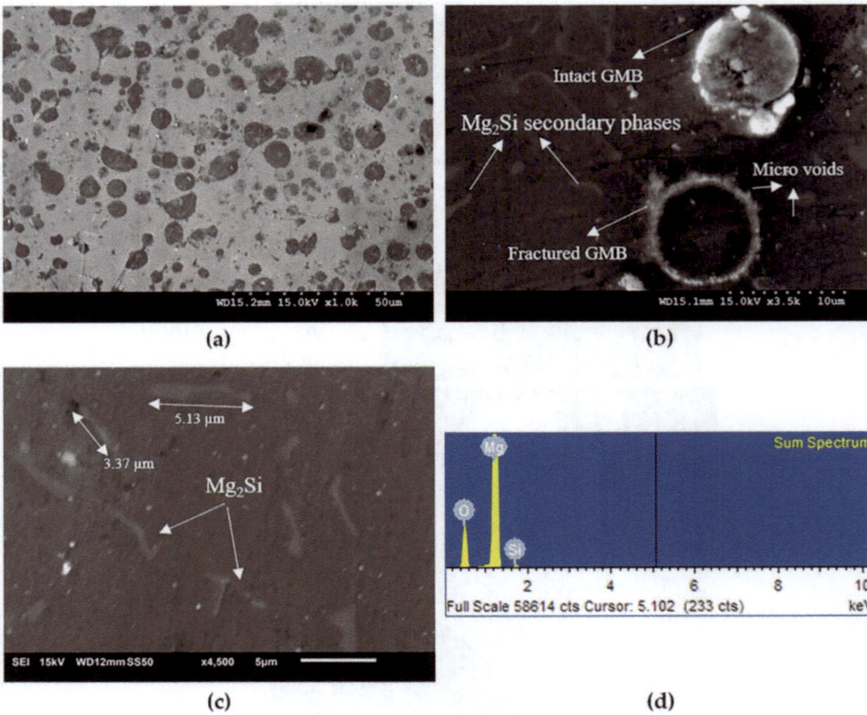

Fig. 7.11 (**a**) Representative microstructure of Mg-15 foam; (**b**) GMB showing uniform wall thickness (**c**) Mg_2Si dendrites present in the Mg matrix and (**d**) Energy dispersive spectrometer (EDS) result indicating the presence of Mg_2Si as secondary phases [15]

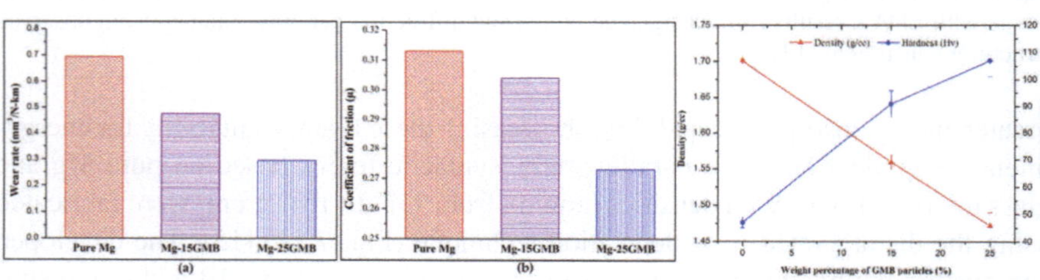

Fig. 7.12 Tribological behavior and densities of Mg/GMB syntactic composites developed by Manakari et al. [16]

Magnesium-lithium alloys reinforced with hollow glass micro-balloons were also developed by the DMD method [17]. The developed composites exhibited densities ranging from 1.45 to 1.66 g/cm^3, representing a significant weight reduction compared to traditional magnesium alloys. The inclusion of lithium improved ductility, while GMB reinforcement enhanced specific strength, thermal stability, and energy absorption properties. Damping behavior analysis revealed that the foams

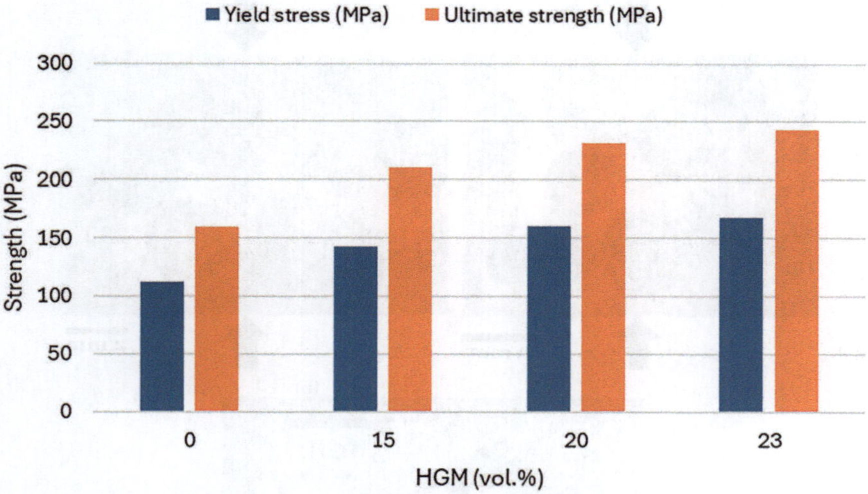

Fig. 7.13 Compressive properties of AZ91D/HGM Syntactic foams (from [19])

could effectively dissipate vibrational energy, making them ideal for aerospace and automotive applications requiring lightweight and energy-efficient materials.

Mg-based syntactic foams based on AZ91D alloy and hollow glass microspheres (HGMs) were also produced via powder metallurgy [18]. This study emphasized the influence of reinforcement content and process parameters on the mechanical properties of syntactic composites. It was found in this study that the increasing amounts of HGM content reduced density to as low as 0.912 g/cm^3 while maintaining high compressive strength and energy absorption capabilities. The authors also used melt stirring and vacuum die casting methods to produce AZ91 alloy-based syntactic composites reinforced with HGM particles [19]. Here, the addition of 15–23% HGMs required the optimization of process parameters such as particle size, stirring speed, and reinforcement content. The microstructural analysis confirmed uniform dispersion of HGMs with good wettability and interfacial bonding along with the presence of $Mg_{17}Al_{12}$ intermetallic phases which were attributed to the improved load transfer and fracture resistance resulting in a compressive strength of 280 MPa and hardness of 148 HV (Fig. 7.13). Here, the authors also clearly highlighted the role of sintering process in achieving the uniform microstructures and strong interfacial bonding [20].

Anantharaman et al. [21] explored the dynamic behavior and damping properties of AZ91D magnesium alloy syntactic foams reinforced with silicon carbide hollow particles (SiCHP). The foams, fabricated using a sub-atmospheric pressure infiltration technique, featured uniform particle dispersion, minimal porosity, and ultra-low densities as low as 0.97 g/cm^3, significantly enhancing weight reduction compared to conventional metallic materials (Fig. 7.14). Despite their low density, the foams maintained excellent mechanical properties, combining lightweight

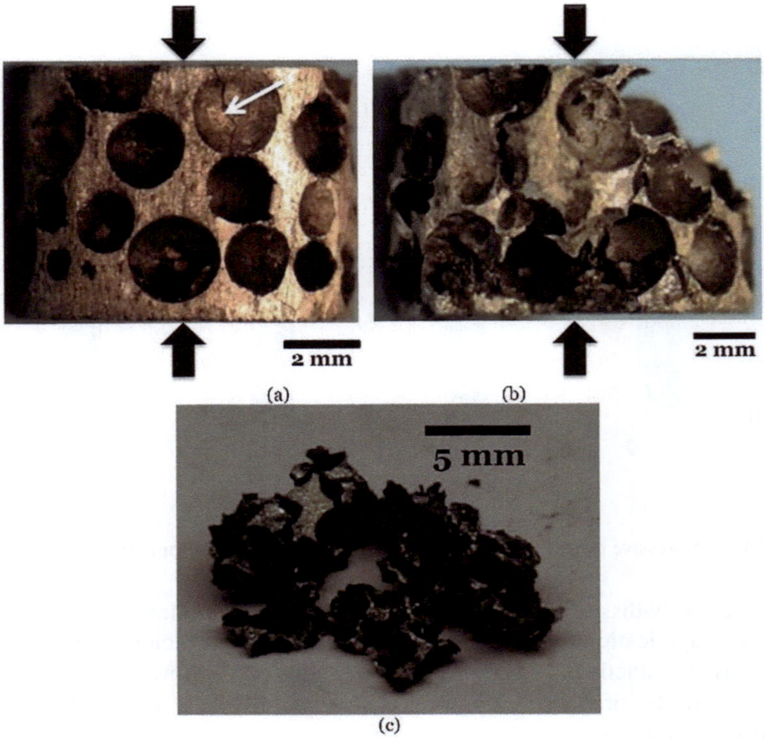

Fig. 7.14 Images showing a specimen of AZ91D/SiCHP syntactic foam at different stages of quasi-static compression [21]

advantages with superior strength and temperature resistance. Dynamic mechanical testing revealed strain-rate sensitivity, with peak compressive stresses and elastic energy absorption up to 1.5 times greater at high strain rates (1330–2300/s) compared to quasi-static loading. Deformation mechanisms included SiC particle fracture, matrix plasticity, and precipitate failure along grain boundaries, which facilitated energy dissipation and delayed catastrophic failure. Dynamic mechanical analysis (DMA) highlighted exceptional vibration damping across a temperature range of 35°C–450°C, with a consistently higher damping parameter (tan δ) than the base alloy. While stiffness decreased with temperature, vibration attenuation improved, as confirmed by master curves constructed using the time-temperature superposition (TTS) principle.

Kubelka et al. [22] studied the effect of interfacial characteristics on the properties of syntactic foams with hollow ceramic spheres and found that they are directly influenced by the cooling conditions. For example, quench cooling resulted in the highest peak strength of 84.30 MPa and stiffness of 4.28 GPa, while reducing the stiffness to 1.93 GPa due to the formation of micropores. On the other hand, the machine-cooled samples displayed intermediate properties. These results underscored the critical role of interface modifications in controlling strength, stiffness, and ductility, highlighting the need for precise control of thermal and mechanical processing parameters (Fig. 7.15).

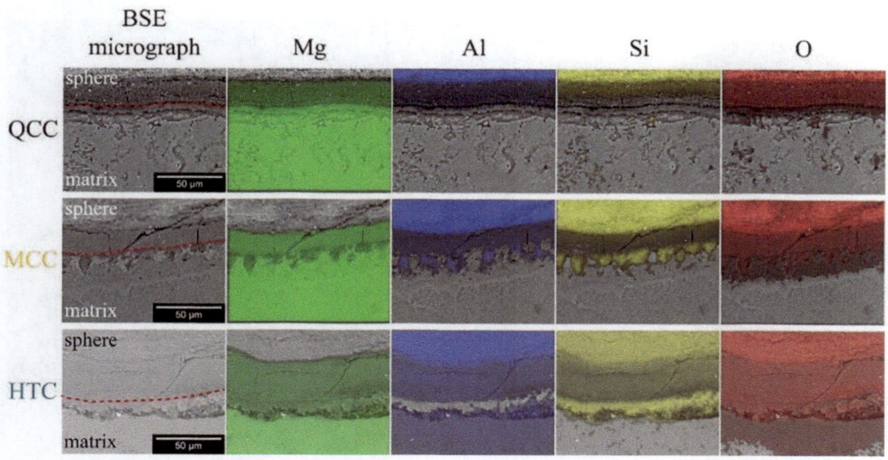

Fig. 7.15 EDX mappings of the respective interfaces of the syntactic foams with the three different material conditions quench cooling, machine cooling, and heat-treated condition with the four present elements Mg, Al, Si, and O. The interface itself is indicated with a red-dotted line in the BSE micrographs [22]

Movahedi et al. [23] developed AZ91 magnesium alloy syntactic foams reinforced with activated carbon (AC) particles using a counter-gravity infiltration method. The foams achieved ultra-low densities between 1.12 and 1.18 g/cm^3, with AC particles demonstrating chemical stability and no significant reactions with the matrix. Microstructural analysis confirmed a clean interface and uniform particle distribution, with MgO formation but no magnesium carbides. Mechanical testing revealed compressive strengths ranging from 32.45 to 51.43 MPa, plateau stresses averaging 12.71 MPa, and volumetric energy absorption up to 13.49 MJ/m^3. Deformation was characterized by brittle shear fracture due to the β-$Mg_{17}Al_{12}$ phase in the matrix (Fig. 7.16).

Dong et al. [24] developed Mg-based syntactic foams using two types of volcanic rock particles, high-density volcanic rock (HD-VR), and low-density volcanic rock (LD-VR), as reinforcements. The foams were fabricated using a pressure infiltration technique, resulting in densities of 1.67 g/cm^3 (HD-VR/Mg) and 0.89 g/cm^3 (LD-VR/Mg) with porosities of 25.5% and 49.5%, respectively. HD-VR particles, characterized by thicker walls and higher density, imparted superior compressive and plateau strengths, reaching 37.8 MPa at room temperature. In contrast, LD-VR particles, with their tubular pore structure and higher porosity, produced foams with more stable deformation and better energy absorption efficiency. Compressive testing revealed temperature-dependent behavior, with both foam types showing reduced strength and energy absorption at elevated temperatures (200°C and 300°C). For instance, HD-VR/Mg′s compressive strength dropped by 32.8% at 300°C, while LD-VR/Mg saw a reduction of 26.5% (Fig. 7.17). The deformation mechanisms also shifted; at room temperature, HD-VR/Mg exhibited brittle shear failure, whereas LD-VR/Mg showed mixed-mode failure. At 300 °C, both transitioned to a more stable layer-by-layer deformation due to increased magnesium matrix plasticity.

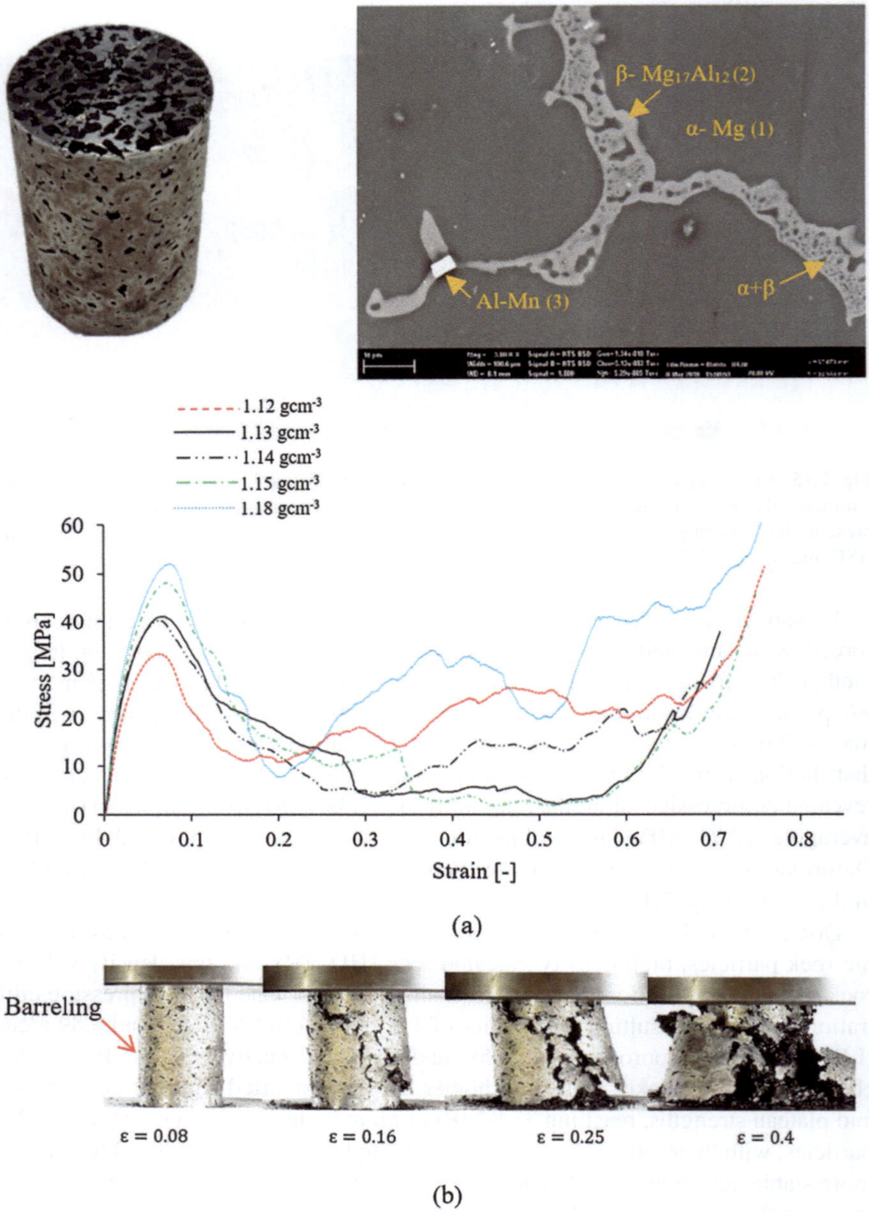

Fig. 7.16 Microstructural and mechanical properties of AZ91-activated carbon (AC) syntactic foam

The corrosion behavior and mechanical properties of magnesium syntactic foams reinforced with hollow silica nanospheres (HSNS) were investigated for potential biomedical applications [25, 26]. Synthesized using the DMD technique, the foams incorporated 0.5–2.0 vol.% HSNS and were tested across simulated body fluids (PBS, HBSS, ABPS, and ASS) and it was found that the 1% HSNS addition was favorable to achieve the best corrosion resistance in PBS, with a corrosion rate of 0.137 mm/year, while 2% addition achieved the lowest corrosion rate (0.062 mm/year) in HBSS due to enhanced passive layer formation (Fig. 7.18). Mechanical testing revealed that Mg-2.0% HSNS also provided significant improvements in

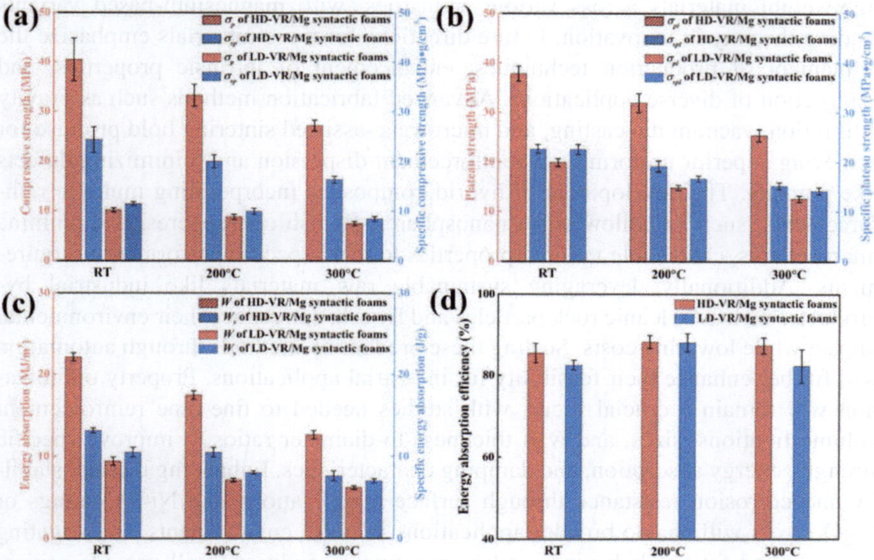

Fig. 7.17 (**a**) The (specific) compressive strength, (**b**) (specific) plateau strength, (**c**) (specific) energy absorption, and (**d**) energy absorption efficiency of HD-VR/Mg and LD-VR/Mg syntactic foams at different temperatures [24]

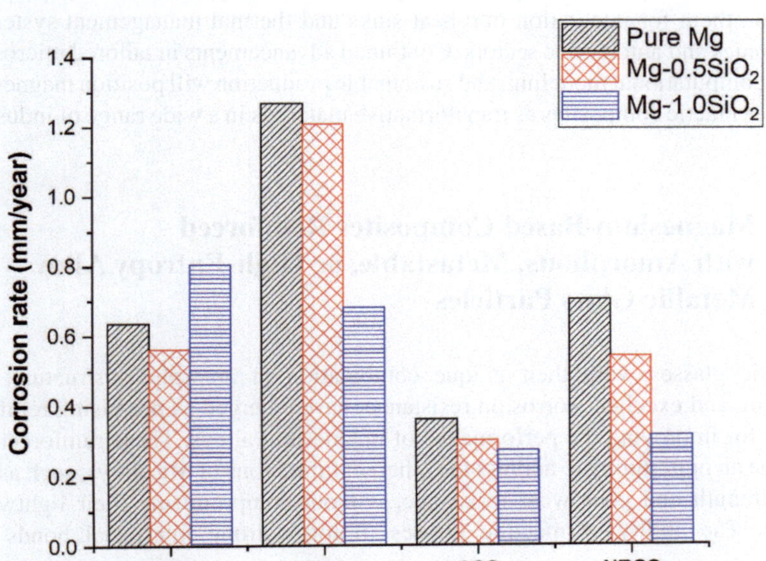

Fig. 7.18 Corrosion rate in Mg-SiO_2 syntactic foams in SBFs [26]

tensile yield strength (167 MPa) and ultimate tensile strength (217 MPa), representing increases of 62% and 46% compared to pure magnesium. The enhanced performance was attributed to the uniform dispersion of HSNS, which facilitated grain refinement and robust passive layer formation.

In summary, the syntactic composites offer significant potential for advancing lightweight materials across various industries, with magnesium-based variants leading the way in innovation. Future directions for these materials emphasize the refinement of production techniques, enhancement of intrinsic properties, and exploration of diverse applications. Advanced fabrication methods such as gravity infiltration, vacuum die casting, and microwave-assisted sintering hold promise for achieving superior uniformity in reinforcement dispersion and minimizing defects like porosity. The development of hybrid composites incorporating multiple reinforcements, such as hollow silica nanospheres, fly ash cenospheres, and alumina microspheres, can enable tailored properties to meet specific performance requirements. Additionally, leveraging sustainable raw materials like industrial by-products, such as volcanic rock particles and fly ash, can reduce their environmental impact while lowering costs. Scaling these production methods through automation will further enhance their feasibility for industrial applications. Property optimization will remain a crucial focus, with studies needed to fine-tune reinforcement volume fractions, sizes, and wall thickness-to-diameter ratios to improve specific strength, energy absorption, and damping characteristics. Enhancing thermal stability and corrosion resistance through surface modifications like Ni-P coatings or MgO layers will enable broader applications in harsh environments. Investigating dynamic and fatigue behaviors under operational conditions will provide critical insights for their use in high-stress applications. These developments align with promising applications in biodegradable implants, lightweight aerospace and automotive structures, marine buoyancy devices, and energy-absorbing systems such as impact-resistant panels and protective gear. Additionally, their thermal properties position them for integration into heat sinks and thermal management systems in electronics and automotive sectors. Continued advancements in tailored microstructures, computational modeling, and sustainable production will position magnesium-based syntactic composites as transformative materials in a wide range of industries.

7.3 Magnesium-Based Composites Reinforced with Amorphous, Metastable, or High-Entropy Alloy Metallic Glass Particles

Metallic glasses, with their unique combination of amorphous structure, high strength, and excellent corrosion resistance, have emerged as promising reinforcements for improving the performance of magnesium alloys. These reinforcements provide an opportunity to address the inherent limitations of Mg alloys, such as their low strength and poor wear resistance, without compromising their lightweight nature. The ability of metallic glasses to form strong interfacial bonds with magnesium matrices further enhances their potential as reinforcements, enabling superior mechanical and thermal properties in the resulting composites.

Recent advancements in magnesium composites include the integration of metallic glass reinforcements for superior mechanical properties. A notable study by Dudina et al. [27] explored the development of magnesium alloy matrix composites reinforced with Zr-based $Zr_{57}Nb_5Cu_{15.4}Ni_{12.6}Al_{10}$ metallic glass particles. The process involved sintering a powder mixture of magnesium alloy and metallic glass at a temperature slightly above the glass transition temperature (Tg) of the reinforcement. During sintering, the metallic glass behaved as a soft liquid-like binder, facilitating compaction, but transformed into a hard reinforcement upon cooling. The XRD analysis confirmed the preservation of the amorphous nature of the metallic glass after sintering, while SEM micrographs revealed a uniform distribution of the reinforcement particles in the matrix. The composite demonstrated no visible reaction layers or porosity, indicating excellent densification (Fig. 7.19). Mechanical testing showed that the composite exhibited significant improvements in hardness, yield strength, and fracture strength compared to the unreinforced magnesium alloy. Specifically, hardness increased from 68 to 123 HV, and yield strength rose from 143 to 325 MPa. The uniaxial compression tests also revealed enhanced strength without a critical loss of ductility, making these composites a promising alternative to traditional metal-ceramic systems.

Jayalakshmi et al. [28] developed $Mg/Ni_{60}Nb_{40}$ composites using the microwave sintering assisted powder metallurgy method followed by hot extrusion. In this study, the $Ni_{60}Nb_{40}$ amorphous alloy powder was produced through mechanical alloying, with XRD analysis of the ball-milled $Ni_{60}Nb_{40}$ powder indicating amorphous characteristics, evidenced by broad humps in the pattern rather than sharp peaks. The thermal properties showed crystallization onset and peak temperatures (Tx-onset and Tx-peak) at 605 °C and 657 °C, respectively. FESEM imaging revealed a nearly spherical morphology of the N $Ni_{60}Nb_{40}$ alloy powder (Fig. 7.20). Optical micrographs of the extruded $Mg/Ni_{60}Nb_{40}$ composites demonstrated a significant refinement in grain size compared to pure magnesium. The grain size reduction was quantified as 2–3 times smaller, emphasizing the effectiveness of metallic

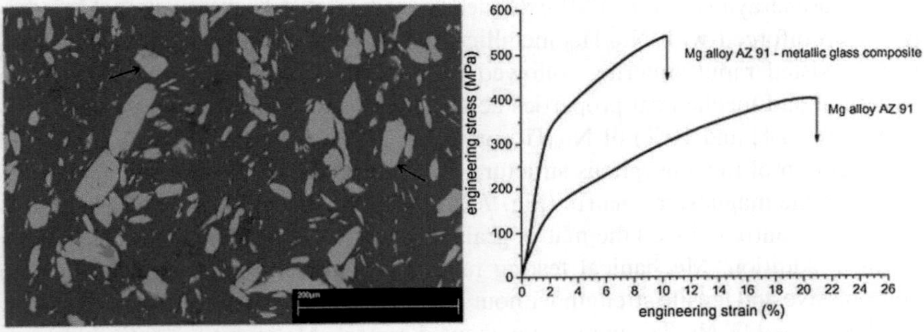

Fig. 7.19 Microstructure and mechanical properties of $Zr_{57}Nb_5Cu_{15.4}Ni_{12.6}Al_{10}$ metallic glass reinforced Mg composite [27]

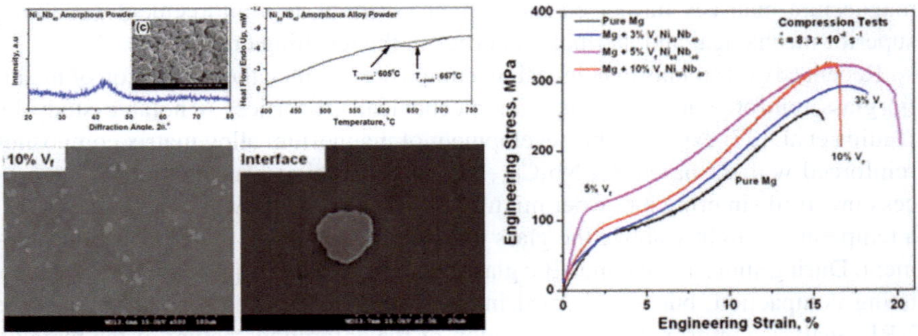

Fig. 7.20 Microstructure and compressive properties of $Mg/Ni_{60}Nb_{40}$ composites from Jayalakshmi et al. [28]

glass reinforcement in grain refinement. The compressive stress-strain curves of $Mg/Ni_{60}Nb_{40}$ composites revealed notable improvements in mechanical properties. With 3 vol.% $Ni_{60}Nb_{40}$ reinforcement, the compressive yield strength and ultimate strength increased by approximately 20%. For 5 vol.% reinforcement, the improvements were ~85% in yield strength and ~30% in ultimate strength (Fig. 7.20). This enhancement is attributed to uniform particle distribution, matrix crystal orientation, and the reinforcing effects of the amorphous phase.

In addition, a comprehensive study by Nafar Dastgerdi et al. [29] examined the high cycle fatigue behavior of $Mg/Ni_{60}Nb_{40}$ composites and reported that the fatigue crack initiation occurred predominantly at the particle–matrix interfaces, while propagation was highly localized and influenced by the distribution and microstructure of reinforcement particles (Fig. 7.21). Specimens with uniformly dispersed particles exhibited superior fatigue resistance compared to those with clustered reinforcements, as crack coalescence was less prevalent in the former. The study also revealed that fatigue crack propagation through regions with well-dispersed particles followed predictable paths governed by established crack growth laws. However, cracks propagating through areas of particle clusters showed significantly faster growth rates due to increased tortuosity.

Sankaranarayanan et al. [30] extended this research by investigating Mg composites reinforced with $Ni_{50}Ti_{50}$ metallic glass particles, synthesized using microwave-assisted rapid sintering followed by hot extrusion. This study focused on structural and mechanical properties across composites with varying volume fractions (3%, 6%, and 10%) of $Ni_{50}Ti_{50}$ particles. Microstructural analysis confirmed the retention of the amorphous structure of the reinforcements and their fair distribution in the magnesium matrix (Fig. 7.22). Notably, the addition of $Ni_{50}Ti_{50}$ particles significantly reduced the matrix grain size, with a 60% refinement observed at 6 vol.% addition. Mechanical testing revealed significant enhancements in both compressive and tensile strength without major compromises in compressive ductility. The 10 vol.% $Ni_{50}Ti_{50}$ composite showed remarkable improvements: compressive yield strength and ultimate compressive strength increased by 79% and 71%, respectively, while tensile yield strength and ultimate tensile strength increased by 98% and 50%, respectively.

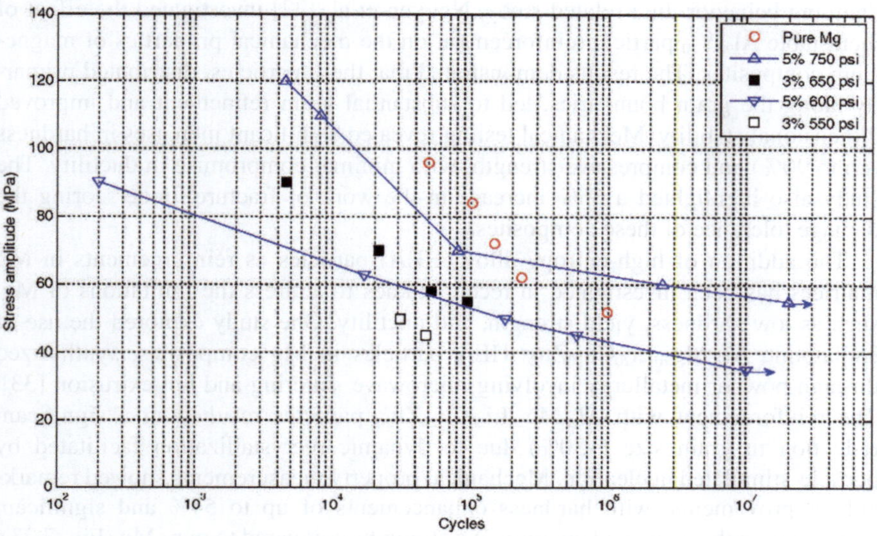

Fig. 7.21 S–N curves of different composition of $Mg/Ni_{60}Nb_{40}$ composite [29]

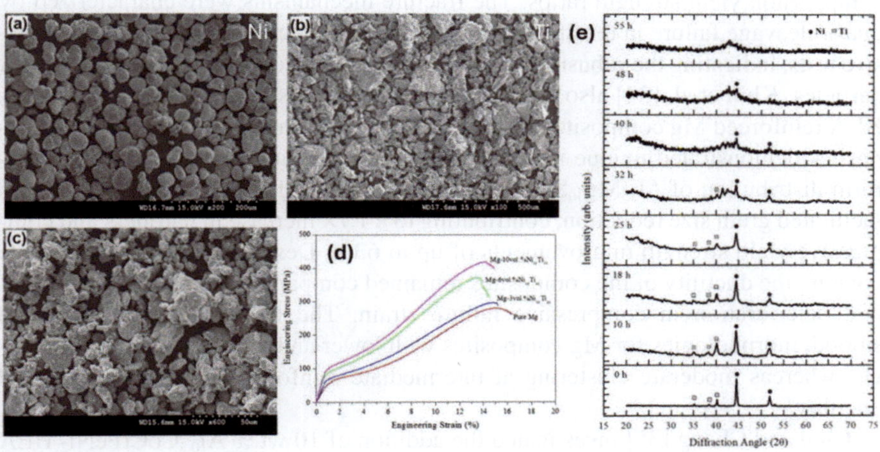

Fig. 7.22 Microstructure and compressive properties of $Mg/Ni_{50}Ti_{50}$ composites [30]

Nguyen et al. [31] further explored the microstructural and damping properties of Mg composites reinforced with metastable $Al_{85}Ti_{15}$ particles. Synthesized through a combination of microwave-assisted sintering and hot extrusion, these composites demonstrated a fair distribution of particles along the grain boundaries and minimal microvoids. The presence of metastable $Al_{8}5Ti_{15}$ particles significantly improved the damping capacity and loss rate, enhancing vibration resistance while maintaining strong interfacial bonding. Additionally, the composites showed equiaxed grain structures and a marked increase in twin formation, contributing to the improved damping behavior. In a related study, Nguyen et al. [32] investigated the effect of metastable $Al_{90}Y_{10}$ particle reinforcement on the mechanical properties of magnesium composites. The results demonstrated that these particles, distributed primarily along the grain boundaries, led to substantial grain refinement and improved dimensional stability. Mechanical testing revealed significant increases in hardness (up to 90%) and compressive strength, with minimal compromise in ductility. The study also highlighted a 66% increase in the work of fracture, underscoring the damage tolerance of these composites.

The addition of high-entropy alloy (HEA) particles as reinforcements in Mg matrices has been investigated in recent studies to address the limitations of Mg, such as low hardness, yield strength, and ductility. One study explored the use of equiatomic $Al_{20}Mg_{20}Li_{20}Cu_{20}Zn_{20}$ HEA particles in Mg composites, synthesized through powder metallurgy involving microwave sintering and hot extrusion [33]. The reinforcement with $Al_{20}Mg_{20}Li_{20}Cu_{20}Zn_{20}$ particles resulted in a significant reduction in grain size (~50%) due to dynamic recrystallization facilitated by particle-stimulated nucleation. Mechanical property measurements showed remarkable improvements, with hardness enhancements of up to 54% and significant increase in both tensile and compressive strengths compared to pure Mg (Fig. 7.23). Furthermore, the presence of $Al_{20}Mg_{20}Li_{20}Cu_{20}Zn_{20}$ particles weakened the fiber texture of Mg, leading to reduced yield asymmetry and achieving near-unity tension/compression yield strength ratios. The fracture mechanisms were characterized by quasi-cleavage failure in tensile tests and limited interfacial debonding in compressive tests, indicating the robust interfacial bonding between the Mg matrix and HEA particles. Khin et al. [34] also investigated the properties of $Al_{35}Mg_{30}Si_{13}Zn_{10}Y_7Ca_5$ HEA reinforced Mg composites. Fabricated using the similar PM method, the composites demonstrated unique microstructural and mechanical properties. The uniform distribution of $Al_{35}Mg_{30}Si_{13}Zn_{10}Y_7Ca_5$ particles in lower reinforcement levels facilitated grain size reduction, contributing to a 1.7× increase in hardness and compressive yield strength improvements of up to 63%. Despite the increased particle content, the ductility of the composites remained comparable to pure Mg, with limited deterioration in compressive failure strain. The fracture surfaces revealed smooth morphologies for Mg composites with lower and higher reinforcement levels, whereas moderate clustering at intermediate reinforcement levels influenced ductility.

Chiu and Chang [35] investigated the addition of 10 wt.% $Al_{0.5}CoCrFeNi_2$ HEA particles on the microstructural and mechanical properties of AZ91 alloy. Here, the composites were fabricated by the PM-based spark plasma sintering technique which was intentionally selected owing to its capability to achieve high-density composites with minimal grain growth and reduced processing time. The properties were compared with that of AZ91 composites reinforced with 10 wt% SiC particles. Microstructural analyses revealed the uniform distribution of HEA particles within the AZ91 matrix, with no interfacial reaction layers observed, a factor attributed to the metallic nature of the HEA and its compatibility with the α-Mg matrix and the primary phases include α-Mg, $Mg_{17}Al_{12}$, and FCC HEA (Fig. 7.24). The porosity of the AZ91–HEA composite was notably low (0.8%) compared to the AZ91–SiC

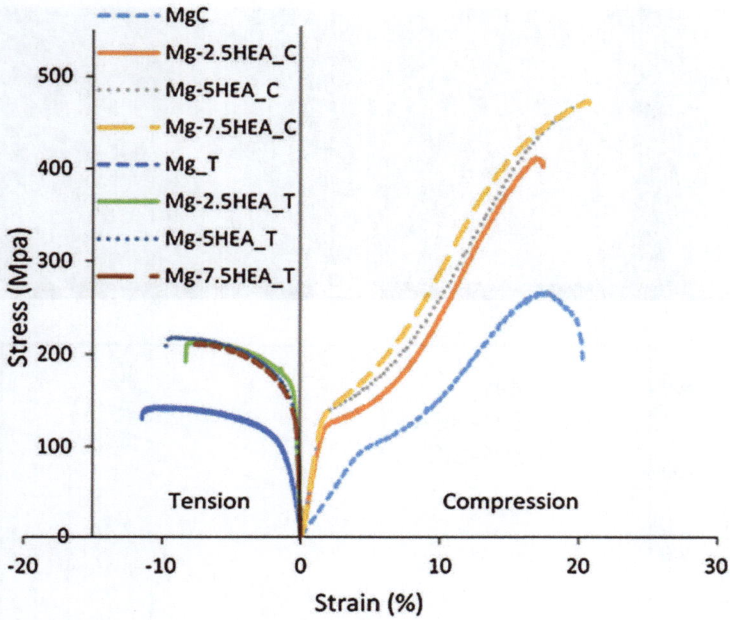

Fig. 7.23 Engineering stress–strain curves of Mg and Mg-$Al_{20}Mg_{20}Li_{20}Cu_{20}Zn_{20}$ HEA composites [33]

composite (3.8%). In terms of mechanical properties, the AZ91–HEA composite exhibited a 48% increase in hardness (from 93 to 138 HV), outperforming the AZ91–SiC composite, which showed a hardness of 123 HV. The compressive yield strength of the AZ91–HEA composite reached 209 MPa, slightly higher than the 204 MPa observed in the AZ91–SiC composite and a 17% improvement over the unreinforced AZ91 alloy (178 MPa). The failure strain of the AZ91–HEA composite slightly increased from 12.2% (AZ91) to 13.7%, demonstrating that the addition of HEA particles maintained the material's ductility (Fig. 7.24). The study concluded that the thermal mismatch between the matrix and HEA particles was the predominant strengthening mechanism, contributing 69% to the yield strength enhancement, followed by load transfer (21%) and grain refinement (7%). The Orowan effect contributed minimally (0.7%) due to the coarse particle size of the HEA reinforcement.

Meng et al. [36] employed the laser melt injection method to introduce AlCoCrCuFeNi HEA particles onto the surface of AZ91D magnesium alloy under controlled conditions. The significant difference in melting points between the HEA particles and the Mg substrate, along with the high mixing enthalpy of the HEA's elements, minimized interfacial reactions, ensuring stability and compatibility. HEA particles were added at volume fractions of 0.1 and 0.4, with the latter exhibiting a more uniform dispersion in the Mg matrix. Microstructural analysis revealed uniformly embedded particles without interfacial reaction products, resulting in

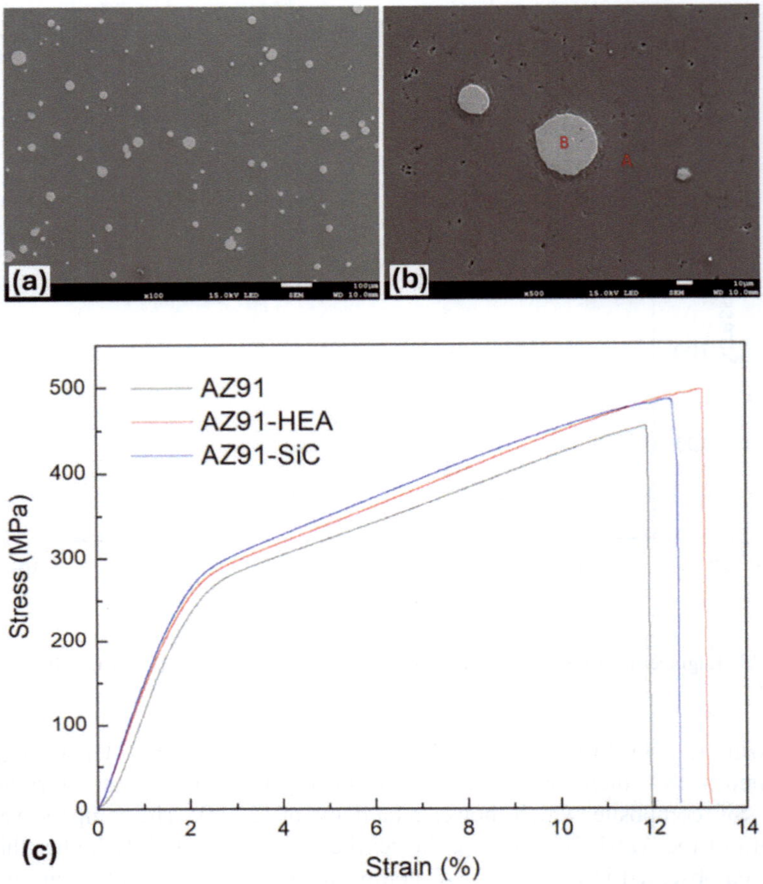

Fig. 7.24 Microstructure and mechanical properties of AZ91/$Al_{0.5}CoCrFeNi_2$ composites [35]

enhanced surface properties. The composite's microhardness increased from 65 HV in the AZ91D substrate to a peak of 515 HV, while tribological tests showed a sevenfold reduction in wear volume loss for the 0.4% HEA composite compared to the unreinforced substrate (Fig. 7.25). Further analysis revealed that Cu rejection from HEA particles into the Mg substrate formed Cu-enriched phases like $CuMg_2$ and complex eutectics, depending on the residual Cu content [37]. These phases, along with HEA particle agglomeration, influenced the composite's microhardness distribution (95–535 HV) and wear performance, with brittle deformation of Cu-rich phases limiting wear resistance despite a lower friction coefficient.

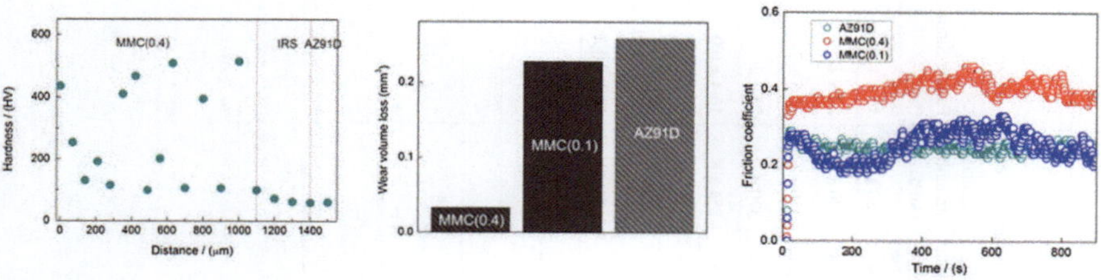

Fig. 7.25 Hardness and tribological properties of AlCoCrCuFeNi reinforced AZ91 alloy composites [36]

The development of high-entropy alloy (HEA) particle-reinforced magnesium (Mg) composites holds immense promise for addressing the inherent limitations of Mg alloys, such as low hardness, limited strength, and poor wear resistance. Future research directions emphasize optimizing HEA compositions to further enhance mechanical and thermal properties while maintaining ductility. Tailored compositions, such as $Al_{20}Mg_{20}Li_{20}Cu_{20}Zn_{20}$ and $Al_{35}Mg_{30}Si_{13}Zn_{10}Y_7Ca_5$, have shown potential for significant performance improvements. Additionally, microstructural refinements, achieved through advanced sintering and extrusion techniques, are critical for ensuring uniform particle distribution, reducing clustering, and refining grain sizes, all of which contribute to superior composite performance.

Interface engineering is another critical focus area, aimed at enhancing the bonding mechanisms between Mg matrices and HEA particles to achieve greater interfacial stability under various loading conditions. Surface engineering methods, such as laser melt injection, offer promising avenues to enhance surface properties while minimizing interfacial reactions and controlling phase formations. Moreover, expanding the evaluation of dynamic properties, including fatigue resistance and damping behavior under high cycle and varying environmental conditions, will be essential to validate these materials for demanding applications in automotive and aerospace industries.

Despite these advancements, challenges remain in achieving cost-effective manufacturing processes that ensure the sustainable and scalable production of Mg-HEA composites. High-entropy alloy particle reinforcements have demonstrated remarkable advancements in mechanical and thermal properties, including significant grain refinement, improved hardness, and enhanced yield strength due to dynamic recrystallization and robust particle-matrix interfaces. These composites also exhibit superior fatigue resistance, vibration damping, and wear performance, making them ideal for lightweight structural applications. However, addressing the remaining challenges in scalability, interface engineering, and dynamic property assessments is crucial for unlocking the full potential of Mg-HEA composites in advanced engineering applications.

Table 7.1 highlights the properties of a few Mg composites reinforced with novel particle reinforcements including hollow micro-balloons, high-entropy alloys, metallic glass, and metastable particles.

Table 7.1 Properties of a few Mg composites reinforced with novel particle reinforcements including hollow micro-balloons, high-entropy alloys, metallic glass, and metastable particles

References	Matrix	Reinforcement			Process	Density	Porosity	Average Grain Size (μm)	Hardness	Tensile properties			Compressive properties		
		Type	Vol.%	Wt.%						Yield Strength (MPa)	Ultimate strength (MPa)	Failure Strain (%)	Yield Strength (MPa)	Ultimate strength (MPa)	Failure Strain (%)
Nguyen et al. [8]	Mg	Ceno		0		1.7361	–	–	47 ± 2	115 ± 5	170 ± 8	6.5 ± 1.1	70 ± 4	180 ± 9	16 ± 1
				5		1.6406	–	–	98 ± 2	180 ± 7	230 ± 10	5.1 ± 0.9	100 ± 8	330 ± 7	22 ± 1.5
				10		1.4905	–	–	110 ± 6	150 ± 8	215 ± 9	3.1 ± 1.0	130 ± 12	350 ± 6	22.5 ± 2
				15		1.4203	–	–	112 ± 7	130 ± 7	180 ± 8	1.5 ± 0.9	150 ± 4	370 ± 6	23 ± 2
Sankaranarayanan et al. [9]	Mg	Ceno	0		DMD	1.7364	–	28 ± 6	49 ± 4	96 ± 8	148 ± 10	5.9 ± 1.3	52 ± 8	254 ± 8	16.4 ± 0.9
			13.24			1.5912	–	22 ± 7	61 ± 3	107 ± 7	158 ± 11	4.4 ± 0.6	64 ± 3	218 ± 4	18.8 ± 0.4
			26.48			1.4707	–	21 ± 11	66 ± 1	95 ± 9	148 ± 13	4.7 ± 0.5	61 ± 6	226 ± 11	21.2 ± 1.2
			39.72			1.4189	–	25 ± 8	54 ± 2	88 ± 12	137 ± 17	4.1 ± 0.2	46 ± 7	221 ± 7	18.7 ± 0.4
Huang et al. [10]	AZ91D	FA Cenosphere	0		Melt stir technique	1.81	–	–	65	–	–	–	–	–	–
			4			1.85	–	–	65	–	–	–	–	–	–
			6			1.87	–	–	72.4	–	–	–	–	–	–
			8			1.93	–	–	76.3	–	–	–	–	–	–
			10			1.92	–	–	74.7	–	–	–	–	–	–
Akinwekomi et al. [11]	AZ61	FA microspheres	0		Microwave sintering techniques	1.58	–	–	–	–	–	–	–	–	–
			20			1.21	23.18	–	–	–	–	–	–	–	–
			30			1.30	17.83	–	–	–	–	–	–	–	–
			40			1.26	20.28	–	–	–	–	–	–	–	–

Akinwekomi et al. [12]		HS	20		Microwave sintering techniques	0.90	–	–	–	–	–	–	–	25.24	–
		HS	30			0.98	–	–	–	–	–	–	–	30.13	–
		HS	40			1.10	–	–	–	–	–	–	–	16.11	–
		CB	20			1.02	–	–	–	–	–	–	–	28.77	–
		CB	30			0.90	–	–	–	–	–	–	–	22.59	–
		CB	40			0.79	–	–	–	–	–	–	–	16.44	–
Padnuru et al. [14]	Mg	GMB	0		MWS	1.73	–	47 ± 4	65 ± 1	–	–	–	98 ± 2	254 ± 6	15.8 ± 0.5
			5			1.59	–	30 ± 4	76 ± 1	–	–	–	91 ± 2	321 ± 7	19.5 ± 0.6
			10			1.46	–	24 ± 2	86 ± 1	–	–	–	88 ± 1	287 ± 6	19.6 ± 0.7
			20			1.28	–	19 ± 2	114 ± 1	–	–	–	85 ± 1	280 ± 4	22 ± 2
Manakari et al. [15]	Mg		0		DMD	1.701	–	–	47 ± 2	–	–	–	66 ± 3.5	194 ± 8	15 ± 1
			5			1.674	–	–	82 ± 4	–	–	–	77 ± 3	232 ± 7	17.2 ± 0.6
			15			1.559	–	–	91 ± 5	–	–	–	102 ± 5	231 ± 6	19.1 ± 0.7
			25			1.472	–	–	107 ± 6	–	–	–	161 ± 4	216 ± 6	37.7 ± 2
Parande et al. [17]	Pure Mg		0		DMD	1.701	–	–	47 ± 2	–	–	–	66 ± 3.5	194 ± 8	15
	Mg-13.1Li		0			1.4495	–	–	94 ± 3	–	–	–	135.9	947.2	>80
	Mg-8.4Li		5			1.6574	–	–	105 ± 5	–	–	–	191.7	1060.6	>80
	Mg-11.9Li		1			1.4967	–	–	83 ± 5	–	–	–	122.1	825.4	>80
Anbuchezhiyan et al. [18]	AZ91D	HGM	0		Stir casting	1.87	–	–	–	–	–	–	112	160	–
			15			1.6	–	–	–	–	–	–	143	211	–
			20			1.5	–	–	–	–	–	–	161	232	–
			23			1.4	–	–	–	–	–	–	168	243	–

(continued)

Table 7.1 (continued)

References	Matrix	Reinforcement			Process	Density	Porosity	Average Grain Size (μm)	Hardness	Tensile properties			Compressive properties		
		Type	Vol.%	Wt.%						Yield Strength (MPa)	Ultimate strength (MPa)	Failure Strain (%)	Yield Strength (MPa)	Ultimate strength (MPa)	Failure Strain (%)
Manakari et al. [26]	Mg	SiO_2	0		MWS + HE	–	–	27 ± 1	59 ± 1	103 ± 5	148 ± 6	7.9 ± 0.5	–	–	–
			0.5			–	–	23 ± 3	73 ± 2	133 ± 3	181 ± 1	6.7 ± 0.2	–	–	–
			1			–	–	15 ± 3	83 ± 2	145 ± 2	198 ± 7	5.7 ± 0.2	–	–	–
			1.5			–	–	13 ± 4	89 ± 1	152 ± 1	203 ± 3	5.2 ± 0.2	–	–	–
			2			–	–	10 ± 2	92 ± 1	167 ± 4	217 ± 7	4.7 ± 0.3	–	–	–
Dudina et al., [27]	Mg		0		BM + sintering	–	–		68	–	–	–	143	404	21.0
			10			–	–		123	–	–	–	325	542	10.5
Jayalakshmi et al. [28]	Mg	Ni60Nb40	0		BM + MWS + HE	–	–	22.0 ± 6	43 ± 2	–	–	–	70 ± 6	265 ± 8	16.2 ± 0.8
			3			–	–	9.4 ± 4	62 ± 4	–	–	–	85 ± 4	283 ± 10	17.6 ± 1.1
			5			–	–	11.0 ± 6	84 ± 5	–	–	–	130 ± 11	320 ± 11	18.4 ± 1.3
			10			–	–	12.5 ± 7	95 ± 5	–	–	–	90 ± 7	322 ± 10	17.2 ± 1.6
Sankaranarayanan et al. [30]	Mg	$Ni_{50}Ti_{50}$	0		BM + MWS + HE	–	–	32 ± 7	32 ± 6	75 ± 9	119 ± 11	8.6 ± 2.1	58 ± 8	234 ± 8	16.4 ± 0.9
			3			–	–	24 ± 12	49 ± 3	94 ± 5	144 ± 6	8.8 ± 1.7	67 ± 9	291 ± 12	15.9 ± 0.7
			6			–	–	12 ± 6	62 ± 4	127 ± 4	183 ± 6	6.5 ± 0.9	89 ± 3	368 ± 8	15.1 ± 1.5
			10			–	–	18 ± 4	66 ± 8	148 ± 7	178 ± 9	2.0 ± 1.3	102 ± 4	417 ± 6	14.9 ± 2.0
Nguyen et al. [31]	Mg	$Al_{85}Ti_{15}$	0	0	BM + MWS + HE	1.734	0.25	32 ± 4	–	–	–	–	–	–	–
			3	3.53		1.767	0.41	26 ± 3	–	–	–	–	–	–	–
			6	6.92		1.805	0.45	24 ± 3	–	–	–	–	–	–	–
			9	10.17		1.841	0.49	23 ± 4	–	–	–	–	–	–	–
Nguyen et al. [32]	Mg	$Al_{90}Y_{10}$	0		BM + MWS + HE	1.722	0.921	32 ± 6	47.6 ± 2.5	–	–	–	69 ± 5	194 ± 7	17.6 ± 2.7
			3			1.758	1.053	16 ± 2	70.1 ± 3.4	–	–	–	90 ± 6	260 ± 11	17.0 ± 2.5
			6			1.795	1.082	15 ± 2	80.8 ± 4.3	–	–	–	110 ± 5	285 ± 15	16.6 ± 1.9
			9			1.831	1.246	10 ± 1	90.1 ± 3.9	–	–	–	120 ± 8	315 ± 25	16.4 ± 2.1

Tun and Gupta [33]	Mg	$Al_{20}Mg_{20}Li_{20}Cu_{20}Zn_{20}$	0		DMD + MWS + HE	–	–	–	–	103 ± 9	162 ± 15	11.5 ± 0.1	91 ± 8	263 ± 16	12 ± 2
			2.5			–	–	–	–	154 ± 14	199 ± 12	6 ± 2	125 ± 3	412 ± 2	13 ± 1
			5			–	–	–	–	147 ± 7	216 ± 9	8 ± 3	135 ± 4	468 ± 7	15 ± 1
			7.5			–	–	–	–	147 ± 17	220 ± 19	7 ± 2	141 ± 4	471 ± 7	15 ± 1
						–	–	–	–	–	–	–	–	–	–
Tun et al. [34]	Mg	$Al_{35}Mg_{30}Si_{13}Zn_{10}Y_7Ca_5$	0	0	Ball milling + MWS + HE	–	–	34 ± 4	47 ± 2	–	–	–	91 ± 8	263 ± 16	12 ± 2
			1.6	2.5		–	–	14 ± 4	56 ± 6	–	–	–	127 ± 5	414 ± 6	15 ± 1
			3.2	5		–	–	12 ± 5	70 ± 6	–	–	–	143 ± 2	469 ± 18	10 ± 2
			4.9	7.5		–	–	12 ± 4	80 ± 7	–	–	–	148 ± 4	472 ± 19	15 ± 2
Chiu et al. [35]	AZ91	HEA	0		PM	–	–	–	93 ± 2	–	–	–	178 ± 4		12.2 ± 0.3
			10			–	–	–	138 ± 2	–	–	–	209 ± 8		13.7 ± 0.5

7.4 Summary

Magnesium-based composites present transformative potential across various high-demand sectors, leveraging their lightweight nature, excellent mechanical properties, and adaptability to diverse reinforcement materials. Innovations in reinforcement techniques, such as the incorporation of hollow microspheres, cenospheres, and high-entropy alloys, have significantly enhanced the structural integrity, energy absorption, and thermal stability of magnesium composites. These advancements enable their application in critical fields like automotive structures, aerospace systems, and biomedical implants. Critical to their success are the optimized fabrication processes like microwave-assisted sintering and melt stir casting, which achieve uniform particle distribution and robust interfacial bonding. The detailed examination of mechanical properties highlights how tailored reinforcement dimensions and compositions improve compressive strength, wear resistance, and energy dissipation while maintaining desirable densities. However, challenges persist in achieving cost-effective, scalable manufacturing solutions. Research must prioritize refining interfacial engineering techniques, such as Ni-P coatings and MgO layers, to improve corrosion resistance and thermal stability under operational conditions. Investigating the dynamic and fatigue behaviors of these materials is equally crucial to validating their reliability in high-stress environments. Furthermore, the exploration of sustainable materials like fly ash cenospheres and volcanic rock particles not only reduces environmental impact but also provides economical alternatives for large-scale production. Future research must also integrate advanced computational modeling to fine-tune composite properties for specific applications, ensuring their adaptability across industrial needs. Magnesium-based composites, with their versatile properties and expanding fabrication techniques, stand poised to redefine the standards for lightweight materials, offering unparalleled opportunities for innovation and industrial application.

References

1. M. Hartmann, I. Crössmann, K. Reindel, and R. Singer, Microstructure and mechanical properties of cellular magnesium matrix composites. Presented at the Metall Foams and Porous Metal Structures, International Conference on Metal Foams and Porous Metal Structures, 1999
2. D.B. Newsome, B.F. Schultz, J.B. Ferguson, P.K. Rohatgi, Synthesis and quasi-static compressive properties of Mg-AZ91D-Al_2O_3 syntactic foams. Materials **8**(9), 6085–6095 (2015). https://doi.org/10.3390/ma8095292
3. C. Li, E. Yang, L. Tang, Y. Li, L. Xu, Compression properties and fabrication of closed-cell metal matrix syntactic foams Al_2O_3hs/AZ91D. Materials **15**(19), 6873 (2022). https://doi.org/10.3390/ma15196873
4. H. Qu et al., Mg-matrix syntactic foam filled with alumina hollow spheres coated by MgO synthesized with solution coating-sintering. J. Mater. Res. Technol. **24**, 2357–2371 (2023). https://doi.org/10.1016/j.jmrt.2023.03.160
5. S. Kannan, S. Pervaiz, R.J. Klassen, D. Huo, M. Haghshenas, An energy-based analysis for machining novel AZ91 magnesium composite foam dispersed with ceramic microspheres. J. Manuf. Sci. Eng. **143**, 031007 (2020). https://doi.org/10.1115/1.4048438

6. S. Kannan, S. Pervaiz, A. Alhourani, R.J. Klassen, R. Selvam, M. Haghshenas, On the role of hollow aluminium oxide microballoons during machining of AZ31 magnesium syntactic foam. Materials **13**(16), 3534 (2020). https://doi.org/10.3390/ma13163534
7. S. Kannan, S. Pervaiz, M.P. Jahan, D. Venkatraghaven, Cryogenic drilling of AZ31 magnesium syntactic foams. Materials **13**(18), 4094 (2020). https://doi.org/10.3390/ma13184094
8. Q.B. Nguyen, M.L. Sharon Nai, A.S. Nguyen, S. Seetharaman, E.W. Wai Leong, M. Gupta, Synthesis and properties of light weight magnesium–cenosphere composite. Mater. Sci. Technol. **32**(9), 923–929 (2016). https://doi.org/10.1080/02670836.2015.1104017
9. S. Sankaranarayanan, Q.B. Nguyen, R. Shabadi, A.H. Almajid, M. Gupta, Powder metallurgy hollow fly ash cenospheres' particles reinforced magnesium composites. Powder Metall. **59**(3), 188–196 (2016). https://doi.org/10.1080/00325899.2016.1139339
10. Z. Huang, S. Yu, M. Li, Microstructures and compressive properties of AZ91D/fly-ash cenospheres composites. Trans. Nonferrous Met. Soc. China **20**, S458–S462 (2010). https://doi.org/10.1016/S1003-6326(10)60518-3
11. A.D. Akinwekomi, Microstructural characterisation and corrosion behaviour of microwave-sintered magnesium alloy AZ61/fly ash microspheres syntactic foams. Heliyon **5**(4), e01531 (2019). https://doi.org/10.1016/j.heliyon.2019.e01531
12. A.D. Akinwekomi et al., Synthesis and characterisation of floatable magnesium alloy syntactic foams with hybridised cell morphology. Mater. Des. **160**, 591–600 (2018). https://doi.org/10.1016/j.matdes.2018.10.004
13. K.N. Braszczyńska-Malik, J. Kamieniak, Analysis of interface between components in AZ91 magnesium alloy foam composite with Ni-P coated fly ash cenospheres. J. Alloys Compd. **720**, 352–359 (2017). https://doi.org/10.1016/j.jallcom.2017.05.285
14. A. Padnuru Sripathy, C. Handjaja, V. Manakari, G. Parande, M. Gupta, Development of lightweight magnesium/glass micro balloon syntactic foams using microwave approach with superior thermal and mechanical properties. Metals **11**(5), 827 (2021). https://doi.org/10.3390/met11050827
15. V. Manakari, G. Parande, M. Doddamani, M. Gupta, Enhancing the ignition, hardness and compressive response of magnesium by reinforcing with hollow glass microballoons. Materials **10**(9), 997 (2017). https://doi.org/10.3390/ma10090997
16. V. Manakari, G. Parande, M. Doddamani, M. Gupta, Evaluation of wear resistance of magnesium/glass microballoon syntactic foams for engineering/biomedical applications. Ceram. Int. **45**(7, Part A), 9302–9305 (2019). https://doi.org/10.1016/j.ceramint.2019.01.207
17. G. Parande, D.J.E. Chow, A.P. Sripathy, V. Manakari, M. Doddamani, M. Gupta, An insight into microstructure and mechanical response of Mg-Li-based composites prepared using glass microballoons as starting reinforcement. J. Mater. Eng. Perform. **33**(22), 12160–12170 (2024). https://doi.org/10.1007/s11665-023-08858-6
18. G. Anbuchezhiyan, B. Mohan, D. Sathianarayanan, T. Muthuramalingam, Synthesis and characterization of hollow glass microspheres reinforced magnesium alloy matrix syntactic foam. J. Alloys Compd. **719**, 125–132 (2017). https://doi.org/10.1016/j.jallcom.2017.05.153
19. G. Anbuchezhiyan, T. Muthuramalingam, B. Mohan, Effect of process parameters on mechanical properties of hollow glass microsphere reinforced magnesium alloy syntactic foams under vacuum die casting. Arch. Civ. Mech. Eng. **18**(4), 1645–1650 (2018). https://doi.org/10.1016/j.acme.2018.07.008
20. G. Anbuchezhiyan, B. Mohan, R. Karthikeyan, Development of magnesium matrix syntactic foams processed through powder metallurgy techniques. Appl. Mech. Mater. **766**, 281–286 (2015)
21. H. Anantharaman, V.C. Shunmugasamy, O.M. Strbik, N. Gupta, K. Cho, Dynamic properties of silicon carbide hollow particle filled magnesium alloy (AZ91D) matrix syntactic foams. Met. Foams Dyn. Load. **82**, 14–24 (2015). https://doi.org/10.1016/j.ijimpeng.2015.04.008
22. P. Kubelka, C. Kádár, N. Jost, Effect of the interface on the compressive properties of magnesium syntactic foams. Mater. Lett. **287**, 129293 (2021). https://doi.org/10.1016/j.matlet.2020.129293

23. N. Movahedi, M. Taherishargh, I.V. Belova, G.E. Murch, T. Fiedler, Mechanical and microstructural characterization of an AZ91–activated carbon syntactic foam. Materials **12**(1), 3 (2019). https://doi.org/10.3390/ma12010003
24. Z. Dong, D. Song, W. Sun, J. Wang, J. Liu, Compressive properties of the magnesium matrix syntactic foams using different types of porous volcanic rock particles. Materialia **25**, 101535 (2022). https://doi.org/10.1016/j.mtla.2022.101535
25. W. Qureshi et al., Influence of silica nanospheres on corrosion behavior of magnesium matrix syntactic foam. IOP Conf. Ser. Mater. Sci. Eng. **346**(1), 012012 (2018). https://doi.org/10.1088/1757-899X/346/1/012012
26. V. Manakari et al., In-vitro degradation of hollow silica reinforced magnesium syntactic foams in different simulated body fluids for biomedical applications. Metals **10**(12), 1583 (2020). https://doi.org/10.3390/met10121583
27. D.V. Dudina et al., A magnesium alloy matrix composite reinforced with metallic glass. Compos. Sci. Technol. **69**(15), 2734–2736 (2009). https://doi.org/10.1016/j.compscitech.2009.08.001
28. S. Jayalakshmi, S. Sahu, S. Sankaranarayanan, S. Gupta, M. Gupta, Development of novel Mg–$Ni_{60}Nb_{40}$ amorphous particle reinforced composites with enhanced hardness and compressive response. Mater. Des. **53**, 849–855 (2014). https://doi.org/10.1016/j.matdes.2013.07.022
29. J. Nafar Dastgerdi, G. Marquis, S. Sankaranarayanan, M. Gupta, Fatigue crack growth behavior of amorphous particulate reinforced composites. Compos. Struct. **153**, 782–790 (2016). https://doi.org/10.1016/j.compstruct.2016.06.071
30. S. Sankaranarayanan, V. Hemanth Shankar, S. Jayalakshmi, N. Quy Bau, M. Gupta, Development of high performance magnesium composites using $Ni_{50}Ti_{50}$ metallic glass reinforcement and microwave sintering approach. J. Alloys Compd. **627**, 192–199 (2015). https://doi.org/10.1016/j.jallcom.2014.12.009
31. Q.B. Nguyen et al., Microstructure and damping characteristics of Mg and its composites containing metastable $Al_{85}Ti_{15}$ particle. J. Compos. Mater. **50**(18), 2565–2573 (2016). https://doi.org/10.1177/0021998315608432
32. Q.B. Nguyen et al., Enhancing hardness, CTE and compressive response of powder metallurgy magnesium reinforced with metastable $Al_{90}Y_{10}$ powder particles. Powder Metall. **59**(3), 209–215 (2016). https://doi.org/10.1080/00325899.2016.1144864
33. K.S. Tun, M. Gupta, Enhanced mechanical properties and near unity yield asymmetry in equiatomic high entropy alloy particles reinforced magnesium composites. J. Alloys Compd. **810**, 151909 (2019). https://doi.org/10.1016/j.jallcom.2019.151909
34. K.S. Tun, Y. Zhang, G. Parande, V. Manakari, M. Gupta, Enhancing the hardness and compressive response of magnesium using complex composition alloy reinforcement. Metals **8**(4), 276 (2018). https://doi.org/10.3390/met8040276
35. C. Chiu, H.-H. Chang, $Al_{0.5}CoCrFeNi_2$ high entropy alloy particle reinforced AZ91 magnesium alloy-based composite processed by spark plasma sintering. Materials **14**(21), 6520 (2021). https://doi.org/10.3390/ma14216520
36. G. Meng, T.M. Yue, X. Lin, H. Yang, H. Xie, X. Ding, Laser surface forming of AlCoCrCuFeNi particle reinforced AZ91D matrix composites. Opt. Laser Technol. **70**, 119–127 (2015). https://doi.org/10.1016/j.optlastec.2015.02.001
37. G. Meng, X. Lin, H. Xie, C. Wang, S. Wang, X. Ding, Reinforcement and substrate interaction in laser surface forming of AlCoCrCuFeNi particle reinforced AZ91D matrix composites. J. Alloys Compd. **672**, 660–667 (2016). https://doi.org/10.1016/j.jallcom.2016.02.191

Index

S. Seetharaman et al., *Magnesium-based Syntactic, Metastable and Nano-composites*, https://doi.org/10.1007/978-3-031-96514-2

MIX
Papier aus verantwortungsvollen Quellen
Paper from responsible sources
FSC® C105338

If you have any concerns about our products,
you can contact us on
ProductSafety@springernature.com

In case Publisher is established outside the EU,
the EU authorized representative is:
Springer Nature Customer Service Center GmbH
Europaplatz 3, 69115 Heidelberg, Germany

Printed by Libri Plureos GmbH
in Hamburg, Germany